T0177014

Statistical Modelling by Exponential Families

This book is a readable, digestible introduction to exponential families, encompassing statistical models based on the most useful distributions in statistical theory, such as the normal, gamma, binomial, Poisson, and negative binomial. Strongly motivated by applications, it presents the essential theory and then demonstrates the theory's practical potential by connecting it with developments in areas such as item response analysis, social network models, conditional independence and latent variable structures, and point process models. Extensions to incomplete data models and generalized linear models are also included. In addition, the author gives a concise account of the philosophy of Per Martin-Löf in order to connect statistical modelling with ideas in statistical physics, such as Boltzmann's law. Written for graduate students and researchers with a background in basic statistical inference, the book includes a vast set of examples demonstrating models for applications and numerous exercises embedded within the text as well as at the ends of chapters.

ROLF SUNDBERG is Professor Emeritus of Statistical Science at Stockholm University. His work embraces both theoretical and applied statistics, including principles of statistics, exponential families, regression, chemometrics, stereology, survey sampling inference, molecular biology, and paleoclimatology. In 2003, he won (with M. Linder) the award for best theoretical paper in the *Journal of Chemometrics* for their work on multivariate calibration, and in 2017 he was named Statistician of the Year by the Swedish Statistical Society.

INSTITUTE OF MATHEMATICAL STATISTICS TEXTBOOKS

IMS Textbooks give introductory accounts of topics of current concern suitable for advanced courses at master's level, for doctoral students and for individual study. They are typically shorter than a fully developed textbook, often arising from material created for a topical course. Lengths of 100–290 pages are envisaged. The books typically contain exercises.

In collaboration with the International Society for Bayesian Analysis (ISBA), selected volumes in the IMS Textbooks series carry the "with ISBA" designation at the recommendation of the ISBA editorial representative.

Other Books in the Series (*with ISBA)

1. *Probability on Graphs*, by Geoffrey Grimmett
2. *Stochastic Networks*, by Frank Kelly and Elena Yudovina
3. *Bayesian Filtering and Smoothing*, by Simo Särkkä
4. *The Surprising Mathematics of Longest Increasing Subsequences*, by Dan Romik
5. *Noise Sensitivity of Boolean Functions and Percolation*, by Christophe Garban and Jeffrey E. Steif
6. *Core Statistics*, by Simon N. Wood
7. *Lectures on the Poisson Process*, by Günter Last and Mathew Penrose
8. *Probability on Graphs (Second Edition)*, by Geoffrey Grimmett
9. *Introduction to Malliavin Calculus*, by David Nualart and Eulàlia Nualart
10. *Applied Stochastic Differential Equations*, by Simo Särkkä and Arno Solin
11. **Computational Bayesian Statistics*, by M. Antónia Amaral Turkman, Carlos Daniel Paulino, and Peter Müller
12. *Statistical Modelling by Exponential Families*, by Rolf Sundberg

Statistical Modelling by Exponential Families

ROLF SUNDBERG
Stockholm University

CAMBRIDGE
UNIVERSITY PRESS

CAMBRIDGE
UNIVERSITY PRESS

University Printing House, Cambridge CB2 8BS, United Kingdom

One Liberty Plaza, 20th Floor, New York, NY 10006, USA

477 Williamstown Road, Port Melbourne, VIC 3207, Australia

314–321, 3rd Floor, Plot 3, Splendor Forum, Jasola District Centre, New Delhi – 110025, India

79 Anson Road, #06–04/06, Singapore 079906

Cambridge University Press is part of the University of Cambridge.

It furthers the University's mission by disseminating knowledge in the pursuit of education, learning, and research at the highest international levels of excellence.

www.cambridge.org
Information on this title: www.cambridge.org/9781108476591
DOI: 10.1017/9781108604574

First published 2019

Printed and bound in Great Britain by Clays Ltd, Elcograf S.p.A.

A catalogue record for this publication is available from the British Library.

Library of Congress Cataloguing-in-Publication data
Names: Sundberg, Rolf, 1942– author.
Title: Statistical modelling by exponential families / Rolf Sundberg (Stockholm University).
Description: Cambridge ; New York, NY : Cambridge University Press, 2019. |
Series: Institute of Mathematical Statistics textbooks ; 12 |
Includes bibliographical references and index.
Identifiers: LCCN 2019009281| ISBN 9781108476591 (hardback : alk. paper) |
ISBN 9781108701112 (pbk. : alk. paper)
Subjects: LCSH: Exponential families (Statistics)–Problems, exercises, etc. |
Distribution (Probability theory)–Problems, exercises, etc.
Classification: LCC QA276.7 .S86 2019 | DDC 519.5–dc23
LC record available at https://lccn.loc.gov/2019009281

ISBN 978-1-108-47659-1 Hardback
ISBN 978-1-108-70111-2 Paperback

To Margareta,
and to Per;
celebrating 50 years

Contents

Examples

Preface

The theoretical importance of exponential families of distributions has long been evident. Every book on general statistical theory includes a few sections introducing exponential families. However, there are three gaps in the literature that I want to fill with the present text:

- It is the first book devoted wholly to exponential families and written at the Master's/PhD level. As a course book, it can stand by itself or supplement other literature on parametric statistical inference.
- It aims to demonstrate not only the essentials of the elegant and powerful general theory but also the extensive development of exponential and related models in a diverse range of applied disciplines.
- It gives an account of the half-century-old but still influential ideas of Per Martin-Löf about such models. Part of this innovative material has previously been available only in my handwritten lecture notes in Swedish from 1969–1970.

Two older books are devoted to the general theory of exponential families: Ole Barndorff-Nielsen's *Information and Exponential Families in Statistical Theory* (Barndorff-Nielsen, 1978) and Larry Brown's *Fundamentals of Statistical Exponential Families* (Brown, 1988). The present text is quite different in intention. The general presentation is less focused on mathematical rigour, and instead the text is much more directed toward applications (even though it does not include much real data). My ambition is not to present a self-contained stringent theory, which would have required more space and distracted the reader from my main messages. Thus I refer to other works for some proofs or details of proofs not short enough or instructive enough to be included.

Instead, in order to show the theory's potential, I make room to describe developments in numerous areas of applied modelling. The topics treated include incomplete data, generalized linear models, conditional independence and latent variable structures, social network models, models for item analysis and spatial point process models. The methodology of generalized linear models will take us a bit beyond exponential family theory. We will also be outside the scope of the standard exponential families when the parameter space or the data are incomplete, but not so far outside that the theory is not beneficial.

The inferential philosophy underlying this text is that of a frequentist statistician. In particular, conditional inference plays a substantial role in the text. This should not preclude potential interest from a Bayesian point of view, however. For both frequentists and Bayesians the likelihood is central, and it plays a dominant role in the present text.

This textbook contains more material than typically covered in a single course. My own proposed selection for a basic course on the topic would cover Chapters/Sections 1, 2.1, 3.1–6, 4.1, 5, 7.1–2, 8.1–3, and 9.1–3. The choice of additional material should be driven by the interests of students and/or the instructor. Instruction on exponential families can also be combined with other aspects of statistical theory. At Stockholm University, a theoretical graduate level course consists of exponential families jointly with more general theory of parametric inference.

As mentioned, the text is written with graduate students primarily in mind. It has a large number of examples and exercises, many of them appearing recurrently. They range from small instructional illustrations to application areas of independent interest and of great methodological or applied utility in their own right, represented by separate sections or chapters. This simplifies the choice of topics by taste. Chapter 2 introduces a large number of common and less common distributions and models. Most of these examples reappear in later chapters. Chapters 10–13 can be regarded as elaborated examples per se.

Exercises are typically found at the ends of sections, but there are also sections entirely devoted to exercises. Solutions to exercises are not included here, but can be found at www.cambridge.org/9781108701112.

As prerequisites, readers should have a general knowledge of the theory of statistics and experience with the use of statistical methods (the more, the better). Appendix A defines some statistical concepts and lists the principles assumed to be in the mind of the reader. Some particularly useful calculus and matrix algebra is found in the other short appendices.

Application of an exponential family model in practice typically requires a statistical computer package. I give some references to the huge, free set of **R** packages, but these references could be regarded as examples, because alternatives often exist in other environments, or even in **R** itself.

My original inspiration to study the theory and methodology of exponential families goes 50 years back in time, in particular to unpublished lectures given by Per Martin-Löf in 1969–1970 (Martin-Löf, 1970). While skipping some technicalities, I try to do justice to many of his original thoughts and deep results, in particular in Chapter 6. Per Martin-Löf's lecture notes were written in Swedish, which of course severely restricted their accessibility and influence outside the Scandinavian languages. Nevertheless, these notes have got about one hundred Google Scholar citations. The Swedish original is available on my department's website:
www.math.su.se/PML1970.

Implicit above is my deep gratitude to Per Martin-Löf for introducing me to the area, for suggesting a fruitful thesis topic and for his sharp mind challenging me to achieve a clear understanding myself.

As mentioned, earlier versions of this text have been used as course texts in my department during their development. I'm particularly grateful to Michael Höhle, who took over the course after me, and whose many questions and comments have led to important corrections and other improvements. I am also grateful to students commenting on previous versions, and to colleagues inside and outside Stockholm University for comments and stimulation.

I appreciate greatly the support of David Cox and Diana Gillooly, who have both been eager to see these notes published in the IMS Textbook series with Cambridge University Press. After many years the goal is now achieved, and I hope they like the result.

Finally, my deep thanks go to my wife Margareta, my wife since 50 years ago, without whose patience and support this project would not have come to completion.

1

What Is an Exponential Family?

We start with a couple of simple, well-known distributions to introduce the common features of the distribution families called *exponential families* or distributions of *exponential type*. First, consider a sample from an *exponential* distribution with intensity parameter λ (expected value $1/\lambda$). A single observation y has the density $\lambda \exp(-\lambda y)$ for $y > 0$, and a sample $\boldsymbol{y} = (y_1, \dots, y_n)$ has the n-dimensional density

$$f(\boldsymbol{y}; \lambda) = \lambda^n e^{-\lambda \sum y_i} = \lambda^n \exp(-\lambda \textstyle\sum y_i), \tag{1.1}$$

for $\boldsymbol{y}$ with all $y_i > 0$. Another basic example is the density for a sample $\boldsymbol{y} = (y_1, \dots, y_n)$ from a two-parameter *normal* (or *Gaussian*) distribution, $\mathrm{N}(\mu, \sigma^2)$. It can be written

$$\begin{aligned} f(\boldsymbol{y}; \mu, \sigma^2) &= (\sigma\sqrt{2\pi})^{-n} e^{-\frac{\sum(y_i-\mu)^2}{2\sigma^2}} \\ &= (\sigma\sqrt{2\pi})^{-n} \exp\left(\tfrac{-1}{2\sigma^2} \textstyle\sum y_i^2 + \tfrac{\mu}{\sigma^2} \sum y_i - \tfrac{n\mu^2}{2\sigma^2}\right), \end{aligned} \tag{1.2}$$

where all dependence on data is found in the two sum type functions of data in the exponent, with a parameter in front of each of them.

As discrete distribution examples, we consider the *binomial* and *Poisson* distribution families. Here is first the binomial probability for y successes in n Bernoulli trials, with success probability π_0 as parameter (Greek letter π is preferred to Roman p for a parameter; π_0 is here used to distinguish from the mathematical constant π)

$$f(y; \pi_0) = \binom{n}{y} \pi_0^y (1-\pi_0)^{n-y} = \binom{n}{y} (1-\pi_0)^n e^{y \log \frac{\pi_0}{1-\pi_0}}, \tag{1.3}$$

for $y = 0, 1, \dots, n$. For a single observation (count) y from a Poisson distribution we analogously write the probability

$$f(y; \lambda) = \frac{\lambda^y}{y!} e^{-\lambda} = \frac{1}{y!} e^{-\lambda} e^{y \log \lambda}, \tag{1.4}$$

for $y = 0, 1, \dots$.

In all four cases we could express the density or probability function as a product of two or three factors, namely one factor depending only on the parameter, another depending only on the data (in some cases not needed), and one factor of exponential form connecting the data with the parameter. The exponent is a product of a (function of the) parameter and a function of data, or more generally a sum of such products.

The second factor was not needed in the normal distribution but was present in the exponential distribution, albeit implicitly, in the form of the characteristic function for the positive real axis (or the product of such functions for a sample).

Many statistical models have these features in common, and are then characterized as being *exponential families*. Generally, let data $\boldsymbol{y}$ be modelled by a continuous or discrete distribution family, with probability density on $\mathbb{R}^n$ or probability (mass) function on $\mathbb{Z}^n$, or on subsets of $\mathbb{R}^n$ or $\mathbb{Z}^n$. The density or probability function (density with respect to the counting measure, or similarly) will be written $f(\boldsymbol{y};\, \boldsymbol{\theta})$ in a parameterization called canonical, where $\boldsymbol{\theta}$ belongs to a k-dimensional parameter space $\boldsymbol{\Theta}$. A *statistic* $\boldsymbol{t}$ is any (measurable) scalar or vector-valued function $\boldsymbol{t}(\boldsymbol{y})$ of data, for example $\boldsymbol{t} = \boldsymbol{t}(\boldsymbol{y}) = (\sum y_i, \sum y_i^2)$ in the Gaussian example (1.2).

Definition 1.1 ***Exponential family***
A parametric statistical model for a data set $\boldsymbol{y}$ is an *exponential family* (or is of exponential type), with *canonical parameter* vector $\boldsymbol{\theta} = (\theta_1, \ldots, \theta_k)$ and *canonical statistic* $\boldsymbol{t}(\boldsymbol{y}) = (t_1(\boldsymbol{y}), \ldots, t_k(\boldsymbol{y}))$, if f has the structure

$$f(\boldsymbol{y};\, \boldsymbol{\theta}) = a(\boldsymbol{\theta})\, h(\boldsymbol{y})\, e^{\boldsymbol{\theta}^T \boldsymbol{t}(\boldsymbol{y})}, \tag{1.5}$$

where $\boldsymbol{\theta}^T \boldsymbol{t}$ is the scalar product of the k-dimensional parameter vector and a k-dimensional statistic $\boldsymbol{t}$, that is,

$$\boldsymbol{\theta}^T \boldsymbol{t} = \sum_{j=1}^{k} \theta_j\, t_j(\boldsymbol{y}),$$

and a and h are two functions, of which h should (of course) be measurable.

It follows immediately that $1/a(\boldsymbol{\theta})$, to be denoted $C(\boldsymbol{\theta})$, can be interpreted as a normalizing constant, that makes the density integrate to 1,

$$C(\boldsymbol{\theta}) = \int h(\boldsymbol{y})\, e^{\boldsymbol{\theta}^T \boldsymbol{t}(\boldsymbol{y})}\, \mathrm{d}\boldsymbol{y}, \tag{1.6}$$

or the analogous sum over all possible outcomes in the discrete case. Of course $C(\boldsymbol{\theta})$ or $a(\boldsymbol{\theta})$ are well-defined only up to a constant factor, which can be borrowed from or lent to $h(\boldsymbol{y})$.

In some literature, mostly older, the canonical parameterization is called the *natural* parameterization. This is not a good term, however, because the canonical parameters are not necessarily the intuitively natural ones, see for example the Gaussian distribution above.

We think of the vector $\boldsymbol{t}$ and parameter space $\boldsymbol{\Theta}$ as in effect k-dimensional (not $< k$). This demand will later be shown to imply that $\boldsymbol{t}$ is minimal sufficient for $\boldsymbol{\theta}$. That $\boldsymbol{t}$ is really k-dimensional means that none of its components t_j can be written as a linear expression in the others. Unless otherwise explicitly told, $\boldsymbol{\Theta}$ is taken to be maximal, that is, comprising all $\boldsymbol{\theta}$ for which the integral (1.6) or the corresponding sum is finite. This maximal parameter space $\boldsymbol{\Theta}$ is called the *canonical parameter space*. In Section 3.1 we will be more precise about regularity conditions.

Before we go to many more examples in Chapter 2, we look at some simple consequences of the definition. Consider first a *sample*, that is, a set of independent and identically distributed (iid) observations from a distribution of exponential type.

Proposition 1.2 ***Preservation under repeated sampling***
If $\boldsymbol{y} = (y_1, \ldots, y_n)$ is a sample from an exponential family, with distribution

$$f(y_i;\, \boldsymbol{\theta}) = a(\boldsymbol{\theta})\, h(y_i)\, e^{\boldsymbol{\theta}^T \boldsymbol{t}(y_i)},$$

then the sample $\boldsymbol{y}$ follows an exponential family with the same canonical parameter space $\boldsymbol{\Theta}$ and with the sum $\sum \boldsymbol{t}(y_i)$ as canonical statistic,

$$f(\boldsymbol{y};\, \boldsymbol{\theta}) = a(\boldsymbol{\theta})^n\, e^{\boldsymbol{\theta}^T \sum \boldsymbol{t}(y_i)} \prod_i h(y_i)\,. \tag{1.7}$$

Proof Formula (1.7) follows immediately from $f(\boldsymbol{y};\, \boldsymbol{\theta}) = \prod_i f(y_i;\, \boldsymbol{\theta})$. □

Exponential families are preserved not only under repeated sampling from one and the same distribution. It is sufficient in Proposition 1.2 that the observations are independent and have the same canonical parameter. Two important examples are log-linear models and Gaussian linear models, discussed in Chapter 2 (Examples 2.5 and 2.9).

The canonical statistic $\boldsymbol{t}$ itself necessarily also has a distribution of exponential type:

Proposition 1.3 ***Distribution for t***
If $\boldsymbol{y}$ has a distribution of exponential type, as given by (1.5), then the (marginal) distribution of $\boldsymbol{t}$ is also of exponential type. Under certain regularity conditions on the function $\boldsymbol{t}(\boldsymbol{y})$, the distribution of $\boldsymbol{t}$ also has a density or probability function, which can then be written

$$f(\boldsymbol{t};\, \boldsymbol{\theta}) = a(\boldsymbol{\theta})\, g(\boldsymbol{t})\, e^{\boldsymbol{\theta}^T \boldsymbol{t}}\,, \tag{1.8}$$

where the structure function g *in the discrete case can be expressed as*

$$g(\boldsymbol{t}) = \sum_{\boldsymbol{t}(\boldsymbol{y})=\boldsymbol{t}} h(\boldsymbol{y}), \tag{1.9}$$

or an analogous integral in the continuous case, written

$$g(\boldsymbol{t}) = \int_{\boldsymbol{t}(\boldsymbol{y})=\boldsymbol{t}} h(\boldsymbol{y})\, \mathrm{d}\boldsymbol{y}\,. \tag{1.10}$$

Proof In the discrete case, the probability function for $\boldsymbol{t}$ follows immediately by summation over the possible outcomes of $\boldsymbol{y}$ for given $\boldsymbol{t}(\boldsymbol{y}) = \boldsymbol{t}$. (Reader, check this!) In the continuous case, the integral representation of $g(\boldsymbol{t})$ is obtained by a limiting procedure starting from a narrow interval of width $d\boldsymbol{t}$ in $\boldsymbol{t}$, corresponding to a thin shell in $\boldsymbol{y}$-space. See also Section 6.1 for a general formula. □

Example 1.1 ***The structure function for repeated Bernoulli trials***
If the sequence $\boldsymbol{y} = (y_1, \ldots, y_n)$ is the realization of n Bernoulli trials, with common success probability π_0, with $y_i = 1$ representing success, the probability function for the sequence $\boldsymbol{y}$ represents an exponential family,

$$f(\boldsymbol{y}; \pi_0) = \pi_0^t (1-\pi_0)^{n-t} = (1-\pi_0)^n\, e^{t \log \frac{\pi_0}{1-\pi_0}}, \tag{1.11}$$

where $t = t(\boldsymbol{y}) = \sum y_i$ is the number of ones. The structure function $g(t)$ is found by summing over all the equally probable outcome sequences having t ones and $n-t$ zeros. The well-known number of such sequences is $g(t) = \binom{n}{t}$, cf. the binomial example (1.3) above. The distribution for the statistic t, induced by the Bernoulli distribution, is the binomial, $\mathrm{Bin}(n; \pi_0)$. △

The distribution for $\boldsymbol{t}$ in the Gaussian example requires more difficult calculations, involving n-dimensional geometry and left aside here.

The conditional density for data $\boldsymbol{y}$, given the statistic $\boldsymbol{t} = t(\boldsymbol{y})$, is obtained by dividing $f(\boldsymbol{y}; \boldsymbol{\theta})$ by the marginal density $f(\boldsymbol{t}; \boldsymbol{\theta})$. We see from (1.5) and (1.8) that the parameter $\boldsymbol{\theta}$ cancels, so $f(\boldsymbol{y}|\, \boldsymbol{t})$ is free from $\boldsymbol{\theta}$. This is the general definition of $\boldsymbol{t}$ being a *sufficient statistic* for $\boldsymbol{\theta}$, with the interpretation that there is no information about $\boldsymbol{\theta}$ in primary data $\boldsymbol{y}$ that is not already in the statistic $\boldsymbol{t}$. This is formalized in the *Sufficiency Principle* of statistical inference: Provided we trust the model for data, all possible outcomes $\boldsymbol{y}$ with the same value of a sufficient statistic $\boldsymbol{t}$ must lead to the same conclusions about $\boldsymbol{\theta}$.

A sufficient statistic should not be of unnecessarily high dimension, so the reduction of data to a sufficient statistic should aim at a *minimal sufficient* statistic. Typically, the canonical statistic is minimal sufficient, see

Proposition 3.3, where the mild additional regularity condition for this is specified.

In statistical modelling we can go a reverse way, stressed in Chapter 6. We reduce the data set $\boldsymbol{y}$ to a small-dimensional statistic $\boldsymbol{t}(\boldsymbol{y})$ that will take the role of canonical statistic in an exponential family, and thus is all we need to know from data for the inference about the parameter $\boldsymbol{\theta}$.

The corresponding parameter-free distribution for $\boldsymbol{y}$ given $\boldsymbol{t}$ is used to check the model. Is the observed $\boldsymbol{y}$ a plausible outcome in this conditional distribution, or at least with respect to some aspect of it? An example is checking a normal linear model by use of studentized residuals (i.e. variance-normalized residuals), e.g. checking for constant variance, for absence of auto-correlation and time trend, for lack of curvature in the dependence of a linear regressor, or for underlying normality.

The statistical inference in this text about the parameter $\boldsymbol{\theta}$ is frequentistic in character, more precisely meaning that the inference is primarily based on the principle of repeated sampling, involving *sampling distributions* of parameter estimators (typically *maximum likelihood*), hypothesis testing via *p-values*, and confidence regions for parameters with prescribed *degree of confidence*. Appendix A contains a summary of inferential concepts and principles, intended to indicate what is a good background knowledge about frequentistic statistical inference for the present text.

Exercise 1.1 ***Scale factor in*** $h(y)$
Suppose $h(y)$ is changed by a constant factor c, to $c\,h(y)$. What effect does this have on the other constituents of the exponential family? △

Exercise 1.2 ***Structure function for Poisson and exponential samples***
Calculate these structure functions by utilizing well-known distributions for t, and characterize the conditional distribution of $\boldsymbol{y}$ given $\boldsymbol{t}$:
(a) Sample of size n from the Poisson $\text{Po}(\lambda)$. Use the reproducibility property for the Poisson, that $\sum y_i$ is distributed as $\text{Po}(\sum \lambda_i)$.
(b) Sample of size n from the exponential with intensity λ. Use the fact that $t = \sum y_i$ is gamma distributed, with density

$$f(t;\, \lambda) = \frac{\lambda^n\, t^{n-1}}{\Gamma(n)}\, e^{-\lambda t}\,,$$

and $\Gamma(n) = (n-1)!$ (Section B.2.2). See also Example 2.7 below.
(c) Note that the conditional density for $\boldsymbol{y}$ is constant on some set, $\mathcal{Y}_t$ say. Characterize $\mathcal{Y}_t$ for the Poisson and the exponential by specifying its form and its volume or cardinality (number of points). △

2

Examples of Exponential Families

Before we proceed to see the elegant and effective statistical theory that can be derived and formulated for exponential families in general, we devote this chapter to a number of examples of exponential families, many of them of high importance in applied statistics. We will return to most of these examples in later sections, in particular those of Section 2.1. It has been proposed that essentailly all important statistical models are of exponential type, and even if this is an exaggeration, the exponential families and some related model types certainly dominate among models available in statistical toolboxes.

2.1 Examples Important for the Sequel

Example 2.1 ***Bernoulli trials and the binomial***
In a Bernoulli trial with success probability π_0, outcome $y = 1$ representing success, and else $y = 0$, the probability function for y is

$$f(y;\,\pi_0) = \pi_0^y\,(1-\pi_0)^{1-y} = (1-\pi_0)\left(\frac{\pi_0}{1-\pi_0}\right)^y = \frac{1}{1+e^\theta}\,e^{\theta y}, \quad y = 0,\,1, \tag{2.1}$$

for $\theta = \theta(\pi_0) = \log(\frac{\pi_0}{1-\pi_0})$, the so-called *logit* of π_0. This is clearly a simple exponential family with $k = 1$, $h(y) = 1$ (constant), $t(y) = y$, and canonical parameter space $\Theta = \mathbb{R}$. The parameter space Θ is a one-to-one representation of the *open* interval $(0, 1)$ for π_0. Note that the degenerate probabilities $\pi_0 = 0$ and $\pi_0 = 1$ are not represented in Θ.

Now, application of Proposition 1.2 on a sequence of n Bernoulli trials yields the following exponential family probability function for the outcome sequence $\boldsymbol{y} = (y_1, \ldots, y_n)$,

$$f(\boldsymbol{y};\,\pi_0) = \frac{1}{(1+e^\theta)^n}\,e^{\theta \sum y_i}, \quad y_i = 0,\,1, \tag{2.2}$$

with canonical statistic $t(\boldsymbol{y}) = \sum y_i$. Note that this is not the binomial distribution, because it is the distribution for the vector $\boldsymbol{y}$, not for t. However, Proposition 1.3 implies that t also has a distribution of exponential type, obtained from (2.2) through multiplication by the structure function

$$g(t) = \binom{n}{t} = \text{number of sequences } \boldsymbol{y} \text{ with } \sum y_i = t,$$

cf. (1.3). As the reader already knows, this is the binomial, $\text{Bin}(n, \pi_0)$. △

The Bernoulli family (binomial family) is an example of a *linear* exponential family, in having $t(y) = y$. Other such examples are the Poisson and the exponential distribution families, see (1.1) and (1.4), respectively, and the normal distribution family when it has a known σ. Such families play an important role in the theory of *generalized linear models*; see Chapter 9.

Example 2.2 ***Logistic regression models***
Now let the Bernoulli trials have different success probabilities π_i, depending on an explanatory variable (regressor) x in such a way that trial i has

$$\pi_i = \frac{e^{\theta x_i}}{1 + e^{\theta x_i}}. \tag{2.3}$$

This is equivalent to saying that the logit is proportional to x, $\text{logit}(\pi_i) = \theta x_i$. Inserting this expression in (2.1), and forming the product over i, we obtain an exponential family with canonical statistic $\sum x_i y_i$. This is a proportional logistic regression model. The ordinary logistic regression model is obtained by adding an intercept parameter to the linear function, to form $\text{logit}(\pi_i) = \theta_0 + \theta_1 x_i$, with $\sum y_i$ as the corresponding additional canonical component. *(Continued in Example 3.5)* △

Example 2.3 ***Poisson distribution family for counts***
The Poisson distribution for a single observation y, with expected value λ, is of exponential type with canonical statistic y and canonical parameter $\theta = \log \lambda$, as seen from the representation (1.4),

$$f(y; \lambda) = \frac{\lambda^y}{y!} e^{-\lambda} = \frac{1}{y!} e^{-e^\theta} e^{\theta y}, \quad y = 0, 1, 2, \ldots . \tag{2.4}$$

The canonical parameter space Θ is the whole real line, corresponding to the open half-line $\lambda > 0$, $h(y) = 1/y!$, and the norming constant C is

$$C(\theta) = e^{\lambda(\theta)} = e^{e^\theta}.$$

For a sample $\boldsymbol{y} = \{y_1, \ldots, y_n\}$ from this distribution we have the canonical statistic $t(\boldsymbol{y}) = \sum y_i$, with the same canonical parameter θ as in (2.4), see Proposition 1.2. In order to find the distribution for t we need the structure function $g(t)$, the calculation of which might appear somewhat complicated from its definition. It is well-known, however, from the reproductive property of the Poisson distribution, that t has also a Poisson distribution, with expected value $n\lambda$. Hence,

$$f(t;\lambda) = \frac{(n\lambda)^t}{t!} e^{-n\lambda} = \frac{n^t}{t!} e^{-n e^{\theta}} e^{\theta t}, \quad t = 0, 1, 2, \dots . \tag{2.5}$$

Here we can identify the structure function g as the first factor (cf. Exercise 1.2).

For the extension to spatial Poisson data, see Section 13.1.1. △

Example 2.4 ***The multinomial distribution***
Suppose $\boldsymbol{y} = \{y_1, \dots, y_k\}$ is multinomially distributed, with $\sum y_j = n$, that is, the y_j are Bin(n, π_j)-distributed but correlated via the constrained sum. If there are no other restrictions on the probability parameter vector $\boldsymbol{\pi} = (\pi_1, \dots, \pi_k)$ than $\sum \pi_j = 1$, we can write the multinomial probability as

$$f(y; \boldsymbol{\pi}) = \frac{n!}{\prod y_j!} \prod_{j=1}^{k} \pi_j^{y_j} = \frac{n!}{\prod y_j!} e^{\sum_j y_j \log(\pi_j/\pi_k) + n \log \pi_k}. \tag{2.6}$$

Note here that the y_k term of the sum in the exponent vanishes, since $\log(\pi_k/\pi_k) = 0$. This makes the representation in effect $(k-1)$-dimensional, as desired. Hence we have an exponential family for which we can select $(y_1, \dots, y_{k-1})$ as canonical statistic, with the corresponding canonical parameter vector $\boldsymbol{\theta} = (\theta_1, \dots, \theta_{k-1})$ for $\theta_j = \log(\pi_j/\pi_k)$. After some calculations we can express $C(\boldsymbol{\theta}) = \exp(-n \log \pi_k)$ in terms of $\boldsymbol{\theta}$ as

$$C(\boldsymbol{\theta}) = \left(1 + \sum_{j=1}^{k-1} e^{\theta_j}\right)^n . \tag{2.7}$$

By formally introducing $\theta_k = \log(\pi_k/\pi_k) = 0$, we obtain symmetry and can write $\boldsymbol{\theta}^T \boldsymbol{t} = \sum_{j=1}^{k} \theta_j y_j$ and

$$C(\boldsymbol{\theta}) = \left(\sum_{j=1}^{k} e^{\theta_j}\right)^n . \tag{2.8}$$

The index choice k for the special role in the parameterization was of course arbitrary, we could have chosen any other index for that role.

If there is a smaller set of parameters of which each θ_j is a linear function, this smaller model is also an exponential family. A typical example follows next, Example 2.5. For a more intricate application, see Section 9.7.1 on Cox regression. △

Example 2.5 ***Multiplicative Poisson and other log-linear models***
For two-way and multiway contingency tables the standard models are based on either the Poisson or the multinomial distribution. We here primarily discuss Poisson-based models, and restrict consideration to two-way tables. The observation numbers in the different cells are then independent and Poisson distributed. In a *saturated* model for the table, each cell has its own free Poisson parameter. In a statistical analysis we look for more structure, and a fundamental hypothesis is the one of row–column independence (multiplicative Poisson). In a two-way table $\boldsymbol{y} = \{y_{ij}\}$, $i = 1, \ldots, r$; $j = 1, \ldots, s$, the saturated Poisson model has the probability function

$$f(\boldsymbol{y}; \boldsymbol{\lambda}) = \prod_{ij} \frac{\lambda_{ij}^{y_{ij}}}{y_{ij}!} e^{-\lambda_{ij}} = \frac{1}{\prod_{ij} y_{ij}!} e^{-\sum \lambda_{ij}} e^{\sum y_{ij} \log \lambda_{ij}} . \tag{2.9}$$

In analogy with the single Poisson, it is seen that the saturated model is an exponential family with the whole table $\boldsymbol{y}$ as canonical statistic and

$$\boldsymbol{\theta}^T \boldsymbol{t} = \sum_{ij} \theta_{ij} y_{ij} = \sum_{ij} y_{ij} \log \lambda_{ij} . \tag{2.10}$$

The multiplicative model is characterized by the factorization

$$\lambda_{ij} = \alpha_i \beta_j \tag{2.11}$$

for some parameters α_i and β_j, or equivalently

$$\theta_{ij} = \log \alpha_i + \log \beta_j , \tag{2.12}$$

that is, log-additivity (independence). This structure represents an exponential family with an $(r + s - 1)$-dimensional canonical statistic formed by the row sums $y_{i.}$ and column sums $y_{.j}$. This is seen by inserting the representation (2.12) in the exponent (2.10):

$$\sum_{ij} \theta_{ij} y_{ij} = \sum_i \log(\alpha_i) y_{i.} + \sum_j \log(\beta_j) y_{.j} , \tag{2.13}$$

Since the row sums and column sums have the same total sum, the row sums and column sums are linearly dependent, and in order to avoid over-parameterization we may for example set $\alpha_1 = 1$, or impose some other parameter restriction. This explains the statement after (2.12), that the canonical statistic is (r+s-1)-dimensional.

The multinomial case is obtained by conditioning on the total $\sum_i \sum_j y_{ij} = \sum_i y_{i.} = \sum_j y_{.j}$. This introduces another linear constraint on the components of $\boldsymbol{t}$, and thereby reduces the dimension to $rs-1$ in the saturated model (just as described in Example 2.4), and to $r + s - 2$ in the log-additive model of independence on the right-hand side of (2.13). See texts on log-linear models for more details, for example Agresti (2013), and Section 5.2 and other examples below. With multiway data more models are possible in between the saturated model and the completely log-additive model. A particularly nice and useful such model class is formulated in terms of *conditional independence* between two variates given any outcome of the other variates. Such models are called *graphical models*, because a suitable form for describing the conditional independence structure is by a conditional independence graph representation, see Section 10.3, but also Example 2.11 on covariance selection models, which are the analogues within the normal distribution framework. △

Example 2.6 ***'Time to first success' (geometric distribution)***
The number of Bernoulli trials up to and including the first success, with probability π_0 for success, has the probability function

$$f(y; \pi_0) = (1 - \pi_0)^{y-1}\, \pi_0, \quad y = 1, 2, \dots . \tag{2.14}$$

This forms a linear exponential family with canonical parameter $\theta = \log(1 - \pi_0)$ and $t(y) = y$. Alternatively and equivalently, we could let $t(y) = y - 1$, that is, the number of unsuccessful trials, which is said to be *geometrically* distributed. The form of the norming constant $C(\theta)$ depends on the choice of version. In the former case,

$$C(\theta) = \frac{(1 - \pi_0)}{\pi_0} = \frac{e^\theta}{(1 - e^\theta)}, \tag{2.15}$$

and in the geometric representation,

$$C(\theta) = 1/\pi_0 = 1/(1 - e^\theta). \tag{2.16}$$

The canonical parameter space Θ is the open interval $(-\infty, 0)$, corresponding to $0 < \pi_0 < 1$, cf. (1.1).

For samples from this distribution, see Exercise 2.2 and Example 4.2. △

Example 2.7 ***Exponential and gamma distribution models***
The gamma distribution family with two parameters (α for scale and β for shape) has the probability density

$$f(y; \alpha, \beta) = \frac{\alpha^{\beta+1}}{\Gamma(\beta + 1)}\, y^\beta\, e^{-\alpha y}, \quad y > 0. \tag{2.17}$$

This forms an exponential family with canonical statistic $t(y) = (y, \log y)$, canonical parameter $\boldsymbol{\theta} = (\theta_1, \theta_2) = (-\alpha, \beta)$, and the canonical parameter space $\boldsymbol{\Theta}$ being the open set $\theta_1 < 0$, $\theta_2 > -1$. For a sample from this distribution, the canonical statistic is $\boldsymbol{t}(\boldsymbol{y}) = (\sum y_i, \sum \log y_i)$, with the same θ, see Proposition 1.2. The structure function $g(\boldsymbol{t})$ of Proposition 1.3 cannot be expressed in a closed form.

The exponential distribution, formula (1.1), is the one-parameter subfamily obtained when setting $\theta_2 = 0$. It corresponds to a reduction of the minimal sufficient statistic from $(\sum y_i, \sum \log y_i)$ to $\sum y_i$ alone. In this subfamily, the structure function has an explicit form, see Exercise 1.2. △

Example 2.8 ***A sample from the normal distribution***
We have already seen in formula (1.2) that for a sample from a $\mathrm{N}(\mu, \sigma^2)$ distribution we have an exponential family with $\boldsymbol{t}(\boldsymbol{y}) = (\sum y_i, \sum y_i^2)$ and $\boldsymbol{\theta} = (\theta_1, \theta_2) = (\mu/\sigma^2, -1/(2\sigma^2))$. The canonical parameter space is the open half-plane $\theta_2 < 0$ ($\sigma^2 > 0$). As minimal sufficient statistic we can take any one-to-one function of $\boldsymbol{t}$, and $(\bar{y}, s^2)$ or $(\bar{y}, s)$ is often a more convenient choice.

Under the hypothesis of a specified mean-value, $\mu = \mu_0$, the dimension of the model is reduced from 2 to 1, in the following way. First note that θ_1 is then proportional to θ_2, namely $\theta_1 = -2\mu_0\theta_2$, so the hypothesis is linear in the canonical parameters. Inserting this expression for θ_1, we obtain an exponential family with the previous θ_2 as canonical parameter and a linear combination of the components t_1 and t_2 as the corresponding canonical statistic, $-2\mu_0 y + y^2$. A more natural choice is $(y - \mu_0)^2$, obtained by the addition of the known constant μ_0^2. This could (of course) also have been seen directly in the usual form for the density of $\mathrm{N}(\mu_0, \sigma^2)$.

We will not make explicit use of the structure function for the normal distribution, so we do not give the formula for it. However, it will play an implicit role, by underlying for example the density of the χ^2 and Student's t distributions. See also Section 6.1 for a more general type of formula. △

Example 2.9 ***Gaussian linear models***
We now relax the assumption in Example 2.8 of a common mean value. Consider a standard Gaussian linear model, with mean value vector $\boldsymbol{\mu} = A\boldsymbol{\beta}$ for a specified $n \times k$ design matrix A of full rank, $n > k$, under the assumption of a normally distributed error term, mutually independent and of constant variance σ^2; Jørgensen (1993) is one of numerous references. The probability density for the observed data set $\boldsymbol{y}$ is then

$$f(\boldsymbol{y}; \boldsymbol{\beta}, \sigma^2) = \left(\frac{1}{\sigma\sqrt{2\pi}}\right)^n e^{-\frac{|\boldsymbol{y} - A\boldsymbol{\beta}|^2}{2\sigma^2}}. \tag{2.18}$$

Here $|y - A\beta|^2 = (y - A\beta)^T(y - A\beta)$ is the squared length of the vector $y - A\beta$, and it may be expanded as

$$|y - A\beta|^2 = y^T y - 2(A\beta)^T y + (A\beta)^T(A\beta). \tag{2.19}$$

The middle term may be rearranged:

$$(A\beta)^T y = \beta^T(A^T y).$$

Hence, we have an exponential family with canonical statistic t consisting of the vector $A^T y$ and the sum of squares $y^T y$). This t is minimal sufficient, of dimension $\dim(\beta) + 1$. Note that the well-known least squares (also ML) estimator $\hat{\beta} = (A^T A)^{-1}(A^T y)$ of β is a one-to-one function of $A^T y$. △

Example 2.10 ***The multivariate normal distribution***
If y is a k-dimensional multivariate normally distributed (column) vector, with arbitrary mean vector μ and nonsingular but otherwise arbitrary covariance matrix Σ, that is, from $N_k(\mu, \Sigma)$, then y has the density

$$f(y; \mu, \Sigma) = (\det \Sigma)^{-1/2}(2\pi)^{-k/2}\, e^{-\frac{1}{2}(y-\mu)^T \Sigma^{-1} (y-\mu)}. \tag{2.20}$$

The quadratic form of the exponent can be expanded to read (cf. (2.19))

$$(y - \mu)^T \Sigma^{-1} (y - \mu) = y^T \Sigma^{-1} y - 2\mu^T \Sigma^{-1} y + \mu^T \Sigma^{-1} \mu. \tag{2.21}$$

In the first term the data vector y is represented by all its squares and pairwise products, each with a parameter associated with it. For a sample of observations y_i, $i = 1, \ldots, n$, from this distribution, we let $y = y_i$ in (2.21) and sum (2.21) over i. We see that we have an exponential family with canonical statistic t consisting of all sums ($\sum y_i$, from the middle term), and all sums of squares and sums of products (from the first term). The latter can be represented by the elements of the matrix $\sum y_i y_i^T$. The canonical parameter θ is represented by the vector $\mu^T \Sigma^{-1}$ together with the matrix Σ^{-1}, more precisely $-\frac{1}{2}$ its diagonal and minus its upper (or lower) nondiagonal part. The matrix Σ^{-1} is often called the *concentration matrix*. Since there is a one-to-one relationship between the two positive definite matrices Σ and Σ^{-1}, the covariance matrix Σ and the concentration matrix Σ^{-1} yield equivalent parameterizations. The canonical parameter set Θ is an open subset of $\mathbb{R}^{k+k(k+1)/2}$.

Primarily for later use we also consider the conditional density for a subvector y_1 given y_2, in $y^T = (y_1^T\; y_2^T)$. Division of the joint density (2.20) by the marginal (normal) density for y_2 will not change the y_1-dependent terms in the exponent (linear and quadratic in y_1). Thus, the conditional distribution is also normal. Let us write the partitioning of Σ^{-1} as

$$\Sigma^{-1} = \begin{pmatrix} \Sigma^{11} & \Sigma^{12} \\ \Sigma^{21} & \Sigma^{22} \end{pmatrix}$$

The quadratic form in $\boldsymbol{y}_1$ appearing in the exponent (2.21) can then be written $\boldsymbol{y}_1^T \Sigma^{11} \boldsymbol{y}_1$. Completing the quadratic when $\boldsymbol{y}_2$ is given and fixed yields

$$(y_1 - \mu_1 + (\Sigma^{11})^{-1}\Sigma^{12}(y_2 - \mu_2))^T\, \Sigma^{11}\, (y_1 - \mu_1 + (\Sigma^{11})^{-1}\Sigma^{12}(y_2 - \mu_2)). \quad (2.22)$$

(Reader, check, by expanding (2.22)!)

Formula (2.22) immediately tells three interesting facts:

- The conditional variance matrix of $\boldsymbol{y}_1$ is $(\Sigma^{11})^{-1}$, that is, invert Σ, pick out the upper left submatrix (for $\boldsymbol{y}_1$), and invert back.
- The conditional mean value, i.e. the multivariate theoretical regression of $\boldsymbol{y}_1$ on $\boldsymbol{y}_2$ is $\boldsymbol{\mu}_1 - (\Sigma^{11})^{-1}\Sigma^{12}(y_2 - \mu_2)$.
- The variance matrix determinants satisfy $\det(\Sigma^{11}) = \det(\Sigma_{22})/\det(\Sigma)$, where Σ_{22} is the $\boldsymbol{y}_2$ submatrix of Σ.

Particularly useful to note is the conditional variance because it appears repeatedly under asymptotic normality in the theory of statistics, for example, in large sample tests of composite hypotheses. It is also the theoretical residual variance matrix in the multivariate linear regression. The regression of $\boldsymbol{y}_1$ on $\boldsymbol{y}_2$ can be rewritten in a more usual form, because

$$(\Sigma^{11})^{-1}\Sigma^{12} = -\Sigma_{12}\Sigma_{22}^{-1},$$

see (B.4) in Proposition B.1, and the residual variance $(\Sigma^{11})^{-1}$ is also

$$\mathrm{var}(\boldsymbol{y}_1 - \Sigma_{12}\Sigma_{22}^{-1}\boldsymbol{y}_2) = \Sigma_{11} - \Sigma_{12}\Sigma_{22}^{-1}\Sigma_{21}$$

see (B.3). Thus Proposition B.1 can be used to obtain two different expressions for the regression and for the residual variance, but alternatively their mutual identity could be turned into a proof of Proposition B.1.

Multivariate linear models are obtained when the Gaussian linear model of Example 2.9 is given a multivariate response variable. We postpone further discussion of such models to Example 7.4. △

Example 2.11 *Covariance selection models*

Suppose we have a sample from a multivariate normal distribution (with unrestricted mean vector) and want a more parsimonious model by reducing the number of parameters in its covariance matrix Σ. A small observed sample correlation might lead us to replace a covariance in Σ by a zero. Unfortunately, however, this will not in general simplify the model, in the sense of not reducing the minimal sufficient statistic. A model with a minimal statistic $\boldsymbol{t}$ that is of larger dimension than the parameter vector is typically not a simple model. We will discuss such models in Chapter 7 under the name of *curved exponential families*. Is there a better alternative? Yes, we could look for *conditional* independences between components of $\boldsymbol{y}$, given the values of the other components. Such independences are clearly interpretable. They are also easily specified in terms of Σ via its inverse Σ^{-1}, the concentration matrix, because a conditional independence is equivalent to a zero in the corresponding position in Σ^{-1}. This can be seen directly from (2.20), which factorizes as a function of y_i and y_j when $\left(\Sigma^{-1}\right)_{ij} = 0$. More efficiently it follows from the result of Example 2.10 that $\left(\Sigma^{11}\right)^{-1}$ is the conditional variance–covariance matrix of the subvector $\boldsymbol{y}_1$, given the rest of $\boldsymbol{y}$, since Σ^{11} is diagonal precisely when $\left(\Sigma^{11}\right)^{-1}$ is diagonal.

Thus, setting off-diagonal elements of Σ^{-1} to zero specifies simple submodels of the saturated model (2.20). From (2.21) it is seen that a zero in Σ^{-1} corresponds to elimination of the corresponding sum of products from $\boldsymbol{t}$. The resulting type of models, with zeros in Σ^{-1}, was introduced by Dempster (1972) under the name of *covariance selection models*. They are analogous to graphical log-linear models for categorical data, and nowadays usually called *graphical Gaussian models*. For an introduction to this theory, see Chapter 10, in particular Section 10.2. △

Example 2.12 ***Exponential tilting families***
This construction is perhaps of more theoretical than practical importance, used for example in connection with saddlepoint approximations, see Section 4.2. Let $f_0(y)$ be any given distribution density. This distribution will now be *embedded* in an exponential family of distributions by so called *exponential tilting*. We form

$$f(y;\theta) \propto e^{\theta y}\, f_0(y), \tag{2.23}$$

being > 0 where $f_0(y) > 0$, and having the norming constant

$$C(\theta) = \int e^{\theta y} f_0(y)\, \mathrm{d}y\,, \tag{2.24}$$

where this integral (the Laplace transform of f_0) exists. More generally, y in $\exp(\theta y)$ may be replaced by some other statistic $t(y)$, that is,

$$f(y;\theta) \propto e^{\theta\, t(y)}\, f_0(y). \tag{2.25}$$

The parameter space Θ of (2.23) is where the integral (2.24) is finite. Note that $\theta = 0$, that is, $f_0(y)$ itself, always belongs to Θ.

A special case, in Davison (2003) called the *uniform density family*, is when f_0 represents the uniform distribution on the unit interval (0, 1). The family (2.23) then contains all truncated exponential distributions, truncated at $y = 1$ $(\theta < 0)$, but also their mirror images increasing exponentially from 0 to 1 $(\theta > 0)$, with the uniform distribution itself in between $(\theta = 0)$. Using instead in (2.25) the uniform distribution in combination with $t(y) = \mathrm{logit}(y)$, we obtain the beta family, see Exercise 2.4.

The name uniform density family is unfortunate, however, and should better be avoided. It invites the misunderstanding (even seen in textbooks!) that the class of uniform distributions forms an exponential family.

Efron and Tibshirani (1996) used the construction in a multiparameter version, with a canonical statistic vector $t(y)$ instead of only y, and with the function f_0 being a (kernel) density estimate of an arbitrary probability density, This estimate was then refined by ML estimation within the tilted family to fit the mean and other moments of the sample data (Example 3.26). △

2.2 Examples Less Important for the Sequel

Example 2.13 ***Finite Markov chains***

Suppose we observe a single realization of a finite Markov chain, from time 0 to a fixed time n. For simplicity of notation and calculations, let it have only two states, 0 and 1. Let the unknown transition matrix be

$$\Gamma = \begin{pmatrix} \gamma_{00} & \gamma_{01} \\ \gamma_{10} & \gamma_{11} \end{pmatrix}.$$

Thus, the one-step transition probabilities are $Pr\{0 \to 1\} = \gamma_{01} = 1 - \gamma_{00}$ and $Pr\{1 \to 0\} = \gamma_{10} = 1 - \gamma_{11}$, so the parameter dimension of Γ is 2.

The probability for observing a specific realized sequence, say $0 \to 0 \to 1 \to 0 \to 0 \cdots$, is the corresponding product of transition matrix probabilities, $\gamma_{00}\gamma_{01}\gamma_{10}\gamma_{00}\cdots$, or more generally,

$$\prod_{i=1}^{2}\prod_{j=1}^{2} \gamma_{ij}^{n_{ij}} = \exp\left\{\textstyle\sum_i \sum_j (\log \gamma_{ij})\, n_{ij}\right\}, \tag{2.26}$$

where n_{ij} is the observed number of consecutive pairs (i, j) (the four possible transitions $i \to j$, including the case $i = j$).

The four n_{ij}-values satisfy the restriction $\sum_{ij} n_{ij} = n$, and it might appear as if there are three linearly independent statistics but only two parameters. However, there is a near equality $n_{01} \approx n_{10}$, because after one transition the next transition must be in the opposite direction. More precisely, say that the chain starts in state 0, then $n_{01} - n_{10} = 0$ or $= 1$. The outcome probability depends on the two parameters, but intuitively there is very little additional information in that outcome, at least for large n.

In analogy with the binomial we have one set of $n_{0\cdot} = n_{00} + n_{01}$ Bernoulli trials when the chain is in state 0, with transition probability γ_{01} $(= 1 - \gamma_{00})$, and another set of $n_{1\cdot} = n_{10} + n_{11}$ trials with another transition probability γ_{10} $(= 1 - \gamma_{11})$. The main difference from the binomial is that the numbers $n_{0\cdot}$ and $n_{1\cdot} = n - n_{0\cdot}$ of these trials are random, and the distributions of $n_{0.}$ and $n_{1.}$ do in fact depend on the γ-values. Another analogue is the time to first success situation (Example 2.6; or the negative binomial, see Exercise 2.2). Bernoulli trials are carried out until the first success, when a transition to the other state occurs. There the situation is repeated with a different probability for success, until the next transition, which means the chain restarts from the beginning. The inferential complication of the negative binomial when the numbers of successes n_{01} and n_{10} are regarded as fixed is that the number n of repetitions then is random.

Note that the probability (2.26), regarded as a likelihood for Γ, looks the same as it would have done in another model, with fixed numbers $n_{0\cdot}$ and $n_{1\cdot}$ of Bernoulli trials. In particular, the ML estimator of Γ is the same in the two models. △

Example 2.14 ***Von Mises and Fisher distributions for directional data***
The von Mises and Fisher distributions, sometimes referred to together under the name *von Mises–Fisher model*, are distributions for directions in the plane and in space, respectively. Such directions can be represented by vectors of length one, that is, vectors on the unit circle and on the unit sphere, respectively. Therefore they fall outside the main stream here, by neither being counts or k-dimensional vectors in $\mathbb{R}^k$. Applications are found in biology, geology, meteorology and astronomy. Both distributions have a density of type

$$f(\boldsymbol{y}; \boldsymbol{\theta}) = \frac{1}{C(\boldsymbol{\theta})} \exp(\boldsymbol{\theta}^T \boldsymbol{y}),$$

where the exponent is the scalar product between the observed direction vector $\boldsymbol{y}$ on the unit circle or sphere and an arbitrary parameter vector $\boldsymbol{\theta}$ in $\mathbb{R}^2$ or $\mathbb{R}^3$, respectively. The special case $\boldsymbol{\theta} = \boldsymbol{0}$ yields a uniformly distributed direction.

The *von Mises distribution* is a symmetric, unimodal model for a sample of directions in the plane. If we represent both $\boldsymbol{\theta}$ and $\boldsymbol{y}$ in polar coordinates, $\boldsymbol{\theta} = (\rho \cos \psi, \rho \sin \psi)$ and $\boldsymbol{y} = (\cos z, \sin z)$, $0 \leq z < 2\pi$, their scalar product $\boldsymbol{\theta}^T \boldsymbol{y}$ in the exponent may be written

$$\boldsymbol{\theta}^T \boldsymbol{y} = \rho \{\cos z \cos \psi + \sin z \sin \psi\} = \rho \cos(z - \psi)$$

The direction ψ is the mean direction, whereas ρ is a measure of the degree of variability around ψ, high values representing small variability. The norming constant per observation is the integral

$$\int_0^{2\pi} e^{\rho \cos(z-\psi)}\, dz = \int_0^{2\pi} e^{\rho \cos z}\, dz = 2\pi\, I_0(\rho),$$

where I_0 is called the *modified Bessel function of the first kind and order 0*. Note that the norming constant is free from ψ. Except for this simplification, the norming constant and the structure function are analytically complicated. We will not further discuss this model, but the interested reader is referred to Martin-Löf (1970) and Mardia and Jupp (2000).

Fisher's distribution for directions in space (Fisher, 1953), is the corresponding density when $\boldsymbol{y}$ is the direction vector on the sphere, and $\boldsymbol{\theta}$ is a parameter vector in $\mathbb{R}^3$. In this case, the expression in polar coordinates becomes somewhat longer and more complicated, so we abstain from details and refer to Mardia and Jupp (2000), see also Diaconis (1988, Sec. 9B). As in two dimensions, the density is symmetrical around the mean direction. This is the direction of $\boldsymbol{\theta}$, and the length of $\boldsymbol{\theta}$ is a concentration parameter. △

Example 2.15 ***Maxwell–Boltzmann model in statistical physics***
Already in 1859 James Clerk Maxwell gave the distribution of kinetic energy among particles in an ideal gas under thermal equilibrium, later established and extended by Ludwig Boltzmann. This distribution now goes under the name Maxwell–Boltzmann distribution. On the so-called microcanonical scale the particles are assumed to interact and change velocities by collisions but move with constant velocity vector between collisions. The Maxwell–Boltzmann distribution can then be derived as describing on the so-called canonical scale the distribution of velocity vectors v among the particles in the gas. Let $v = (v_1, v_2, v_3) \in \mathbb{R}^3$, where the components are the velocities (with sign) in three orthogonal directions. Then the Maxwell–Boltzmann distribution for the vector v is given by the density

$$f_v(v_1, v_2, v_3;\, T) = \left(\frac{m}{2\pi kT}\right)^{3/2} e^{-\frac{m\,|v|^2}{2kT}}, \tag{2.27}$$

where $|v|^2 = v^T v = v_1^2 + v_2^2 + v_3^2$ is the speed squared, neglecting its direction,

m is the particle mass (assumed known), and k and T are other constants (see the next paragraph for their interpretations).

It is clear that the density (2.27) is a three-dimensional Gaussian distribution for mutually independent components v_1, v_2 and v_3, each being $N(0, \sigma^2)$ with $\sigma^2 = kT/m$. It is also clear that we have an exponential family with $|v|^2$ as canonical statistic. Equivalently we could use the *kinetic energy* $E = m|v|^2/2$ as canonical statistic, with the corresponding canonical parameter $\theta = -1/(kT)$. Here T is the *thermodynamic temperature* of the gas, which is regarded as a parameter, whereas k is the *Boltzmann constant*, whose role is to transform the unit of temperature to the unit of energy. In statistical thermodynamics the notation β for $-\theta = 1/(kT)$ is standard, called the *thermodynamic beta*.

From the normal distribution for v we can for example easily find the mean energy per particle, in usual physics notation $\langle E \rangle$, as

$$\langle E \rangle = \frac{m}{2}\langle |v|^2 \rangle = \frac{m}{2}\, 3\sigma^2 = \frac{3}{2}\, kT\,.$$

Note that in thermodynamics the only available measurements will typically be of macroscopic characteristics such as temperature, thus representing the mean value $\langle E \rangle$. The sample size will be enormous, a laboratory quantity of gas of reasonable density containing say $n = 10^{20}$ particles, so the corresponding sampling error will be negligible.

The whole distribution of $E = m|v|^2/2$ among particles can also easily be deduced from the normal distribution (2.27) of v. In principle it means that we must calculate the structure function $g(E)$, a task that might appear difficult (includes finding the surface area of a sphere with given radius). In the present case, however, we can immediately conclude without calculations that the distribution of E is proportional to a $\chi^2(3)$. It follows as a consequence that the density for E is

$$f_E(E;\, T) = 2\sqrt{\frac{E}{\pi (kT)^3}}\; e^{-E/kT}, \qquad E > 0. \tag{2.28}$$

The density for $E/(kT)$ is illustrated in Figure 2.1. Similarly the density for other related characteristics can be found from (2.27) by variable transformation, for example the density for the speed $|v|$.

The terms *Boltzmann distribution* and *Boltzmann's law* represent a generalization in several directions. For example, the gas particles may be in a gravitational field, and the energy variable may be generalized to allow discrete energy levels E_j, as in the Ising model introduced in the next example. The resulting model is still an exponential family, but instead of

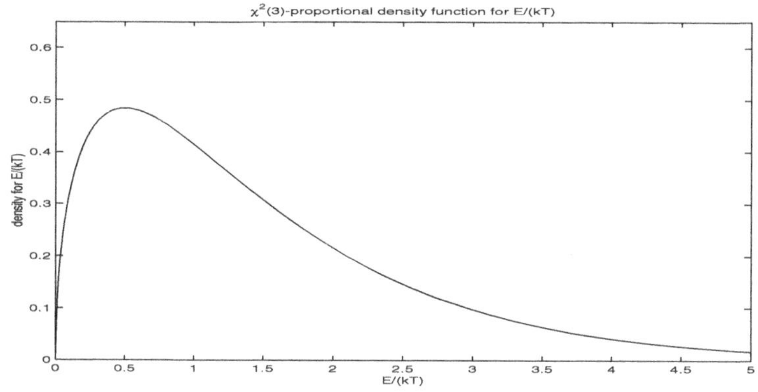

Figure 2.1 Illustration of the $\chi^2(3)$-proportional Maxwell–Boltzmann density for the energy E, normalized by kT.

(2.27) we now have, for state j, the probability

$$f(j;\,T) = \frac{1}{C(-\frac{1}{kT})}\,e^{-\frac{E_j}{kT}}, \tag{2.29}$$

where the norming constant $C(-1/(kT)) = \sum_j e^{-E_j/(kT)}$ is the sum over all possible states. In statistical physics this sum is denoted $Z(T)$ and is called the *partition function*. To calculate the partition function in various types of situations is a central problem in statistical physics. Note that the number of states is typically huge. One difference from the models we have met earlier in this chapter is that there need not be an explicit expression for the sum $Z(T)$. This is a problem that is shared with several statistical model types, for example social network models (Chapter 11) and models for processes in space or time (Chapter 13).

For a standard treatment of Boltzmann distributions in physics, including the Ising model, see for example Huang (1987). △

Example 2.16 ***The Ising model***

The basic Ising model is a model for binary response data, allowing interaction between the particles/items/individuals of a population. Variations of it are widely used in many fields, e.g. for interacting cells in biology or individuals in sociology. Historically, it was proposed by W. Lenz and investigated by Ernst Ising in the 1920s as a model for ferromagnetism, to explain some of the behaviour of magnetic materials. The binary variable is then the spin of an electron, which can have two different directions ('up' or 'down'). The n electrons are assumed to be located in a grid net (lattice), with interactions primarily between neighbours. The model type has inspired quite wide development and usage, not only within physics, material science (e.g. binary alloys), and image analysis, but also to model dynamically so different biological examples as bird flocks, fish shoals, networks of firing neurons and beating heart cells, and in sociology, for example, voting behaviour. Like in Example 2.15, the model probability for the state of the system, now being a spin configuration $\{s_i\}$ under thermal equilibrium, is given by a Boltzmann distribution,

$$f(s_1, \dots, s_n;\, \theta) = \frac{1}{C(\theta)}\, e^{\theta\, E(s_1, \dots, s_n)}. \tag{2.30}$$

Here the canonical statistic E will be specified in (2.31), the canonical parameter θ is (as before) proportional to the inverse temperature $1/T$, and the norming constant (partition function) is the sum over all the 2^n possible spin configurations,

$$C(\theta) = \sum_{s_1=\pm 1, \dots, s_n=\pm 1} e^{\theta\, E(s_1, \dots, s_n)}.$$

The energy E in (2.30), corresponding to the kinetic energy in Example 2.15, is specified as

$$E = \sum_{\text{neighbours i, j}} E_{i,j}, \tag{2.31}$$

with

$$E_{i,j} = \begin{cases} -J\, s_i s_j & \text{when } i \text{ and } j \text{ are neighbours} \\ 0 & \text{otherwise} \end{cases}$$

The product $s_i s_j = \pm 1$, the sign depending on whether the two spins are the same or opposite, and J is a constant ($J > 0$ for ferromagnetic materials). By assuming the energy measured in units of J, we can and will forget J.

This basic Ising model turns out to behave quite differently depending on the number of neighbours, corresponding to the grid net dimension. In a one-dimensional model, with the electrons consecutively ordered along a line, each grid point has only two neighbours, and we may write

$$E = -\sum_i s_i s_{i+1}.$$

In a two-dimensional model, each position has four neighbours, and with explicit double index on s we can write

$$E = -\sum_{i,j} \left\{ s_{i,j} s_{i+1,j} + s_{i,j} s_{i,j+1} \right\}$$

The two-dimensional Ising model is much more intricate that its one-dimensional analogue. Not until twenty years after Ising's analysis of the one-dimensional model, it was understood by Onsager that the two-dimensional model has a phase transition temperature, implying that if the temperature is changed near the critical value, the spin configuration appearance will change considerably from unordered to ordered or vice versa. Nowadays, there is a vast literature about the phase transition properties of the Ising model.

The model can be extended and modified in various ways. One extension is to allow an external magnetic field, represented by an additional component $\sum_i s_i$ of the canonical statistic. This, and other statistics linear in $\{s_i\}$, can be natural to try in future applications. △

2.3 Exercises

Exercise 2.1 ***Weighted multiplicative Poisson***
The standard multiplicative Poisson model has Poisson parameters $\lambda_{ij} = \alpha_i \beta_j$. In many applications there is a need to include a known weights factor N_{ij} (often a population size), and write

$$\lambda_{ij} = N_{ij} \alpha_i \beta_j.$$

The oldest uses of the model were in Swedish actuarial statistics, but here we refer to E.B. Andersen (1977), who developed inference in terms of exponential families. He applied the model on an epidemiological set of data, representing death numbers in lung cancer in four Danish cities, divided in six agegroups. Death numbers are expected to be proportional to population size, and the model was used with N_{ij} being the registered number of inhabitants in the corresponding city and agegroup.

Characterize this family in the same way as the standard family in Example 2.5, that is, find its canonical statistic and canonical parameter, and consider the parameter dimensionality. *(Continued in Exercise 3.21)* △

Exercise 2.2 ***Special negative binomial***
Consider the number of Bernoulli trial failures until the k-th success. This has a negative binomial distribution with success parameter π_0, say. Characterize this one-parameter distribution as an exponential family. *(Continued in Exercise 3.20)* △

Exercise 2.3 ***Logarithmic distribution***
Consider the family with probability function

$$f(y;\,\psi) \propto \frac{\psi^y}{y}, \qquad y = 1, 2, 3, \dots,$$

Find the canonical parameter and its parameter space, and the norming constant. Plot the probability function for some parameter value, at least schematically. See also Example 8.7. △

Exercise 2.4 ***The beta distribution family***
The beta distribution family has two parameters and its density is usually expressed as proportional to

$$y^{\alpha-1}\,(1-y)^{\beta-1}\,,\ 0 < y < 1.$$

The norming constant is

$$B(\alpha,\beta) = \frac{\Gamma(\alpha)\,\Gamma(\beta)}{\Gamma(\alpha+\beta)}\,.$$

Find a version of the canonical statistic and the corresponding canonical parameter and parameter space. △

Exercise 2.5 ***Rayleigh distribution***
If y has the density

$$f(y;\,\sigma^2) = \frac{y}{\sigma^2}\,\exp\left(-\frac{y^2}{2\sigma^2}\right) \qquad y \geq 0,$$

it is said to be Rayleigh distributed. This distribution is an analogue in only two dimensions to the Maxwell–Boltzmann distribution (2.27), and it has some applications in physics. Characterize the distribution type as an exponential family. Find out how it is related to a particular χ^2 distribution. △

Exercise 2.6 ***Inverse Gaussian***
The *inverse Gaussian* (IG) distribution family will repeatedly reappear (Exercises 3.4, 3.25, 4.6, 9.7). The family is not only of theoretical interest, but also of some practical ditto. The reason for the name 'inverse' is that this distribution represents the time required for a Brownian motion with positive drift to reach a certain fixed (> 0) level, in contrast to the ordinary Gaussian for the level after a fixed time. The two-parameter density is

$$f(y;\,\mu,\alpha) = \sqrt{\frac{\alpha}{2\pi y^3}}\,\exp\left(-\frac{\alpha(y-\mu)^2}{2\mu^2 y}\right) \qquad y, \mu, \alpha > 0, \tag{2.32}$$

where μ is in fact the mean value of the distribution (to be shown in Exercise 3.25).

(a) Suppose data $(y_1, ..., y_n)$ are modelled as a sample from this distribution, the IG(μ, α). Characterize the model as an exponential family by finding its canonical statistic, canonical parameter vector, and canonical parameter space. For later use, find an expression for the norming constant $C(\theta)$.

(b) In the absence of drift in the Brownian motion, a one-parameter distribution family appears, the *Lévy distribution*. The parameter μ is inversely proportional to the drift, so it follows that the special case is obtained by letting $\mu \to \infty$ in (2.32). Carry out this limiting procedure and characterize the resulting distribution family.

Three IG densities are illustrated in Figure 2.2: IG(1, 1), IG(3, 1) and IG(1, 3). Note how insensitive the mode (i.e maximum point) is to the increase in the mean value parameter μ alone, whereas a change in α alone moves the mode of the density (but not the mean value). △

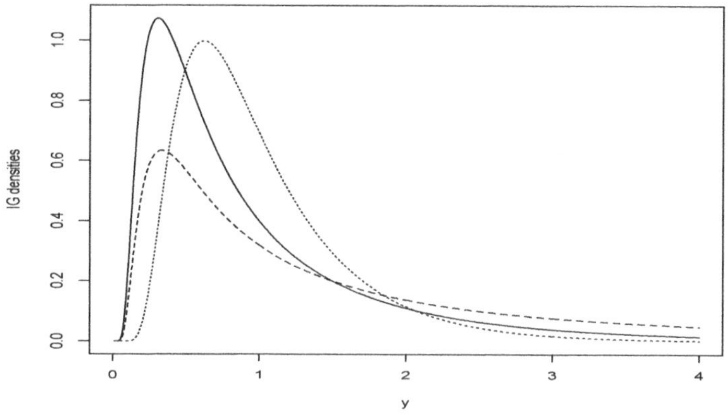

Figure 2.2 Densities for three IG distributions, IG(μ, α). Solid curve: IG(1, 1); Dashed: IG(3, 1); Dotted: IG(1, 3)

3

Regularity Conditions and Basic Properties

In this chapter, we introduce some regularity conditions and other concepts, and prove a number of basic results, in particular about ML estimation. We consider alternatives to the canonical parameterization, and argue that inference about canonical parameter components should be conditional inference. Other topics, taken up toward the end of the chapter, are Basu's theorem and the Cramér–Rao inequality.

3.1 Regularity and Analytical Properties

In Chapter 1, we assumed that $\boldsymbol{t}$ and $\boldsymbol{\theta}$ are 'in effect k-dimensional', and not less, provided they have k components. We make this statement precise by the following definition.

Definition 3.1 ***Order and minimality***
The *order* of an exponential family is the lowest dimension of the canonical statistic for which the family of distributions can be represented in the form (1.5) of Definition 1.1, with $\boldsymbol{\theta}$ in some set of parameters. The representation is said to be *minimal* if its canonical statistic t has this lowest dimension.

What can it look like when the representation is not minimal? One possibility is that the components of $\boldsymbol{t}$ are linearly (affinely[1]) dependent. An example is if we write the probability function for the multinomial model (see Example 2.4) on the form

$$f(y; \boldsymbol{\pi}) = \frac{n!}{\prod y_j!} \prod_{j=1}^{k} \pi_j^{y_j} = \frac{n!}{\prod y_j!} e^{\sum_{j=1}^{k} y_j \log \pi_j}. \tag{3.1}$$

and take $\boldsymbol{t} = (y_1, \ldots, y_k)$, $\boldsymbol{\theta} = (\log \pi_1, \ldots, \log \pi_k)$. Since $\sum y_j = n$ (given), the components of $\boldsymbol{t}$ vary in a $(k-1)$-dimensional hyperplane, so we do

[1] An affine transformation is a linear transformation in combination with a translation, so affinely independent means that there is no linear combination that is a constant.

not need more than $k-1$ of the k components to represent the distribution family. Correspondingly, the components of $\boldsymbol{\theta}$ are functionally dependent, $\sum_{j=1}^{k} e^{\theta_j} = 1$, so $\boldsymbol{\theta}$ is also in effect $(k-1)$-dimensional. Such situations can always be avoided by reduction of $\boldsymbol{t}$ in combination with a reparameterization of $\boldsymbol{\theta}$, see Example 2.4 as a typical illustration. An extension is when in a two-dimensional multiplicative Poisson table (Example 2.5) we reduce the table to the set of all row sums and the set of all column sums. These two sets are linearly connected by both of them summing to the table total. For symmetry reasons we might want, in some cases, to retain the redundant dimension in these examples.

Another possibility is that the components of $\boldsymbol{\theta}$ are linearly (affinely) dependent, $\sum c_j\theta_j = c_0$. After reparameterization we can assume that θ_k has a fixed value. Then t_k is redundant and should simply be omitted from $\boldsymbol{t}$, whose dimension is reduced. As soon as we want to test a hypothesis that is linear in the canonical parameters, we have this kind of situation, with a hypothesis model of reduced dimension, see Example 2.8 for an illustration. Exceptionally, however, this situation may occur unintended, in spite of a maximal $\boldsymbol{\Theta}$. As an example, admittedly artificial, let

$$f(y;\boldsymbol{\theta}) \propto \exp(\theta_1 y + \theta_2 y^2 + \theta_3 y^3) \tag{3.2}$$

for y on the whole real line. For $\theta_3 = 0$ and $\theta_2 < 0$ we have a Gaussian density, cf. Example 2.8, but for $\theta_3 \neq 0$ the density is not integrable. Hence the maximal canonical parameter space $\boldsymbol{\Theta}$ is a two-dimensional subspace of $\mathbb{R}^3$, and the component y^3 is meaningless and should be omitted from $\boldsymbol{t}$.

Hence, in order to check that a given representation of an exponential family is minimal,

- check that there is no linear (affine) dependency between the components of t, and
- check that the parameter space is not such that there is a linear (affine) dependency between the components of $\boldsymbol{\theta}$.

These two requirements together are precisely what is needed for minimality. It is usually a trivial task to carry out the checking.

Although symmetry can often be a motivation to allow redundant components, for example in the multinomial model, it is simpler from a general theoretical point of view to assume that we have a minimal representation. As described after (3.1), this can always be achieved. Therefore we take all representations of exponential families in this chapter to be minimal.

We also assumed in Chapter 1 that the parameter space for $\boldsymbol{\theta}$ was maximal (= canonical $\boldsymbol{\Theta}$), that is, comprising all $\boldsymbol{\theta}$ for which the norming constant $C(\boldsymbol{\theta})$ was finite. We combine this demand with the minimality of Definition 3.1 in the important concept of a *full family*:

Definition 3.2 ***Full families***
A minimal exponential family is called *full* if its parameter space is maximal, that is, equals the canonical space $\boldsymbol{\Theta}$.

Most of the theory to follow will only hold for full exponential families. When the family is not full, it is typically because the parameter space is a lower-dimensional 'curved' subset of the maximal, canonical space $\boldsymbol{\Theta}$ (mathematically a lower-dimensional manifold). Such families are called *curved exponential families*. Their treatment will be postponed to Chapter 7, because the statistical theory for curved families is much more complicated than for the full families. As a simple example, let the coefficient of variation of a normal distribution be assumed known, equal to 1 say. The model distribution is then $\mathrm{N}(\mu, \mu^2)$, $\mu > 0$, and its parameter space is represented by a curve in the canonical parameter set, the curve $\theta_2 = -\theta_1^2/2$ in the notations of Example 2.8. Even though this is a one-dimensional subspace of $\mathbb{R}^2$, the minimal sufficient statistic for this model is the same two-dimensional one as in the full model, as a consequence of Proposition 3.3.

Proposition 3.3 ***Minimal sufficiency of t***
If $\boldsymbol{y}$ *has the exponential family model (1.5), and the representation is minimal, then the statistic* $\boldsymbol{t} = \boldsymbol{t}(\boldsymbol{y})$ *is minimal sufficient for* $\boldsymbol{\theta}$.

Proof That $\boldsymbol{t}$ is sufficient follows immediately from the *factorization criterion*, see e.g. Pawitan (2001, Sec. 3.1), since the density f is the product of the parameter-free factor $h(\boldsymbol{y})$ and a factor that depends on data through $\boldsymbol{t}$ alone. For the minimality we must show that a ratio of the density in two points $\boldsymbol{y}$ and $\boldsymbol{z}$, that is, $f(\boldsymbol{y};\boldsymbol{\theta})/f(\boldsymbol{z};\boldsymbol{\theta})$, cannot be independent of $\boldsymbol{\theta}$ unless $\boldsymbol{t}(\boldsymbol{y}) = \boldsymbol{t}(\boldsymbol{z})$. This ratio is

$$\frac{f(\boldsymbol{y};\boldsymbol{\theta})}{f(\boldsymbol{z};\boldsymbol{\theta})} = \frac{h(\boldsymbol{y})}{h(\boldsymbol{z})}\, e^{\boldsymbol{\theta}^T\{\boldsymbol{t}(\boldsymbol{y}) - \boldsymbol{t}(\boldsymbol{z})\}}.$$

The minimality of the exponential family representation guarantees that there is no affine dependence between the components of $\boldsymbol{\theta}$, and hence the ratio cannot be independent of $\boldsymbol{\theta}$ unless $\boldsymbol{t}(\boldsymbol{y}) = \boldsymbol{t}(\boldsymbol{z})$, which was to be proved. □

A converse of Proposition 3.3 also holds, actually. We only sketch the theory. The property was conjectured by Fisher 1935 and soon afterwards proved independently by Darmois, Koopman and Pitman, whose names have therefore sometimes been used to name the exponential families. Suppose we have a sample of size n from a distribution family only known to satisfy a number of regularity conditions. The most important condition is that all distributions have the same support. This demand is of course satisfied by an exponential family – its support is the parameter-independent set where $h(y) > 0$.

Proposition 3.4 ***Dimension reduction implies exponential family***
Under parameter-free support and other regularity conditions, if there is a sufficient statistic of a dimension less than the sample size, the distribution family must be an exponential family (but not necessarily a full family).

For a precise formulation and a proof, see e.g. Kagan et al. (1973, Th. 8.4.1). Compare also with Boltzmann's law in Chapter 6.

Examples of distribution families of nonexponential type with parameter-free support are the location–scale Cauchy distribution and the Weibull distribution (with exponent of type y^α instead of $t(y)$). By Proposition 3.4, they both have the whole (ordered) sample as minimal sufficient statistic. The uniform distribution with location (or location and scale) as parameter(s) has a merely two-dimensional minimal sufficient statistic (minimum and maximum observations) even though it is not of exponential type – this is possible because it has a parameter-dependent support.

The next results will concern the canonical parameter space $\boldsymbol{\Theta}$, so we restrict considerations to full exponential families.

Proposition 3.5 ***Convexity of*** $\boldsymbol{\Theta}$
The canonical parameter space $\boldsymbol{\Theta}$ *is a convex set in* $\mathbb{R}^k$.

Proof We show that if the points $\boldsymbol{\theta}'$ and $\boldsymbol{\theta}''$ are in $\boldsymbol{\Theta}$, then any point on the line segment between $\boldsymbol{\theta}'$ and $\boldsymbol{\theta}''$ must also be in $\boldsymbol{\Theta}$, that is, $h(\boldsymbol{y})\exp(\boldsymbol{\theta}^T\boldsymbol{t}(\boldsymbol{y}))$ has finite integral or sum. Let $\boldsymbol{\theta} = w\boldsymbol{\theta}' + (1-w)\boldsymbol{\theta}''$ with $0 \leq w \leq 1$. The desired result follows from Hölder's inequality (which is essentially only the inequality between the geometric and arithmetic averages):

$$\int e^{\boldsymbol{\theta}^T \boldsymbol{t}(\boldsymbol{y})}\, h(\boldsymbol{y})\, d\boldsymbol{y} = \int \left(e^{\boldsymbol{\theta}'^T \boldsymbol{t}(\boldsymbol{y})}\, h(\boldsymbol{y})\right)^w \left(e^{\boldsymbol{\theta}''^T \boldsymbol{t}(\boldsymbol{y})}\, h(\boldsymbol{y})\right)^{1-w} d\boldsymbol{y}$$
$$\leq \left(\int e^{\boldsymbol{\theta}'^T \boldsymbol{t}(\boldsymbol{y})}\, h(\boldsymbol{y})\, d\boldsymbol{y}\right)^w \left(\int e^{\boldsymbol{\theta}''^T \boldsymbol{t}(\boldsymbol{y})}\, h(\boldsymbol{y})\, d\boldsymbol{y}\right)^{1-w} < \infty, \tag{3.3}$$

with the integrals exchanged for sums in the discrete case. □

The canonical parameter space $\boldsymbol{\Theta}$ is usually an *open* convex set. This was true for almost all examples of the previous chapter (the inverse Gaussian was an exception, see Exercise 3.4). The Bernoulli trials Example 2.1 is typical for understanding this feature: In the conventional parameterization by the probability parameter π_0 we could think of the distribution as defined on the closed interval $0 \leq \pi_0 \leq 1$, but the boundary points then represent degenerate, one-point distributions, not belonging to the exponential family as defined. When we transform π_0 to the canonical parameterization by the logit, they are mapped on the infinity points, and thus left out.

In the rare cases when $\boldsymbol{\Theta}$ is not an open set in $\mathbb{R}^k$, the maximum likelihood theory will be more complicated, so we will need a name for the nice models with an open canonical parameter set. They will be called *regular*.

Definition 3.6 ***Regular families***
An exponential family of order k is called *regular* if its canonical parameter space $\boldsymbol{\Theta}$ is an open set in $\mathbb{R}^k$.

Here is an example showing that exponential families are not necessarily regular. Let us construct an exponential family by taking $t(y) = y$ and

$$h(y) = 1/y^2, \; y > 1. \tag{3.4}$$

Since $h(y)$ is integrable by itself, the parameter space will include the boundary point zero, $\Theta = (-\infty, 0]$. In Section 3.2.1 we will define a concept *steep* to include the more well-behaving nonregular models.

On the other hand, here is a sufficient condition for regularity.

Proposition 3.7 ***Finite support in t implies regularity***
If the set of possible $\boldsymbol{t}$*-values is finite, a full family is necessarily regular, and the canonical parameter space is* $\boldsymbol{\Theta} = \mathbb{R}^{\dim t}$ *(unless* $\boldsymbol{\Theta}$ *is empty).*

Proof This result is quite obvious, since for each $\boldsymbol{\theta}$ the norming constant is just a finite sum of finite terms, unless the structure function happens to be infinite for some t. However, since $e^{\theta^T t} > 0$ whatever is $\boldsymbol{\theta}$, the latter case would imply that the norming constant did not exist for any $\boldsymbol{\theta}$. □

For an application of Proposition 3.7, see the comments following Section 14.3. Note that it is not sufficient for the result of Proposition 3.7 that the set of possible values of $\boldsymbol{t}$ is bounded, because in the continuous case the structure function $g(\boldsymbol{t})$ itself need not be integrable. But if it is, then the norming constant integral is finite for any $\boldsymbol{\theta}$, and hence the family is regular.

Most of the following results do not hold on the boundary of $\boldsymbol{\Theta}$, so by restricting interest to regular families, for which $\boldsymbol{\Theta}$ has no boundary, the presentation is simplified. However, there will also be results about maximum likelihood for which assumptions about the boundary are really needed, and for which a regular family is a sufficient condition.

Now we will see that there is much information contained in the form of the norming constant as a function of $\boldsymbol{\theta}$,

$$C(\boldsymbol{\theta}) = \int e^{\boldsymbol{\theta}^T t(\boldsymbol{y})}\, h(\boldsymbol{y})\, \mathrm{d}\boldsymbol{y}.$$

Proposition 3.8 ***Properties of*** $\log C(\boldsymbol{\theta})$
The function $\log C(\boldsymbol{\theta})$ *is strictly convex, and for a regular family arbitrarily many times differentiable, with gradient vector* $D \log C(\boldsymbol{\theta})$ *and second derivatives (=Hessian) matrix* $D^2 \log C(\boldsymbol{\theta})$ *given by*

$$D \log C(\boldsymbol{\theta}) = E_{\theta}(\boldsymbol{t}) = \boldsymbol{\mu}_t(\boldsymbol{\theta}) \tag{3.5}$$

$$D^2 \log C(\boldsymbol{\theta}) = \mathrm{var}_{\theta}(\boldsymbol{t}) = V_t(\boldsymbol{\theta}), \tag{3.6}$$

i.e. by the mean vector and the variance–covariance matrix, respectively, in two different notational variants.

Proof That $\log C(\boldsymbol{\theta})$ is convex is seen directly from inequality (3.3) by taking the logarithm. For the strict convexity, note that equality in (3.3) for some $0 < w < 1$ would require $(\boldsymbol{\theta}' - \boldsymbol{\theta}'')^T \boldsymbol{t} = 0$ (with probability 1). This is not possible for a family in minimal representation, however, unless $\boldsymbol{\theta}' = \boldsymbol{\theta}''$, so the inequality is strict, and hence the convexity is strict. This result will also follow from (3.6), as soon as this formula has been proved.

The function $\exp(\boldsymbol{\theta}^T \boldsymbol{t})$ of $\boldsymbol{\theta}$ is differentiable arbitrarily many times, and so smooth that we are allowed to differentiate $C(\boldsymbol{\theta})$ under the integral (or sum) sign. The result is immediate:

$$\frac{\partial C(\boldsymbol{\theta})}{\partial \theta_j} = E_{\theta}(t_j)\, C(\boldsymbol{\theta}), \tag{3.7}$$

$$\frac{\partial^2 C(\boldsymbol{\theta})}{\partial \theta_i\, \partial \theta_j} = E_{\theta}(t_i t_j)\, C(\boldsymbol{\theta}). \tag{3.8}$$

Hence

$$\frac{\partial \log C(\boldsymbol{\theta})}{\partial \theta_j} = \frac{\partial C(\boldsymbol{\theta})}{\partial \theta_j} \frac{1}{C(\boldsymbol{\theta})} = E_{\theta}(t_j),$$

$$\frac{\partial^2 \log C(\boldsymbol{\theta})}{\partial \theta_i\, \partial \theta_j} = E_{\theta}(t_i\, t_j) - E_{\theta}(t_i)\, E_{\theta}(t_j) = \mathrm{cov}_{\theta}(t_i,\, t_j).$$

Expressed in matrix notation these are the desired formulae (3.5) and (3.6).

□

The fact that $\log C(\boldsymbol{\theta})$ is infinitely differentiable within $\boldsymbol{\Theta}$ reflects the fact that this function is an analytic function in the set $\boldsymbol{\Theta}$. However, this will not be proved here, and we will not exploit the analyticity property below.

Remark 3.9. ***Extension to higher-order moments.*** Repeated differentiation of $C(\boldsymbol{\theta})$ yields higher-order raw moments, after normalization by $C(\boldsymbol{\theta})$ itself, generalizing (3.7) and (3.8). Alternatively they can be obtained from the *moment-generating function* (the Laplace transform) for the distribution of $\boldsymbol{t}$, with argument $\boldsymbol{\psi}$,

$$E_{\theta}(e^{\boldsymbol{\psi}^T \boldsymbol{t}}) = \frac{C(\boldsymbol{\theta} + \boldsymbol{\psi})}{C(\boldsymbol{\theta})}, \tag{3.9}$$

or, if preferable, from the *characteristic function*, that can be written

$$E_{\theta}(e^{i\boldsymbol{\psi}^T \boldsymbol{t}}) = \frac{C(\boldsymbol{\theta} + i\,\boldsymbol{\psi})}{C(\boldsymbol{\theta})}.$$

For example, for component t_j we have

$$E_{\theta}(t_j^r) = \frac{\partial^r C(\boldsymbol{\theta})}{\partial \theta_j^r} \frac{1}{C(\boldsymbol{\theta})}. \tag{3.10}$$

Repeated differentiation of $\log C(\boldsymbol{\theta})$ yields other expressions in terms of higher-order moments. They can be precisely characterized. It is often simpler to work with $\log C(\boldsymbol{\theta})$ than with $C(\boldsymbol{\theta})$. The corresponding logarithm of the moment-generating function is generally called the *cumulant function*, and the moment expressions derived by differentiating the cumulant function are called the *cumulants* (or *semi-invariants*). The first cumulants, up to order three, are the mean, the central second-order moments (that is the variance–covariance matrix) and the central third-order moments. For higher cumulants, see for example Cramér (1946, Sec. 15.10). Now, for exponential families, differentiation of the cumulant function $\log C(\boldsymbol{\theta} + \boldsymbol{\psi}) - \log C(\boldsymbol{\theta})$ with respect to $\boldsymbol{\psi}$ in the point $\boldsymbol{\psi} = \boldsymbol{0}$ evidently yields the same result as differentiation of $\log C(\boldsymbol{\theta})$ in the point $\boldsymbol{\theta}$. Hence, $\log C(\boldsymbol{\theta})$ can be interpreted as the cumulant function for the distribution of $\boldsymbol{t}$. △

Exercise 3.1 ***A recursive moment formula***

Let u be a component of $\boldsymbol{t}$, with the corresponding canonical parameter component ψ. Show that

$$\frac{\partial E_{\theta}(u)}{\partial \psi} = \mathrm{var}_{\theta}(u),$$

and more generally that

$$\frac{\partial E_\theta(u^r)}{\partial \psi} = E_\theta(u^{r+1}) - E_\theta(u)\, E_\theta(u^r).$$

△

Exercise 3.2 ***Skew-logistic distribution***
Consider the family of densities

$$f(y;\,\alpha) = \frac{\alpha\, e^{-y}}{(1+e^{-y})^{\alpha+1}}, \qquad y \in \mathbb{R}$$

indexed by shape parameter $\alpha \in (0, \infty)$. This is the location–scale standardized version of the so-called skew-logistic distribution or the Type I generalized logistic distribution. First show and characterize this as an exponential family, and next show that $E(\log(1+e^{-y})) = 1/\alpha$ and $\mathrm{var}(\log(1+e^{-y})) = 1/\alpha^2$. Note that a brute force calculation of these moments would seem impossible. △

Exercise 3.3 ***Deviation from uniform model***
The family of densities

$$f(y;\,\theta) = \begin{cases} \theta\,(2y)^{\theta-1} & \text{for } 0 \le y \le 1/2 \\ \theta\,(2-2y)^{\theta-1} & \text{for } 1/2 \le y \le 1 \end{cases}$$

has been suggested as a parametric model for deviations from the standard uniform distribution (Bernardo and Smith, 1994, Ex. 5.13). Show that this is a regular exponential family. △

3.2 Likelihood and Maximum Likelihood

We now turn to the likelihood function $L(\boldsymbol{\theta})$, that is, the density or probability $f(\boldsymbol{y};\boldsymbol{\theta})$ for the data $\boldsymbol{y}$ (or $\boldsymbol{t}$, since $\boldsymbol{t}$ is sufficient), regarded as a function of the canonical $\boldsymbol{\theta}$. The importance of the concepts of likelihood and maximum likelihood in statistical inference theory should be known to the reader, but see also Appendix A. There is not enough space here for a comprehensive introduction, but some basic statements follow, additionally making our terminology and notation clear. More conveniently than the likelihood itself, we will study $\log L(\boldsymbol{\theta})$, given by (3.11). Not only for exponential families, but generally under some weak smoothness conditions, automatically satisfied for exponential families, the first derivative (the gradient) $U(\boldsymbol{\theta}) = D \log L(\boldsymbol{\theta})$ of the log-likelihood function exists and has nice and useful properties. It is called the (Fisher) *score function* or *score vector*. The expected score vector is zero, if the score and the density in this expected value use the same $\boldsymbol{\theta}$, that is, $E_{\boldsymbol{\theta}}\{U(\boldsymbol{\theta})\} = \mathbf{0}$. Setting the actual score vector to zero yields the likelihood equation system, to which the maximum likelihood (ML) estimator $\hat{\boldsymbol{\theta}}$ is usually a root. The variance–covariance matrix for the score vector is the *Fisher information matrix*, $I(\boldsymbol{\theta})$. The Fisher information can alternatively be calculated as the expected value of $-D^2 \log L(\boldsymbol{\theta})$, that is, minus the second-order derivative (the Hessian) of the log-likelihood. This function, $J(\boldsymbol{\theta}) = -D^2 \log L(\boldsymbol{\theta})$ is called the *observed information*. Hence, the Fisher information is the expected value of the observed information. The observed information $J(\boldsymbol{\theta})$ tells the curvature of the likelihood function, and is of particular interest in the ML point $\boldsymbol{\theta} = \hat{\boldsymbol{\theta}}$, where the inverse of the (observed or expected) information matrix asymptotically equals the variance–covariance matrix of the maximum likelihood estimator, see Chapter 4. For the general theory of likelihood in more detail, see for example Cox and Hinkley (1974), Pawitan (2001) and Davison (2003, Ch. 4).

Proposition 3.10 ***Log-likelihood properties in canonical parameters***
For a full exponential family in canonical parameterization, the log-likelihood function,

$$\log L(\boldsymbol{\theta}) = \boldsymbol{\theta}^T \boldsymbol{t} - \log C(\boldsymbol{\theta}) + constant, \tag{3.11}$$

is a smooth and strictly concave function of $\boldsymbol{\theta}$, the score function is

$$U(\boldsymbol{\theta}) = \boldsymbol{t} - \boldsymbol{\mu}_t(\boldsymbol{\theta}), \tag{3.12}$$

and the observed information $J(\boldsymbol{\theta})$ equals the expected (Fisher) information $I(\boldsymbol{\theta})$, and they are both given by the variance of $\boldsymbol{t}$,

$$I(\boldsymbol{\theta}) = J(\boldsymbol{\theta}) = V_t(\boldsymbol{\theta}). \tag{3.13}$$

Proof The form of the log-likelihood function is seen from the form (1.5) of $f(y; \boldsymbol{\theta})$ or (1.8) of $f(t; \boldsymbol{\theta})$. Since the log-likelihood differs from $-\log C(\boldsymbol{\theta})$ by only a term linear in $\boldsymbol{\theta}$, all the stated properties of $\log L(\boldsymbol{\theta})$ follow almost immediately from Proposition 3.8. □

In the next proposition we look more precisely at the consequences for likelihood maximization.

Proposition 3.11 ***Maximum likelihood***
In a regular exponential family, $\boldsymbol{\mu}_t(\boldsymbol{\theta})$ is a one-to-one function of $\boldsymbol{\theta}$ for $\boldsymbol{\theta} \in \boldsymbol{\Theta}$, with $\boldsymbol{\mu}_t(\boldsymbol{\Theta})$ also open, and the log-likelihood function has a unique

maximum in $\boldsymbol{\Theta}$ *if and only if* $\boldsymbol{t} \in \boldsymbol{\mu}_t(\boldsymbol{\Theta})$, *and then the ML estimate* $\hat{\boldsymbol{\theta}}(\boldsymbol{t})$ *is the unique root of the likelihood equations*

$$\boldsymbol{t} = \boldsymbol{\mu}_t(\boldsymbol{\theta}), \tag{3.14}$$

i.e. the functions $\boldsymbol{\mu}_t$ *and* $\hat{\boldsymbol{\theta}}$ *are mutual inverses,* $\hat{\boldsymbol{\theta}}(\boldsymbol{t}) = \boldsymbol{\mu}_t^{-1}(\boldsymbol{t})$.

Proof Since $\boldsymbol{\mu}_t$ is the gradient vector of the strictly convex $\log C(\boldsymbol{\theta})$, $\boldsymbol{\mu}_t$ is a one-to-one mapping of $\boldsymbol{\Theta}$ on $\boldsymbol{\mu}_t(\boldsymbol{\Theta})$, both open sets in $\mathbb{R}^k$ (since the family is regular and both $\boldsymbol{\mu}_t$ and its inverse are continuous functions; even analytic functions).

The strict concavity of the log-likelihood function implies that if a root of (3.14) is found within the open $\boldsymbol{\Theta}$, it must be unique and global. Vice versa, by the differentiability of the likelihood function and since $\boldsymbol{\Theta}$ is open, a global maximum must have a zero value of the score function, i.e. be a root of (3.14).

Hence, the requirement that $\boldsymbol{t} \in \boldsymbol{\mu}_t(\boldsymbol{\Theta})$ is precisely what is needed for the existence of a root of (3.14) in $\boldsymbol{\Theta}$. □

Remark 3.12. ***Extension.*** For a nonregular family the result of this proposition holds if we exclude the boundary points by restricting consideration to the interior of $\boldsymbol{\Theta}$ and the corresponding image under $\boldsymbol{\mu}_t$. This follows from the proof given. △

An observed $\boldsymbol{t}$ in a regular family will usually be found to satisfy the requirement $\boldsymbol{t} \in \boldsymbol{\mu}_t(\boldsymbol{\Theta})$ of Proposition 3.11. When it does not, the observed $\boldsymbol{t}$ typically suggests some degenerate distribution as having a higher likelihood, a distribution which falls outside the exponential family but corresponds to a boundary value of the parameter in some alternative parameterization. An example is provided by the binomial distribution family: If we observe $t = 0$ successes, which has a positive probability for any n and θ, the ML estimate of the success probability π_0, $0 \leq \pi_0 \leq 1$, is $\hat{\pi}_0 = 0$. The value $\pi_0 = 0$ has no corresponding canonical θ-value in $\Theta = \mathbb{R}$. It is generally possible (at least for discrete families) to extend the definition of the model, so that the ML estimation always has a solution (Lauritzen, 1975; Barndorff-Nielsen, 1978, Sec. 9.3). This is perhaps not a very fruitful approach, however, since we would lose some of the nice regularity properties possessed by regular exponential families.

Suppose we have a sample of scalar $\boldsymbol{y}$-values from a distribution in a family with a p-dimensional parameter. The classic *moment method of estimation* equates the sample sums of $y, y^2, \ldots, y^p$ to their expected values. The form of the likelihood equation for exponential families shows that the

ML estimator in exponential families can be regarded as a sort of a generalized, more sophisticated moment estimator. Proposition 1.2 namely tells that the j-th component of the t-vector is a sum of a function $t_j(y)$ over the sample y-values, so the function $t_j(y)$ replace the simple moment y^j.

The set of possible outcomes of t, for which the ML estimate exists, can be characterized in a simple way for a regular exponential family. This result is due to Barndorff-Nielsen:

Proposition 3.13 ***MLE existence, continued***
In a regular exponential family the maximum likelihood estimate (MLE) $\hat{\theta}(t) = \mu_t^{-1}(t)$ exists precisely when **t** *belongs to the interior of the closed convex hull of the support of the family of distributions for* **t***, and, as a consequence, this set of* **t***-values equals $\mu_t(\Theta)$.*

Proof This result will not be proved here, only illustrated. For a proof, see Barndorff-Nielsen (1978, Ch. 9) or Johansen (1979). □

As a simple illustration of Proposition 3.13, we return to the binomial model, Bin(n, π_0). Let t be the relative success frequency. The possible outcomes of t form a set of rational numbers: $\{0/n = 0, 1/n, 2/n, \ldots, n/n = 1\}$. The convex hull of this set is the closed unit interval [0, 1]. The interior of the unit interval is the open interval (0, 1). For any t-value in this interval, the MLE of $\theta = \text{logit}(\pi_0)$ is defined through the formula $\hat{\theta} = \text{logit}(t)$. Proposition 3.13 tells that the formula for $\hat{\theta}$ is valid for any t-value in (the open) (0, 1), even though only a finite number of t-values are possible for each fixed n. Proposition 3.13 also tells that the MLE for θ does not exist when t falls at the boundary points 0 and 1 of the unit interval.

For other models, with more parameters, in particular non-iid-models, it can be a bit more difficult to characterize when the MLE exists, cf. Exercise 3.17. For discrete models, where the y-vector takes its values in a finite or countable set, we refer to Jacobsen (1989) for an alternative criterion, simpler to use. He demonstrated its use for logistic regression, Rasch models (Chapter 12), etc., and even for Cox partial likelihoods (Section 9.7.1), which are only similar to exponential family likelihoods.

3.2.1 The Steepness Criterion

For the truth of the result of Proposition 3.13, it is not quite necessary that the exponential family is regular. A *necessary and sufficient condition* for Proposition 3.13 to hold is that the model is *steep* (Barndorff-Nielsen). Steepness is a requirement on the log-likelihood or equivalently on $\log C(\boldsymbol{\theta})$, that it must have infinite derivatives in all boundary points belonging to the canonical parameter space $\boldsymbol{\Theta}$. Thus, the gradient $\boldsymbol{\mu}_t(\boldsymbol{\theta})$ does not exist in boundary points. In other words, there cannot be an observation $\boldsymbol{t}$ satisfying $\boldsymbol{t} = \boldsymbol{\mu}_t(\boldsymbol{\theta})$ for a steep family. Conversely, suppose the family is not steep, and furthermore suppose $\dim \boldsymbol{\theta} = 1$, for simplicity of argument. Since the family is not steep, there is a boundary point where $\mu_t(\theta)$ is finite. Then, there must be possible t-values on both sides of this μ_t (otherwise we had a one-point distribution in that point). Thus there are possible t-values on the outside of the boundary value of μ_t. This is inconsistent with the conclusion of Proposition 3.13, so steepness is required.

Note that regular families have no boundary points in $\boldsymbol{\Theta}$, so for them the steepness requirement is vacuous. A precise definition of steepness must involve derivatives along different directions within $\boldsymbol{\Theta}$ towards a boundary point of $\boldsymbol{\Theta}$, but we abstain from formulating a more precise version.

Examples of models that are steep but not regular are rare in applications. A simple theoretical construction is based on the h-function (3.4), a linear (i.e. $t(y) = y$) exponential family with $h(y) = 1/y^2$ for $y > 1$. The family is not regular, because the norming constant $C(\theta)$ is finite for θ in the *closed* interval $\theta \leq 0$ but it is steep. The left derivative of $C(\theta)$ (hence also of $\log C(\theta)$) at zero is infinite, because $1/y$ is not integrable on $(1, \infty)$. If $h(y)$ is changed to $h(y) = 1/y^3$, the derivative of $C(\theta)$ approaches a bounded constant as θ approaches 0, since $1/y^2$ is integrable. Hence, $h(y) = 1/y^3$ generates a nonregular linear exponential family that is not even steep. The inverse Gaussian distribution is a steep model of some practical interest, see Exercise 3.4. The Strauss model for spatial variation, Section 13.1.2, has also been of applied concern; it is a nonregular model that is not even steep.

For models that are steep but not regular, boundary points of $\boldsymbol{\Theta}$ do not satisfy the same regularity conditions as the interior points, but for Proposition 3.11 to hold we need only exclude possible boundary points. In Barndorff-Nielsen and Cox (1994), such families, steep families with canonical parameter space restricted to the interior of the maximal $\boldsymbol{\Theta}$, are called *prime exponential models*. Note that for outcomes of $\hat{\boldsymbol{\theta}}(\boldsymbol{t})$ in Proposition 3.13, it is the interior of $\boldsymbol{\Theta}$ that is in a one-to-one correspondence with the interior of the convex hull.

Exercise 3.4 ***Inverse Gaussian, continued from Exercise 2.6***
Note first that the inverse Gaussian, see Exercise 2.6, is not regular, more precisely because the Lévy model is located on its boundary. Next, show that the family is steep. △

Exercise 3.5 ***A nonregular family, not even steep***
Exponential tilting (Example 2.12) applied to a Pareto type distribution yields an exponential family with density

$$f(y;\,\theta) \propto y^{-a-1}\, e^{\theta y}, \qquad y > 1,$$

where $a > 1$ is assumed given, and the Pareto distribution is

$$f(y;\,0) = a y^{-a-1}, \qquad y > 1.$$

(a) Show that the canonical parameter space is the closed half-line, $\theta \leq 0$.
(b) $E_\theta(t) = \mu_t(\theta)$ is not explicit. Nevertheless, show that its maximum must be $a/(a-1) > 1$.
(c) Thus conclude that the likelihood equation for a single observation y has no root if $y > a/(a-1)$. However, note that the likelihood actually has a maximum in Θ for any such y, namely $\hat{\theta} = 0$. △

Exercise 3.6 ***Legendre transform***
The Legendre transform (or convex conjugate) of $\log C(\theta)$ has a role in understanding convexity-related properties of canonical exponential families, see Barndorff-Nielsen (1978). For a convex function $f(x)$ the transform $f^\star(x^\star)$ is defined by $f^\star(x^\star) = x^T x^\star - f(x)$, for the maximizing x, which must (of course) satisfy $Df(x) = x^\star$.
As technical training, show for $f(\theta) = \log C(\theta)$ that
(a) the transform is $f^\star(\mu_t) = \theta^T \mu_t - \log C(\theta)$, where $\theta = \hat{\theta}(\mu_t)$;
(b) $Df^\star(\mu_t) = \hat{\theta}(\mu_t)$ and $D^2 f^\star(\mu_t) = V_t(\hat{\theta}(\mu_t))^{-1}$;
(c) The Legendre transform of $f^\star(\mu_t)$ brings back $\log C(\hat{\theta}(\mu_t)) = \log C(\theta)$. △

3.3 Alternative Parameterizations

The canonical parameterization yields some useful and elegant results, as shown in Section 3.2. However, sometimes other parameterizations are theoretically or intuitively more convenient, or of particular interest in an application. Before we introduce some such alternatives to the canonical parameterization, it is useful to have general formulas for how score vectors and information matrices are changed when we make a transformation of the parameters. After this ‘Reparameterization lemma’, we introduce two alternative types of parameterization of full exponential families: the mean value parameterization and the mixed parameterization. These parameters are often at least partially more ‘natural’ than the canonical parameters, but this is not the main reason for their importance.

3.3.1 Reparameterization and Restricted Parameterization Formulae

Suppose we change the parameterization of a model from a parameter $\boldsymbol{\theta}$ to a different parameter $\boldsymbol{\psi}$. When we talk of reparameterization we usually have a one-to-one function in mind, so $\boldsymbol{\theta} = \boldsymbol{\theta}(\boldsymbol{\psi})$ and $\boldsymbol{\psi} = \boldsymbol{\psi}(\boldsymbol{\theta})$, but the formulas below hold also in the case when the model with $\boldsymbol{\theta} \in \boldsymbol{\Theta}$ is restricted to a smaller-dimensional submodel parameterized by some $\boldsymbol{\psi} \in \boldsymbol{\Psi}$ (typically a curved family; Chapter 7). Each $\boldsymbol{\psi}$ then corresponds to a particular $\boldsymbol{\theta} = \boldsymbol{\theta}(\boldsymbol{\psi})$ in $\boldsymbol{\Theta}$. To express the relation between the two score functions and the two information matrices we only need the Jacobian matrix $\left(\frac{\partial \boldsymbol{\theta}}{\partial \boldsymbol{\psi}}\right)$ of the function $\boldsymbol{\theta}(\boldsymbol{\psi})$ (assumed to exist).

Proposition 3.14 ***Reparameterization lemma***
If $\boldsymbol{\psi}$ and $\boldsymbol{\theta} = \theta(\boldsymbol{\psi})$ are two equivalent parameterizations of the same model, or if parameterization by $\boldsymbol{\psi}$ represents a lower-dimensional subfamily of a family parameterized by $\boldsymbol{\theta}$, via $\boldsymbol{\theta} = \boldsymbol{\theta}(\boldsymbol{\psi})$, then the score functions are related by

$$U_{\psi}(\boldsymbol{\psi}; \boldsymbol{y}) = \left(\frac{\partial \boldsymbol{\theta}}{\partial \boldsymbol{\psi}}\right)^T U_{\theta}(\boldsymbol{\theta}(\boldsymbol{\psi}); \boldsymbol{y}), \tag{3.15}$$

and the corresponding information matrices are related by the equations

$$I_{\psi}(\boldsymbol{\psi}) = \left(\frac{\partial \boldsymbol{\theta}}{\partial \boldsymbol{\psi}}\right)^T I_{\theta}(\boldsymbol{\theta}(\boldsymbol{\psi})) \left(\frac{\partial \boldsymbol{\theta}}{\partial \boldsymbol{\psi}}\right), \tag{3.16}$$

$$J_{\psi}(\hat{\boldsymbol{\psi}}) = \left(\frac{\partial \boldsymbol{\theta}}{\partial \boldsymbol{\psi}}\right)^T J_{\theta}(\boldsymbol{\theta}(\hat{\boldsymbol{\psi}})) \left(\frac{\partial \boldsymbol{\theta}}{\partial \boldsymbol{\psi}}\right), \tag{3.17}$$

where the information matrices refer to their respective parameterizations and are calculated in points $\boldsymbol{\psi}$ and $\boldsymbol{\theta}(\boldsymbol{\psi})$, respectively, and where $\left(\frac{\partial \boldsymbol{\theta}}{\partial \boldsymbol{\psi}}\right)$ is the Jacobian transformation matrix for the reparameterization, calculated in the same point as the information matrix.

Note that it is generally crucial to distinguish the Jacobian matrix from its transpose. The definition of the Jacobian is such that the (i, j) element of $\left(\frac{\partial \boldsymbol{\theta}}{\partial \boldsymbol{\psi}}\right)$ is the partial derivative of θ_i with respect to ψ_j.

When $\boldsymbol{\psi}$ and $\boldsymbol{\theta}$ are two equivalent parameterizations, it is sometimes easier to derive $\left(\frac{\partial \boldsymbol{\psi}}{\partial \boldsymbol{\theta}}\right)$ than $\left(\frac{\partial \boldsymbol{\theta}}{\partial \boldsymbol{\psi}}\right)$ for use in the reparameterization expressions. Note then that they are the inverses of each other, cf. Section B.2.

Note also that Proposition 3.14 does not allow $\boldsymbol{\psi}$ to be a function of $\boldsymbol{\theta}$, so for example not a component or subvector of $\boldsymbol{\theta}$. Reducing the score to a subvector of itself, and the information matrix to the corresponding submatrix, is legitimate only when the excluded parameter subvector is regarded as known (cf. Proposition 4.3).

Note finally that the relationship (3.16) between the two Fisher informations I holds for any ψ, whereas the corresponding relation (3.17) for the observed information J only holds in the ML point. The reason for the latter fact will be clear from the following proof.

Proof Repeated use of the chain rule for differentiation yields first

$$D_\psi \log L = \left(\frac{\partial \boldsymbol{\theta}}{\partial \psi}\right)^T D_\theta \log L, \tag{3.18}$$

and next

$$D_\psi^2 \log L = \left(\frac{\partial \boldsymbol{\theta}}{\partial \psi}\right)^T D_\theta^2 \log L \left(\frac{\partial \boldsymbol{\theta}}{\partial \psi}\right) + \left(\frac{\partial^2 \boldsymbol{\theta}}{\partial \psi^2}\right)^T D_\theta \log L. \tag{3.19}$$

where the last term represents a matrix obtained when the three-dimensional array $\left(\frac{\partial^2 \boldsymbol{\theta}}{\partial \psi^2}\right)^T$ is multiplied by a vector. In the MLE point this last term of (3.19) vanishes, since $D_\theta \log L = 0$, and (3.17) is obtained. Also, if we take expected values, the same term vanishes because the score function has expected value zero, and the corresponding relation between the expected informations follows. □

3.3.2 Mean Value Parameterization

For a regular exponential family, a parameterization by the mean value vector $\boldsymbol{\mu}_t$ on $\boldsymbol{\mu}_t(\boldsymbol{\Theta})$ is equivalent to the canonical parameterization by $\boldsymbol{\theta}$ on $\boldsymbol{\Theta}$. This was part of Proposition 3.11 and is due to the strict convexity of $\log C(\boldsymbol{\theta})$, whose gradient vector is $\boldsymbol{\mu}_t$. Proposition 3.13 gave an alternative characterization of the parameter space $\boldsymbol{\mu}_t(\boldsymbol{\Theta})$ for regular families.

Proposition 3.15 ***Mean value parameter estimation***
The mean value parameter $\boldsymbol{\mu}_t = \boldsymbol{\mu}_t(\boldsymbol{\theta})$ is unbiasedly estimated by its MLE, which is $\hat{\boldsymbol{\mu}}_t = \boldsymbol{t}$, with variance $\mathrm{var}(\hat{\boldsymbol{\mu}}_t) = \mathrm{var}(\boldsymbol{t})$. *This is also exactly the inverse of the Fisher information matrix under parameterization with $\boldsymbol{\mu}_t$,*

$$I(\boldsymbol{\mu}_t) = \mathrm{var}(\boldsymbol{t})^{-1}, \tag{3.20}$$

and in the MLE, $J(\hat{\boldsymbol{\mu}}_t) = I(\hat{\boldsymbol{\mu}}_t)$.

Proof Since the likelihood equations are $\boldsymbol{\mu}_t(\boldsymbol{\theta}) = \boldsymbol{t}$, it is evident that $\hat{\boldsymbol{\mu}}_t =$

t, and consequently $\text{var}(\hat{\mu}_t) = \text{var}(t)$. Formula (3.20) and the equality between observed and expected information in the MLE point follow by application of the Reparameterization lemma (Proposition 3.14), telling generally how the score function and the information matrix are changed under reparameterization. From Proposition 3.8 we know both that $I(\theta) = \text{var}(t)$, and that the Jacobian matrix for the function $\mu_t(\theta)$ is $\text{var}(t)$. For application of Proposition 3.14 we need the inverse of this Jacobian, that is $\text{var}(t)^{-1}$.

The equality $J(\hat{\mu}_t) = I(\hat{\mu}_t)$ is valid much more generally, see Exercise 3.7 below, but away from the point $\hat{\mu}_t$, it is typically not satisfied. □

Exercise 3.7 ***Expected and observed information in general parameter***
Use the Reparameterization lemma for any (noncanonical) choice of parameter ψ, with $\theta = \theta(\psi)$, to show that in a full exponential family, $I(\hat{\psi}) = J(\hat{\psi})$. △

Exercise 3.8 ***Normal distribution in mean value parameterization***
Find the mean value parameter space $\mu_t(\Theta)$ for the simple Gaussian two-parameter model.
Hint: It might simplify to use μ and σ^2 as intermediate parameters. △

Exercise 3.9 ***Linear exponential family in mean value parameterization***
Consider a linear family (that is, having $t(y) = y$). Derive $\frac{\partial \log C}{\partial \mu}$ as a function of μ (use chain rule). Conclude that if two linear families have the same variance function $V_y(\mu)$ for y as function of μ, and the same range of μ-values, $\mu_t(\Theta)$, they must have the same moment-generating function (3.9), or equivalently the same cumulant function. Thus the families are the same. In other words, $V_y(\mu)$ characterizes the linear family uniquely (Morris, 1982). See also Sections 9.4 and 9.5. △

Even though many mean value parameter spaces $\mu_t(\Theta)$ share the convexity with the corresponding canonical parameter spaces Θ, this is not a general property. A counter-example can be found in Efron (1978, p. 364).

3.3.3 Mixed Parameterizations. Marginal and Conditional Families

Another parameterization that also allows important theoretical conclusions, in particular in connection with conditional inference, is the *mixed parameterization*. If the canonical statistic t is partitioned into two components u and v, and correspondingly the canonical parameter vector θ in the two components θ_u and θ_v, a mixed parameterization is obtained by using the partial canonical parameter θ_u in combination with the mean value vector μ_v, or vice versa, θ_v with μ_u. Before we show that this is actually a parameterization, and that it has nice properties, we take a brief look at marginal and conditional distribution families generated from an exponential family. This will not only be needed for the proof of the subsequent proposition on the mixed parameters, but will also motivate the interest in this parameterization.

Proposition 3.16 ***Marginal and conditional families***
In a regular exponential family, consider a partitioning of $\boldsymbol{t}$ *in* $\boldsymbol{u}$ *and* $\boldsymbol{v}$*, and the corresponding partitioning of* $\boldsymbol{\theta}$ *in* $\boldsymbol{\theta}_u$ *and* $\boldsymbol{\theta}_v$*.*

- *The marginal model for* $\boldsymbol{u}$ *is a regular exponential family for each given* $\boldsymbol{\theta}_v$*, depending on* $\boldsymbol{\theta}_v$ *but with one and the same parameter space for its mean value parameter* $\boldsymbol{\mu}_u$*.*
- *The conditional model for* $\boldsymbol{y}$*, given* $\boldsymbol{u}$*, and thus also for* $\boldsymbol{v}$*, given* $\boldsymbol{u}$*, is a regular exponential family with canonical statistic* $\boldsymbol{v}$*. The conditional model depends on* $\boldsymbol{u}$*, but with one and the same canonical parameter* $\boldsymbol{\theta}_v$ *as in the joint model.*

Corollary 3.17 ***Mixed parameterization***
The mixed parameterization is a valid parameterization.

Proof of Corollary 3.17 From Proposition 3.16 it follows that the conditional density can be parameterized by $\boldsymbol{\theta}_v$. When $\boldsymbol{\theta}_v$ has been specified, Proposition 3.16 additionally tells that the marginal for $\boldsymbol{u}$ is an exponential family that is for example parameterized by its mean value $\boldsymbol{\mu}_u$. Finally, since the joint density for $\boldsymbol{u}$ and $\boldsymbol{v}$ is the product of these marginal and conditional densities, it follows that the mixed parameterization with $(\boldsymbol{\mu}_u, \boldsymbol{\theta}_v)$ is also a valid parameterization of the family for $\boldsymbol{t}$. □

Proof of Proposition 3.16 The marginal density for $\boldsymbol{u}$ is obtained by integrating the density (1.8) for $\boldsymbol{t}$ with respect to $\boldsymbol{v}$, for fixed $\boldsymbol{u}$:

$$\begin{aligned} f(\boldsymbol{u};\boldsymbol{\theta}) &= \int a(\boldsymbol{\theta})\, g(\boldsymbol{u},\boldsymbol{v})\, e^{\boldsymbol{\theta}_u^T \boldsymbol{u} + \boldsymbol{\theta}_v^T \boldsymbol{v}}\, \mathrm{d}\boldsymbol{v} \\ &= a(\boldsymbol{\theta})\, e^{\boldsymbol{\theta}_u^T \boldsymbol{u}} \int g(\boldsymbol{u},\boldsymbol{v})\, e^{\boldsymbol{\theta}_v^T \boldsymbol{v}}\, \mathrm{d}\boldsymbol{v}. \end{aligned} \tag{3.21}$$

The integral factor on the second line is generally a function of both data and the parameter ($\boldsymbol{\theta}_v$), which destroys the exponential family property. However, for each given $\boldsymbol{\theta}_v$, this factor is only a function of data, and then the family is exponential. Its canonical parameter is $\boldsymbol{\theta}_u$, but *the parameter space for* $\boldsymbol{\theta}_u$ *may depend on* $\boldsymbol{\theta}_v$, being the intersection of $\boldsymbol{\Theta}$ with the hyperplane or affine subspace $\boldsymbol{\theta}_v$ = fixed. Since this exponential family is automatically regular in $\boldsymbol{\theta}_u$, the range for its mean value $\boldsymbol{\mu}_u$, that is, its mean value parameter space, is identical with the interior of the closed convex hull of the support of the family of distributions for $\boldsymbol{u}$, according to Proposition 3.13. However, since this set is independent of $\boldsymbol{\theta}_v$, the parameter space for $\boldsymbol{\mu}_u$ is also independent of $\boldsymbol{\theta}_v$, as was to be shown.

For the conditional model we do not bother about possible mathematical technicalities connected with conditional densities, but simply write as for conditional probabilities, using the marginal density (3.21) to simplify for $\boldsymbol{u} = \boldsymbol{u}(\boldsymbol{y})$ the expression:

$$f(\boldsymbol{y}|\boldsymbol{u};\boldsymbol{\theta}) = f(\boldsymbol{u},\boldsymbol{y};\boldsymbol{\theta})/f(\boldsymbol{u};\boldsymbol{\theta}) = \frac{e^{\boldsymbol{\theta}_v^T \boldsymbol{v}(\boldsymbol{y})}\, h(\boldsymbol{y})}{\int e^{\boldsymbol{\theta}_v^T \boldsymbol{v}}\, g(\boldsymbol{u},\boldsymbol{v})\, \mathrm{d}\boldsymbol{v}}. \tag{3.22}$$

To obtain $f(\boldsymbol{v}|\boldsymbol{u};\boldsymbol{\theta})$ we only substitute $g(\boldsymbol{u},\boldsymbol{v})$ for $h(\boldsymbol{y})$ in the numerator of the exponential family density (3.22). Note that $f(\boldsymbol{v}|\boldsymbol{u};\boldsymbol{\theta})$ is defined only for those $\boldsymbol{v}$-values which are possible for the given $\boldsymbol{u}$, and the denominator is the integral over this set of values. This makes the model depend on $\boldsymbol{u}$, even though the canonical parameter is the same, $\boldsymbol{\theta}_v$, and in particular the conditional mean and variance of $\boldsymbol{v}$ will usually depend on u. □

Example 3.1 ***Marginality and conditionality for Gaussian sample***
As an illustration of Proposition 3.16, consider $u = \sum y_i^2$ for a sample from the normal distribution, with $v = \sum y_i$. First, the marginal distribution of $\sum y_i^2$ is proportional to a noncentral χ^2 and does not in general form an exponential family. Next, consider instead its conditional distribution, given $\sum y_i = n\bar{y}$. This might appear quite complicated, but given $\bar{y}$, $\sum y_i^2$ differs by only an additive, known constant from $\sum y_i^2 - n\bar{y}^2 = (n-1)s^2$. Thus, it is enough to characterize the distribution of $(n-1)s^2$, and it is well-known that s^2 is independent of $\bar{y}$ (see also Section 3.7), and that the distribution of $(n-1)s^2$ is proportional (by σ^2) to a (central) $\chi^2(n-1)$. From the explicit form of a χ^2 it is easily seen that the conditional distribution forms an exponential family (Reader, check!). △

The canonical parameter space for the conditional family is at least as large as the maximal set of $\boldsymbol{\theta}_v$-values in $\boldsymbol{\Theta}$, that is, the set of $\boldsymbol{\theta}_v$ such that $(\boldsymbol{\theta}_u, \boldsymbol{\theta}_v) \in \boldsymbol{\Theta}$ for some $\boldsymbol{\theta}_u$. Typically the two sets are the same, but there are models in which the conditional canonical parameter space is actually larger and includes parameter values $\boldsymbol{\theta}_v$ that lack interpretation in the joint model, see for example the Strauss model in Section 13.1.2 and the model in Section 14.3(b).

Returning now to the mixed parameterization, note that for a given partitioning of $\boldsymbol{\theta} = (\boldsymbol{\theta}_u, \boldsymbol{\theta}_v)$, there are two different mixed parameterizations possible, $(\boldsymbol{\mu}_u, \boldsymbol{\theta}_v)$ and $(\boldsymbol{\theta}_u, \boldsymbol{\mu}_v)$. So choosing the normal distribution as an example, with the usual canonical $(\mu/\sigma^2, -1/2\sigma^2)$, we find the mixed ones $(\mu, -1/2\sigma^2)$ and $(\mu/\sigma^2, \mu^2 + \sigma^2)$. Both parameterizations have a property that is obvious only for the first of them; they partition the parameter vector in components which are *variation independent*.

Definition 3.18 ***Variation independent parameter components***
The components of a partitioning (ψ, λ) are said to be *variation independent* if the parameter space for (ψ, λ) is the (Cartesian) product space $\Psi \times \Lambda$ of the sets Ψ and Λ of possible values of ψ and λ, respectively.

With $(\mu, -1/2\sigma^2)$ as example, the joint parameter space (lower half plane) is the 'rectangle' formed as the product space of the whole real line for μ and the negative halfline for $-1/2\sigma^2$.

Remark 3.19 ***Likelihood orthogonality.*** When a likelihood factorizes in one factor for ψ and another factor for λ, we want to state that inference for ψ and λ are totally separated (*likelihood orthogonality*). The factorization is not enough, however, to guarantee that whatever knowledge is provided about λ (for example an estimate of λ), there is no information in it whatsoever about ψ. We also need variation independence, implying that the set of possible ψ-values is the same for each fixed λ. It is easy to forget stating or checking the demand on variation independent parameters (happens even in quite good books, e.g. Pawitan (2001, Sec. 10.2)), because it is so often satisfied automatically by natural parameterizations. However, 'often' is not 'always'. For example, the mean value parameters of a normal distribution, $\psi = \mu$ and $\lambda = \mu^2 + \sigma^2$, are not variation independent (check!). $\triangle$

The following proposition has important consequences, in particular in the next section. It states that mixed parameterizations are variation independent by construction and that the corresponding information matrix is block-diagonal, a particularly nice property for large sample methodology. The latter property is generally called *information orthogonality*.

Proposition 3.20 ***Properties of mixed parameterizations***
If a regular exponential family is given a mixed parameterization by $(\boldsymbol{\mu}_u, \boldsymbol{\theta}_v)$, *the two parameter components are variation independent and information*

orthogonal, and the Fisher information for (μ_u, θ_v) *is*

$$I(\mu_u, \theta_v) = \begin{pmatrix} \mathrm{var}(u)^{-1} & 0 \\ 0 & (\mathrm{var}(t)^{vv})^{-1} \end{pmatrix} \tag{3.23}$$

where $\mathrm{var}(t)^{vv}$ *stands for the* $v v$*-block of* $\mathrm{var}(t)^{-1}$. *The same formula holds for the observed information in the MLE,* $J(\hat{\mu}_u, \hat{\theta}_v)$.

The lower right item of (3.23) can be interpreted as the (theoretical) residual variance matrix for a multivariate linear regression of v on u. In the special case of normally distributed (u, v), this is identically the same as the conditional variance of v given u, see Example 2.10. The Gaussian point of view is particularly relevant in a large sample setting, when t satisfies the central limit theorem.

Information orthogonality means that the information matrix is block-diagonal. Also its inverse will then be block-diagonal in the same way, implying that the ML estimators of components μ_u and θ_v are asymptotically uncorrelated, see further Section 4.1. Note also that since (3.23) is a local property, this large sample result holds also for nonregular families, provided the parameter is restricted to the interior of Θ.

Proof The variation independence of μ_u and θ_v was essentially already contained in Proposition 3.16. There it was found that the mean value parameter vector μ_u for u has a range of possible values which is independent of θ_v. This means that the two components are variation independent.

That the information matrix is correspondingly block diagonal and that it has the form given in (3.23) is shown by use of the Reparameterization lemma, Proposition 3.14 from Section 3.3.1, see Exercise 3.10. That it holds also for the observed information in the MLE point is seen in the same way, or by reference to the result of Exercise 3.7. □

Exercise 3.10 ***Fisher information for mixed parameters***
Use the Reparameterization lemma, Proposition 3.14, to prove the form (3.23) of the Fisher information under a mixed parameterization. △

3.3.4 Profile Likelihood

Likelihood maximization with several parameters can always be carried out in stages, typically by first maximizing over parameters without direct interest (nuisance parameters), before the final likelihood consideration and maximization is done over the parameter(s) of particular interest. The intermediate result is called a *profile likelihood*. It is not a proper likelihood, but has many such properties. As a general definition, suppose a model parameter vector is split in components $(\boldsymbol{\lambda}, \boldsymbol{\psi})$. Then the profile likelihood function $L_p(\boldsymbol{\psi})$ for $\boldsymbol{\psi}$ is formed as

$$L_p(\boldsymbol{\psi}) = L\left(\hat{\boldsymbol{\lambda}}(\boldsymbol{\psi}), \boldsymbol{\psi}\right), \tag{3.24}$$

where $\hat{\boldsymbol{\lambda}}(\boldsymbol{\psi})$ is the MLE of $\boldsymbol{\lambda}$ when $\boldsymbol{\psi}$ is regarded as given.

Generally, $L_p(\boldsymbol{\psi})$ can be used as likelihood for $\boldsymbol{\psi}$ in large sample hypothesis testing and likelihood-based confidence regions, where $-D^2 \log L_p$ will yield the proper precision of $\hat{\boldsymbol{\psi}}$, as if it were the observed information in a model with only $\boldsymbol{\psi}$. For smaller samples, profile likelihoods do not eliminate problems with nuisance parameters, and there is an extensive literature on 'modified profile likelihoods'. Profile likelihoods are particularly simple and natural for an exponential family when $\boldsymbol{\psi}$ is a component vector of the canonical parameter and $\boldsymbol{\lambda}$ is chosen according to a mixed parameterization, $(\boldsymbol{\lambda}, \boldsymbol{\psi}) = (\boldsymbol{\mu}_u, \boldsymbol{\theta}_v)$. In this case, $\hat{\boldsymbol{\lambda}}(\boldsymbol{\psi})$ does not depend on $\boldsymbol{\psi}$, because whatever value of $\boldsymbol{\theta}_v$, the MLE of $\boldsymbol{\mu}_u$ is $\boldsymbol{\mu}_u = \boldsymbol{u}$ (Proposition 3.11). Thus, the profile likelihood for $\boldsymbol{\theta}_v$ is

$$L_p(\boldsymbol{\theta}_v) = L(\boldsymbol{u}, \boldsymbol{\theta}_v) = f(\boldsymbol{u}, \boldsymbol{v}; \boldsymbol{\mu}_u = \boldsymbol{u}, \boldsymbol{\theta}_v).$$

The Hessian (second derivatives) of $-\log L_p(\boldsymbol{\theta}_v)$ follows from Proposition 3.20:

$$-D^2 \log L_p(\boldsymbol{\theta}_v) = (\mathrm{var}(\boldsymbol{t})^{vv})^{-1}. \tag{3.25}$$

Thus, invert $\mathrm{var}(\boldsymbol{t})$, pick out the vv corner, and invert back. We will see this formula again in later chapters, in particular in Section 5.5. Note that we do not need to specify the mean value component $\boldsymbol{\mu}_u$, the role here of the mixed parameterization is purely theoretical.

Exercise 3.11 ***Profile likelihoods for normal sample***
Let data be a sample of size n from $\mathrm{N}(\mu, \sigma^2)$. Find the profile likelihood functions for σ^2 and for μ. △

Exercise 3.12 ***Profile likelihood equation***
Show that the equation constructed by differentiating the profile likelihood is identical with the equation obtained by eliminating $\boldsymbol{\lambda}$ from the likelihood equation system for $(\boldsymbol{\lambda}, \boldsymbol{\psi})$. △

3.4 Solving Likelihood Equations Numerically: Newton–Raphson, Fisher Scoring and Other Algorithms

For a full family in mean value parameterization the MLE is evidently explicit, $\hat{\mu}_t = t$. For other parameterizations, by some ψ say, we must invert the equation system $\mu_t(\hat{\psi}) = t$, and this need not have an explicit solution.

A classic and presumably well-known algorithm for iterative solution of equation systems is the *Newton–Raphson method*. Its basis here is the following linearization of the score U, which is adequate for ψ close enough to the unknown $\hat{\psi}$ (note that $-J$ is the Hessian matrix of partial derivatives of U):

$$0 = U_\psi(\hat{\psi}) \approx U_\psi(\psi) - J_\psi(\psi)(\hat{\psi} - \psi). \tag{3.26}$$

Iteratively solving for $\hat{\psi}$ in this linearized system yields successive updates of ψ_k to ψ_{k+1}, $k = 0, 1, 2, \ldots$:

$$\psi_{k+1} = \psi_k + J_\psi(\psi_k)^{-1} U_\psi(\psi_k). \tag{3.27}$$

This is a locally fast algorithm (quadratic convergence), provided that it converges. The method can be quite sensitive to the choice of starting point ψ_0. Generally, if $J_\psi(\psi_0)$ is not positive definite or close to singular, it is not likely to converge, or it might converge to a minimum of $\log L$, so there is no guarantee that the resulting root represents even a local likelihood maximum. For a full exponential family, however, the odds are better. In particular, there is not more than one root, and if the parameter is the canonical, θ, we know that $J_\theta = I_\theta = \mathrm{var}(t)$, a necessarily positive definite matrix.

Quasi-Newton methods are modifications of (3.27), where the J matrix is approximated, e.g. by the secant method, typically because an explicit formula for J is not available.

In *Fisher's scoring method* we make a statistically motivated replacement of the observed information J in (3.27) by the expected (Fisher) information $E(J) = I$. The scoring method is much less sensitive to the choice of starting point. In particular, I is everywhere positive definite (or at least nonnegative definite). This implies that it will never approach a minimum or saddlepoint of $\log L$. The convergence rate can be characterized as follows. The scoring method is of type $\psi_{k+1} = g(\psi_k)$. If $\dim \psi = 1$, and the starting point is close enough to $\hat{\psi}$, the method converges if the derivative (the Hessian) satisfies $|g'(\hat{\psi})| < 1$, and it does not converge to $\hat{\psi}$ if $|g'(\hat{\psi})| > 1$. The deviation from $\hat{\psi}$ is approximately reduced by the factor $g'(\hat{\psi})$ for each iteration. For larger $\dim \psi$, the largest eigenvalue of the Hessian of g determines the rate of convergence. The Hessian for the scoring method can be written $I^{-1}(I - J)$. Typically, $I - J$ is much smaller than I

(and J), and the convergence is then fast, albeit linear. In full exponential models, when $I_\psi(\hat{\psi}) = J_\psi(\hat{\psi})$ (Exercise 3.7), the convergence is of course particularly fast, and for the canonical parameter, $\psi = \theta$, $I = J$ everywhere.

There are further aspects when comparing these methods. Which is the simpler method to use? Sometimes J, sometimes I is simpler to express and to compute than the other. Even if the observed information is optimal, the expected information might be easier to deal with.

There are models for which both J and I are difficult to find, because the log-likelihood is so complicated. To solve the likelihood equations then, we could try differences as numerical estimates of the score function derivatives, or the estimate

$$\hat{J} = \frac{1}{n} \sum_{i=1}^{n} U(\psi_k; y_i) U(\psi_k; y_i)^T$$

when this kind of expression is relevant. However, these replacements for J or I are less reliable, in particular when the likelihood is relatively flat, and when the parameter dimension is relatively large and n is not much larger.

Some more special algorithms work well for special classes of situations. A method described by Song et al. (2005) assumes that the likelihood function may be decomposed in one factor that has a relatively simple likelihood equation and another factor that contributes the complicated part of the information, and the method iterates between the two score function components. The TM algorithm finds the MLE when the likelihood can be truly or artificially regarded as a conditional likelihood, but the full likelihood is more easily maximized (Edwards and Lauritzen, 2001; Sundberg, 2002). The applications are mostly in graphical chain models, related to the models of Chapter 10.

The EM algorithm with variations, see Section 8.3, is applicable when observed data may be regarded as incomplete data from a model in which complete data would have been more easily handled.

An even more difficult situation is when there is no closed form for $C(\theta)$ and its derivatives. Examples are social network models (Chapter 11) and spatial point process models (Chapter 13). Tools developed for such situations include Monte Carlo and MCMC techniques and so-called pseudo-likelihood methods, see the chapters referred to.

3.5 Conditional Inference for Canonical Parameter

Suppose ψ is a *parameter of interest*, either being itself a subvector θ_v of θ, perhaps after a linear transformation of the canonical statistic t, or being some other one-to-one function of $\boldsymbol{\theta}_v$. Corresponding to the partitioning $\boldsymbol{t} = (\boldsymbol{u}, \boldsymbol{v})$, we may write $\boldsymbol{\theta}$ as $(\boldsymbol{\lambda}, \boldsymbol{\psi})$. Here $\boldsymbol{\lambda}$, represented by $\boldsymbol{\theta}_u$ or $\boldsymbol{\mu}_u$, is regarded as a *nuisance parameter*, supplementing $\boldsymbol{\psi}$. As shown in Section 3.3.3 above, $\boldsymbol{\lambda} = \boldsymbol{\mu}_u = E_\theta(\boldsymbol{u})$ is the preferable nuisance parameter, at least in principle, since $\boldsymbol{\psi}$ and $\boldsymbol{\mu}_u$ are variation independent and information orthogonal (Proposition 3.20).

Proposition 3.21 ***Conditionality principle for full families***
Statistical inference about the canonical parameter component $\boldsymbol{\psi}$ in presence of the nuisance parameter $\boldsymbol{\lambda}$ or $\boldsymbol{\mu}_u = E_\theta(\boldsymbol{u})$ should be made conditional on $\boldsymbol{u}$, that is, in the conditional model for $\boldsymbol{y}$ or $\boldsymbol{v}$ given $\boldsymbol{u}$.

Motivation. The likelihood for $(\boldsymbol{\mu}_u, \boldsymbol{\psi})$ factorizes as

$$L(\boldsymbol{\mu}_u, \boldsymbol{\psi};\, t) = L_1(\boldsymbol{\mu}_u, \boldsymbol{\psi};\, u)\, L_2(\boldsymbol{\psi};\, \boldsymbol{v} \,|\, \boldsymbol{u}), \tag{3.28}$$

where the two parameters are variation independent. In some cases, exemplified by Example 3.2, L_1 depends only on $\boldsymbol{\mu}_u$,

$$L(\boldsymbol{\mu}_u, \boldsymbol{\psi};\, t) = L_1(\boldsymbol{\mu}_u;\, \boldsymbol{u})\, L_2(\boldsymbol{\psi};\, \boldsymbol{v} \,|\, \boldsymbol{u}). \tag{3.29}$$

Then it is clear that there is no information whatsoever about $\boldsymbol{\psi}$ in the first factor L_1, and the argument for the principle is compelling. The terminology for this situation is that the factorization is called a *cut*, and $\boldsymbol{u}$ is called *S-ancillary* for $\boldsymbol{\psi}$. But also when L_1 depends on $\boldsymbol{\psi}$ (illustrated in Example 3.3), there is really no information about $\boldsymbol{\psi}$ in $\boldsymbol{u}$, as seen by the following argument. Note first that $\boldsymbol{u}$ and $\boldsymbol{\mu}_u$ are of the same dimension (of course), and that $\boldsymbol{u}$ serves as an estimator (the MLE) of $\boldsymbol{\mu}_u$, whatever be $\boldsymbol{\psi}$ (Proposition 3.11). This means that the information in $\boldsymbol{u}$ about $(\boldsymbol{\mu}_u, \boldsymbol{\psi})$ is totally consumed in the estimation of $\boldsymbol{\mu}_u$. Furthermore, the estimated value of $\boldsymbol{\mu}_u$ does not provide any information about $\boldsymbol{\psi}$, and $\boldsymbol{\mu}_u$ would not do so even if it were known, due to the variation independence between $\boldsymbol{\mu}_u$ and $\boldsymbol{\psi}$. Thus, the first factor L_1 contributes only information about $\boldsymbol{\mu}_u$ in (3.28). □

In wider generality, (3.28) and (3.29) correspond to cases 4.(b) and 4.(a), respectively, in the arguments for conditioning formulated by Barndorff-Nielsen and Cox (1994, p. 38). The advantage of the present exponential family setting is that the argument becomes even stronger, in particular by reference to the mixed parameterization. Cf. also Section 7.5 for curved families.

Example 3.2 ***Two Poisson variates***
Suppose we want to make inference about the relative change in a Poisson parameter from one occasion to another, with one observation per occasion.

Specifically, let y_1 and y_2 be independent Poisson distributed variables with mean values e^{λ} and $e^{\psi+\lambda}$, respectively. The model is

$$f(y_1, y_2; \lambda, \psi) = \frac{h(y_1, y_2)}{C(\lambda, \psi)} e^{\lambda(y_1 + y_2) + \psi y_2}$$

The parameter of prime interest is e^{ψ}, or equivalently the canonical ψ. The conditionality principle states that the inference about ψ should be made in the conditional model for $v = y_2$ given the other canonical statistic $u = y_1 + y_2$. Simple but important calculations left as an exercise (Exercise 3.13) show that this model is the binomial distribution with logit parameter ψ, $\text{Bin}\{y_1 + y_2;\, e^{\psi}/(1 + e^{\psi})\}$. The marginal distribution for $u = y_1 + y_2$ is $\text{Po}(\mu_u)$, not additionally involving ψ. Thus we have a cut, and S-ancillarity. Note the crucial role of the mixed parameterization to obtain a cut. If expressed instead in terms of the canonical parameters λ and ψ of the joint model, the Poisson parameter in the marginal for u would have been dependent on both λ and ψ, since $\mu_u = E(y_1 + y_2) = e^{\lambda}(1 + e^{\psi})$. △

Under S-ancillarity, the MLEs will be the same when derived in the conditional or marginal models as in the joint model, and the observed and expected information matrices in the joint model will be block diagonal with blocks representing the conditional and marginal models. This can sometimes be used in the opposite way, by artificially extending a conditionally defined model to a simpler joint model with the same MLEs etc, see Section 5.6. Outside S-ancillarity we cannot count on these properties, but we must instead distinguish joint model and conditional model MLEs. Here is one such example, see Example 3.6 for another one.

Example 3.3 ***Conditional inference for Gaussian sample, cont'd***
Suppose we want to make inference about σ^2, or σ. Then we are led to consider the conditional distribution of $\sum y_i^2$, given $\bar{y}$, that depends on σ alone (via the canonical, being $\propto 1/\sigma^2$). The marginal distribution of $\bar{y}$ depends on both μ and σ^2, so the joint and conditional likelihoods are different functions of σ^2.

As we have seen in Example 3.1, in the conditional approach $n\bar{y}^2$ is a constant, and after subtraction of this empirical constant from $\sum y_i^2$ we are led to the use of the statistic s^2. Now, we already know that $\bar{y}$ and s^2 are independent, so the even simpler result is that the inference should be based on the marginal model for $(n-1)s^2/\sigma^2$, with its $\chi^2(n-1)$ distribution. In particular this leads to the conditional and marginal ML estimator $\hat{\sigma}^2 = s^2$, which differs by the factor $n/(n-1)$ from the MLE in the joint model (with denominator n). This argument for use of s and s^2 is much more compelling than the unbiasedness of $\widehat{\sigma^2} = s^2$ (but not of $\hat{\sigma} = s$).

In passing, note that not only in this example, but more generally – for example in Gaussian linear models (Exercise 3.15) – conditional inference for variance estimation is equivalent to marginal inference based on the residuals alone, after fitting the mean value parameters of the linear model. Restriction to residuals, generally denoted REML (REsidual, REstricted or REduced ML), is a useful procedure in mixed models and variance components models, in particular in unbalanced situations. Smyth and Verbyla (1996) give a sophisticated argument for a conditional interpretation in a class of models much wider than the Gaussian, incorporating models that are exponential families and generalized linear models for fixed variance parameters.

In this example $u = \bar{y}$ is not S-ancillary for $\psi = \sigma^2$, since u is $N(\mu, \sigma^2/n)$, which depends on both σ^2 and μ_u. However, since μ_u is unknown and arbitrary, it is clear that u does not provide any information about σ^2.

The corresponding question for μ does not motivate a conditional inference. There is no linear combination of the canonical parameters μ/σ^2 and $-1/2\sigma^2$ that is a function of μ alone, so the conditionality principle is not applicable. Also, the distribution for the MLE $\bar{y}$ clearly depends on σ^2 (whereas $\bar{y}$ is independent of s^2). We cannot get rid of σ^2 by conditioning. △

Exercise 3.13 ***Conditioning on a Poisson sum***
Show that if y_1 and y_2 are independent and Poisson distributed with mean values μ_1 and μ_2, respectively, then the conditional distribution of y_1 given their sum $y_1 + y_2$ is binomial, $\text{Bin}(y_1 + y_2, \mu_1/(\mu_1 + \mu_2))$. △

Exercise 3.14 ***Two Poisson variates in different clothes***
Suppose y_1 and y_2 are independent Poisson, parameterized as $\text{Po}(\psi\lambda)$ and $\text{Po}((1-\psi)\lambda)$, respectively. Show that inference about ψ should be a conditional inference, and that in a suitable parameterization of the joint model we have a cut. The model reappears in Example 7.11. △

Exercise 3.15 ***Variance parameter of a linear model***
Suppose we have a set of n observations from a linear model with $k < n$ mean value parameters. Use linear models theory to extend the argument of Example 3.3 and to yield the conditional = marginal ML estimator of the error variance σ^2. △

Exercise 3.16 ***Structural and incidental parameters***
An even more extreme variation of Example 3.3 and Exercise 3.15 was discussed by Neyman and Scott (1948). Suppose we have k normal samples each of size $n = 2$, say, and each with its own 'incidental' mean value μ_i, $i = 1, \ldots, k$, but with the 'structural' variance parameter σ^2 in common. This is a frequent type of situation in practice, representing say a measuring instrument used to measure k different objects, each in duplicates. Follow Neyman and Scott by showing that the unconditional MLE of σ^2 is seriously biased and not consistent as $k \to \infty$, whereas the conditionality principle leads to a conditional MLE, given the sample averages observed, that is both unbiased and consistent. △

3.6 Common Models as Examples

Example 3.4 ***Bernoulli trials, continued from Example 2.1***
We know that the canonical parameter is $\theta = \text{logit}(\pi_0)$ and that the norming constant is $C(\theta) = (1 + e^{\theta})^n$. Hence $\Theta = \mathbb{R}$, which is trivially open and convex. The mean value parameter is $n\pi_0$ with $0 < \pi_0 < 1$. Transforming from $t = \sum y_i$ to $t' = t/n = \overline{y}$ changes this parameter to π_0. The likelihood equation is $t = n\pi_0$. The outcomes $t = 0$ and $t = n$ fall outside the image $\mu(\Theta)$ of Θ, and the likelihood equation in θ then has no solution. In the parameterization by π_0 we can write $\hat{\pi}_0 = 0$ and $\hat{\pi}_0 = 1$, respectively, but the corresponding degenerate distributions do not belong to the exponential family (as also discussed in Section 3.2).

The Fisher information for θ can be found by differentiation of $\log C(\theta)$:

$$I(\theta) = \frac{d^2 \log C(\theta)}{d\theta^2} = n\frac{e^{\theta}}{(1 + e^{\theta})^2} = n\pi_0(1 - \pi_0) = \text{var}(t).$$

The Fisher information for the mean value parameter $\mu_t = n\pi_0$ is the inverse of this, see (3.20), from which we get

$$I(\pi_0) = \frac{n}{\pi_0(1 - \pi_0)}.$$

In this case, $\text{var}(\hat{\pi}_0) = I(\pi_0)^{-1}$, with exact equality, cf. Proposition 3.25 in Section 3.8. △

Example 3.5 ***Logistic regression, continued from Example 2.2***
For the two-parameter logistic regression model the canonical statistic consists of $\sum y_i$ and $\sum x_i y_i$. Since

$$E(y_i) = \pi_i = e^{\theta_0 + \theta_1 x_i}/(1 + e^{\theta_0 + \theta_1 x_i}),$$

the likelihood equation system is

$$\begin{cases} \sum y_i = \sum \dfrac{e^{\theta_0+\theta_1 x_i}}{1+e^{\theta_0+\theta_1 x_i}}, \\ \sum x_i y_i = \sum \dfrac{x_i\, e^{\theta_0+\theta_1 x_i}}{1+e^{\theta_0+\theta_1 x_i}}. \end{cases} \tag{3.30}$$

This equation system cannot be solved explicitly, but, except in some degenerate cases, a root exists in the canonical parameter space $\boldsymbol{\Theta} = \mathbb{R}^2$ (cf. Exercise 3.17). In this model the canonical parameter is a more natural parameter than the mean value parameter.

Since $\mathrm{var}(y_i) = \pi_i(1-\pi_i)$, the information matrix is

$$I(\boldsymbol{\theta}) = J(\boldsymbol{\theta}) = \begin{pmatrix} \sum \pi_i(1-\pi_i) & \sum x_i \pi_i(1-\pi_i) \\ \sum x_i \pi_i(1-\pi_i) & \sum x_i^2 \pi_i(1-\pi_i) \end{pmatrix}, \tag{3.31}$$

where $\pi_i(1-\pi_i) = e^{\theta_0+\theta_1 x_i}/(1+e^{\theta_0+\theta_1 x_i})^2$, when expressed in θ.

For multiple logistic regression it is more intricate to specify when the likelihood equations have a unique or partially unique solution in $\boldsymbol{\Theta}$ (Albert and Anderson, 1984; Jacobsen, 1989; Kolassa, 1997).

Continued in Example 3.6. △

Example 3.6 ***Conditional inference in logistic regression***

For inference about the regression parameter θ_1 in the simple logistic regression of Example 3.5, the conditionality principle of Section 3.5 tells us to make the inference conditional on the value of $t_0 = \sum y_i$. The first of the likelihood equations (3.30) gives the mean value parameter μ_0 for $t_0 = \sum y_i$ expressed in θ, but the distribution for $\sum y_i$ is not characterized by μ_0 alone. Thus, t_0 is not S-ancillary for inference about θ_1, and the conditional MLE of θ_0 differs from the joint model MLE (cf. σ^2 in Example 3.3).

Conditional inference is not always used in logistic regression, because the conditional quantities are numerically intractable, and with much data the improvement in relevance should be negligible. In some variations of this example, the situation is different, however. Suppose data represent individuals sampled from different strata, with one intercept θ_{0s} per stratum s, but the same θ_1 for all strata. Then there are as many mean values to condition on as there are strata. In particular, if the strata are pairs, as they are in a *matched pairs* study, the total number of individuals is only twice the number of strata parameters, and the conditional approach is highly motivated. See for example Agresti (2013, Sec. 7.3, 11.2) for more examples and more details (but his statement '$\sum y_i$ is sufficient for θ_0' is inadequate, since the distribution for $\sum y_i$ involves the whole of θ).

The numerical tasks require special software, e.g. `clogit` in **R**. △

Example 3.7 ***Poisson sample, continued from Example 2.3***
For the simple Poisson model, $\mu_t(\theta) = e^\theta = \lambda$, so $\mu(\Theta) = \mathbb{R}_+$, the open positive half-axis. The likelihood equation is $y = e^\theta$ or $y = \lambda$. Provided $y > 0$, the solution represents a proper exponential family. The Fisher (and the observed) information for θ is $\mathrm{var}(y) = \lambda$. Note how the equality of mean and variance for the Poisson connects with the fact that $\log C(\theta) = e^\theta$ does not change under differentiation with respect to θ. △

Example 3.8 ***Multiplicative Poisson model, continued from Example 2.5***
In the Poisson family (2.9), the general likelihood equations (3.14) take the form

$$\begin{cases} y_{i.} = E(y_{i.}) = \sum_{j=1}^{s} \alpha_i \beta_j = \alpha_i \beta_. & i = 1, \dots, r, \\ y_{.j} = E(y_{.j}) = \sum_{i=1}^{r} \alpha_i \beta_j = \alpha_. \beta_j & j = 1, \dots, s, \end{cases} \tag{3.32}$$

well-known from the theory of log-linear models. With, for example, the constraint $\alpha_1 = 1$ in order not to overparameterize, the canonical parameter space is the open convex $\boldsymbol{\Theta} = \mathbb{R}^{r+s-1}$. since the canonical parameters are the $\log \alpha_i$ and $\log \beta_j$. The image of $\boldsymbol{\Theta}$ under the function μ is $(\mathbb{R}_+)^{r+s-1}$.

The likelihood equation system has the explicit solutions

$$\begin{cases} \hat{\alpha}_i = c\, y_{i.}\,, \\ \hat{\beta}_j = y_{.j}/(c y_{..})\,, \end{cases} \tag{3.33}$$

where c is a constant that cancels from the product $\hat{\alpha}_i \hat{\beta}_j$, and which is specified by the parameterization restriction, e.g. $c = 1/y_{1.}$ for $\alpha_1 = 1$.

The norming constant is

$$\log C(\boldsymbol{\theta}) = \sum_{i=1}^{r} \sum_{j=1}^{s} \alpha_i \beta_j = \sum_i \sum_j e^{\log \alpha_i + \log \beta_j}\,,$$

which is evidently a convex function of the log-parameters. Differentiation with respect to the log-parameters yields the right-hand sides of (3.32). Differentiation once more yields the elements of $I_t(\theta) = \mathrm{var}(\boldsymbol{t})$. Alternatively, and simpler, we get them directly from the Poisson properties and the mutual independence of all y_{ij}. In any case, the result is that the nonzero elements of $I_t(\theta)$ are (as expressed in the αs and βs):

$$\mathrm{var}(y_{i.}) = \alpha_i \beta_.\,,$$
$$\mathrm{var}(y_{.j}) = \alpha_. \beta_j\,,$$
$$\mathrm{cov}(y_{i.}, y_{.j}) = \alpha_i \beta_j\,.$$

The matrix is singular, unless the matrix size is reduced (for example by setting $\alpha_1 = 1$ and deleting the first row and first column).

To see the implications of the conditionality principle of Section 3.5, let the parameter constraint instead be chosen as $\sum_i \alpha_i = 1$. Then we see from (3.32) that inference about the individual α_i should be made in the conditional distribution of data (or of the row sums), given the set of column sums. This conditional distribution for the row sums is a multinomial distribution, with multinomial probabilities α_i (with $\sum_i \alpha_i = 1$) and canonical parameters $\log \alpha_i$. On the other hand, the column sums $y_{.j}$ are marginally Poisson distributed with mean value parameters β_j. Thus we have a mixed parameterization, where the column sums are S-ancillary for inference about the canonical $\log \alpha_i$ in the multinomial distribution of the row sums $\{y_{i.}\}$, given the column sums. This is an extension of Example 3.2. Note that $\log \alpha_i$ uniquely determines α_i, and vice versa. △

Example 3.9 ***Exponential and gamma, continued from Example 2.7***
The canonical parameter for the gamma distribution family (2.17) is $\boldsymbol{\theta} = (\theta_1, \theta_2) = (-\alpha, \beta)$, where $\theta_2 = 0$ yields the exponential distribution subfamily. The canonical parameter space $\boldsymbol{\Theta}$ is the open and convex set $\theta_1 < 0,\ \theta_2 > -1$. The log of the norming constant,

$$\log C(\boldsymbol{\theta}) = \log \Gamma(\theta_2 + 1) - (\theta_2 + 1)\log(-\theta_1), \tag{3.34}$$

immediately yields the mean value and variance for y, by differentiation with respect to θ_1:

$$E(y) = -\frac{\theta_2 + 1}{\theta_1} = \frac{\beta + 1}{\alpha}, \tag{3.35}$$

$$\mathrm{var}(y) = \frac{\theta_2 + 1}{\theta_1^2} = E(y)^2/(\beta + 1). \tag{3.36}$$

An alternative and often convenient parameterization of the gamma family is the mixed one, by its mean value $\mu = E(y)$ together with $\theta_2 = \beta$.

The log of the norming constant, (3.34), is a convex function of $\boldsymbol{\theta}$. This is guaranteed by Proposition 3.8. Without this general result, we would have had to show that the special function $\log \Gamma$ is convex, which is not an elementary result.

The function $\log \Gamma$ complicates ML estimation, too. The likelihood equation system for a sample from a gamma distribution can be expressed as

$$\begin{cases} \overline{y} = -(\theta_2 + 1)/\theta_1\,, \\ \overline{\log y} = E(\log y) = \dfrac{\partial \log C(\boldsymbol{\theta})}{\partial \theta_2}\,. \end{cases}$$

Thus, for the likelihood equations we need the derivative of the $\log \Gamma$ function, which is also not an elementary function. The second order derivative of $\log\Gamma$, needed in the information matrix, has a closed form only in the special case $\theta_2 = 0$ (the exponential distribution), when it equals $\pi^2/6$ (see e.g. Cramér (1946, p. 131), a fact that can be utilized in a test statistic for testing if a gamma sample is actually from an exponential distribution, see Exercise 5.13.

The likelihood maximization can be carried out in the following way. First we use the likelihood equation $\overline{y} = \mu$ that yields $\hat{\theta}_1$ explicitly in terms of θ_2 and $\overline{y}$ as

$$\hat{\theta}_1(\theta_2) = -(\theta_2 + 1)/\overline{y}. \tag{3.37}$$

Inserting this expression for θ_1 in $\log L$, we are left with the profile likelihood (Section 3.3.4) for θ_2 to be maximized. This can be done numerically in (for example) **R** or MATLAB by using the built-in $\log\Gamma$ functions. However, with a statistical package available, we rather use the fact that the gamma distribution gives rise to a generalized linear model, see further Example 9.5. Note also that setting $\theta_2 = 0$ in (3.37) yields the exponential distribution MLE of θ_1. △

Example 3.10 ***A Gaussian sample, continued from Example 2.8***
For the normal distribution family, with $\boldsymbol{\theta} = (\mu/\sigma^2, -1/2\sigma^2)$, the canonical parameter space $\boldsymbol{\Theta}$ is an open half-space, hence convex. The joint (unconditional) likelihood equations are most easily found using Proposition 3.10:

$$\begin{cases} \sum_i y_i &= E(\sum y_i) = n\,\mu\,, \\ \sum_i y_i^2 &= E(\sum y_i^2) = n\,(\sigma^2 + \mu^2)\,. \end{cases} \tag{3.38}$$

In this particular model it is seen that the ML method agrees with the *moment method of estimation*, which equates the first two moments to their expected values.

The Fisher information matrix for $\boldsymbol{\theta}$ can be written

$$\begin{aligned} I(\boldsymbol{\theta}) = J(\boldsymbol{\theta}) = V_t(\boldsymbol{\theta}) &= n\begin{pmatrix} \mathrm{var}(y) & \mathrm{cov}(y,\, y^2) \\ \mathrm{cov}(y,\, y^2) & \mathrm{var}(y^2) \end{pmatrix} \\ &= n\begin{pmatrix} \sigma^2 & 2\mu\sigma^2 \\ 2\mu\sigma^2 & 2\sigma^4 + 4\mu^2\sigma^2 \end{pmatrix}, \end{aligned} \tag{3.39}$$

where for simplicity, but perhaps also confusion, the matrix elements are expressed in terms of the moments μ and σ^2. The formulas for $\mathrm{cov}(y,\, y^2)$ and $\mathrm{var}(y^2)$ may be derived for example by differentiation twice with respect to θ_1 and θ_2 of

$$\log C(\boldsymbol{\theta}) = \frac{\mu^2}{2\sigma^2} + \log\sigma = -\frac{\theta_1^2}{4\theta_2} - \frac{1}{2}\log(-\theta_2) + \text{constant}.$$

To do this is left to the reader, see Exercise 3.23.

The equation system (3.38) may be nonlinearly reorganized and expressed in terms of the alternative sufficient statistics $\overline{y}$ and $\sum(y_i - \overline{y})^2$,

$$\begin{cases} \overline{y} &= \mu\,, \\ \sum_i (y_i - \overline{y})^2 &= n\sigma^2\,. \end{cases} \tag{3.40}$$

Note that after this nonlinear transformation, the right-hand side of the second equation, $n\sigma^2$, is no longer the expected value of the left-hand side. Hence, the well-known fact that the ML estimator of σ^2 is not unbiased. However, as we saw in Example 3.1 and Example 3.3, inference about σ^2 should be conditional on $\overline{y}$. This leads to the usual (unbiased) s^2 as being the solution of the conditional likelihood equation for σ^2,

$$\sum_i (y_i - \overline{y})^2 = (n-1)\,\sigma^2\,. \tag{3.41}$$

△

Example 3.11 ***The multivariate normal, continued from Example 2.10***
An immediate extension of the previous univariate results, in particular (3.41), to a multivariate normal sample yields the following likelihood equation system. The observations $\boldsymbol{y}_i$ are now vectors, together with their average $\overline{\boldsymbol{y}}$ and mean $\boldsymbol{\mu}$:

$$\begin{aligned} \hat{\boldsymbol{\mu}} &= \overline{\boldsymbol{y}}\,, \\ \widehat{\Sigma} &= \sum_i (\boldsymbol{y}_i - \overline{\boldsymbol{y}})(\boldsymbol{y}_i - \overline{\boldsymbol{y}})^T / (n-1)\,. \end{aligned} \tag{3.42}$$

Like in the univariate case, the conditional MLE of Σ is unbiased. △

Example 3.12 ***Covariance selection models, continued***
In covariance selection models, as introduced in Example 2.11, we reduce the model dimension by prescribing off-diagonal zeros in Σ^{-1}, interpreted in terms of conditional independence. This corresponds to omitting sums of products from the sufficient statistic. The likelihood equations for the model are obtained by simply deleting the corresponding equations from the covariance part of the equation system (3.42). By Proposition 3.13 the MLE exists as soon as the sample covariance matrix is nonsingular, and this has probability 1 if $n > \dim y$. For the remaining task, to find the elements of $\widehat{\Sigma}$ after inserting zeros in Σ^{-1}, see Section 10.2. △

Exercise 3.17 ***MLE existence in logistic regression, Example 3.5 cont'd***
Assume a logistic regression model of y on a single x, and suppose (for simplicity) we make only three observations; one for each of $x = -1, 0, 1$. Investigate for which outcomes of data the MLE exists, by characterizing $\boldsymbol{\Theta}$, $\mu_t(\boldsymbol{\Theta})$, $\mathcal{T}$ = set of possible outcomes of t, and int(convex closure($\mathcal{T}$)). Make a diagram showing $\mathcal{T}$ and int(convex closure($\mathcal{T}$)) as sets in $\mathbb{R}^2$. △

Exercise 3.18 ***Alternative parameterizations in logistic regression***
An alternative canonical parameterization in logistic regression is obtained like in linear regression models by rewriting $\theta_1 + \theta_2 x_i = \eta_1 + \eta_2(x_i - \overline{x})$. Here (η_1, η_2) is related with (θ_1, θ_2) by a partially trivial linear transformation. In simple linear regression a motivation for this transformation is that it makes the information matrix diagonal. Calculate the information matrix for the parameter vector (η_1, η_2) by use of the Reparameterization lemma. Find that there is no reason why this matrix should be diagonal.

Another parameter often of interest in connection with logistic regression is the ED_{50} (or LD_{50}), which is the x-value for which the logit is zero. If we think of a dose–response experiment, where a random sample of individuals is exposed to individual doses x_i, and each individual has a latent threshold value ξ_i such that it responds (yields $y = 1$) precisely when the dose exceeds the threshold value, then this parameter can be interpreted as the median of the population distribution of threshold values, that is, the 'effective dose' (ED) or 'lethal dose' (LD), to which half of the population responds. Denote ED_{50} by μ and reparameterize according to $\theta_1 + \theta_2 x_i = \beta(x_i - \mu)$. Use the Reparameterization lemma 3.14 to derive the information matrix for the parameterization by (μ, β), and compare with that of (η_1, η_2). △

Exercise 3.19 ***Conditional inference about an odds ratio***
Suppose we have two observations y_0 and y_1, mutually independent and distributed as $\text{Bin}(n_0, \pi_0)$ and $\text{Bin}(n_1, \pi_1)$, respectively, for example representing a study of disease occurrence in two cohorts. We want to make inference about the odds ratio, or equivalently the log odds ratio,

$$\psi = \text{logit}(\pi_1) - \text{logit}(\pi_0) = \log\left(\pi_1(1 - \pi_0)/\pi_0(1 - \pi_1)\right).$$

(a) Express this model in terms of ψ and $\text{logit}(\pi_0)$ as canonical parameters.
(b) For inference about ψ, motivate conditional inference and derive the conditional distribution for y_1 given $y_0 + y_1$. (*Remark:* This is the *generalized* (or noncentral) *hypergeometric distribution*, see Cox (2006, Ch. 4), Agresti (2013, Sec. 7.3.3, 16.6.4); cf. Section 5.2 on testing $\psi = 0$. △

Exercise 3.20 ***Negative binomial (continued from Exercise 2.2)***
Suppose y has a one-parameter negative binomial distribution, with

$$f(y;\pi_0) = \binom{y+k-1}{k-1} \pi_0^k (1-\pi_0)^y = \frac{\Gamma(y+k)}{y!\,\Gamma(k)} \pi_0^k (1-\pi_0)^y.$$

When k is (a known) integer, y can be interpreted as the number of failures in Bernoulli trials until k successes, with success probability p, However, k need not be a positive integer, for this distribution to be a well-defined family, but k can be allowed to be any known positive number.
(a) Find the relationship between π_0 and the canonical θ, and find $C(\theta)$.
(b) Use $C(\theta)$ to derive $\mu = E(y)$ and var(y). Show that $\mathrm{var}(y) = \mu + \mu^2/k$.
(c) Derive the MLE for π_0 from the MLE for μ, given a sample from the distribution (known k).
(continued in Example 4.2) △

Exercise 3.21 ***Weighted multiplicative Poisson, continued***
The weighted multiplicative Poisson model, with Poisson means $\lambda_{ij} = N_{ij}\,\alpha_i\beta_j$ for known N_{ij}-values, was introduced in Exercise 2.1. Details concerning the following tasks are found in Andersen (1977).
(a) Reconsider the canonical statistic and derive the likelihood equations.
(b) Unfortunately, there is only exceptionally an explicit solution to it. Suggest a 'flip-flop' type algorithm for solving the likelihood equations, which is alternatingly fitting the α_is and the β_js.
(c) Show that the likelihood equations can be reexpressed as one set of equations for only the α_is and another set for only the β_js.
(d) Derive the expected and observed information matrices in the parameterization chosen.
(e) For the sequel, suppose the row effects are of particular interest, whereas the column effects are not. Factorize the likelihood by conditioning on the set of column sums. Show that the conditional likelihood depends on only the α-parameters, whereas the marginal likelihood (for the column sums) can be reparameterized to depend on only one nuisance parameter per column. Hence, argue in favour of conditional inference.
(f) Interpret the preceding result in terms of a *mixed* parameterization of the original family.
(g) Show that the conditional model likelihood equations for the α-part parameters are the same as in the original model, cf. (c). △

Exercise 3.22 ***Multinomial distribution***
For y from a multinomial distribution, see Example 2.4, choose the mean value parameterization, and derive and solve the likelihood equations. △

Exercise 3.23 ***A sample from a normal distribution***
For a sample from $\mathrm{N}(\mu, \sigma^2)$,
(a) verify formula (3.39) for $I(\boldsymbol{\theta})$ by differentiation of $\log C(\boldsymbol{\theta})$.
(b) find $I(\mu, \sigma^2)$ from $I(\boldsymbol{\theta})$ by use of the Reparameterization lemma, Proposition 3.14. Note the result that $I(\mu, \sigma^2)$ is diagonal, with diagonal elements n/σ^2 and $n/(2\sigma^4)$. △

Exercise 3.24 ***Gaussian linear models, continued***
For the case of a data vector y from a Gaussian linear model (Example 2.9),
(a) go through and check or interpret the propositions so far in Chapter 3, in analogy with the other examples in Section 3.6;
(b) in particular derive and solve the likelihood equations, and derive the corresponding information matrix. Make for example use of Example 3.10.
△

Exercise 3.25 ***Inverse Gaussian distribution family, continued***
(a) Let y have the inverse Gaussian distribution, see Exercise 2.6. Use the expression for $\log C(\theta)$ from there to derive moments of y by suitable differentiation. In this way, show that $E(y) = \mu$ and $\mathrm{var}(y) = \mu^3/\alpha$.
(b) Suppose we have a sample from this inverse Gaussian. Derive the likelihood equations and show that they have the explicit solution $\hat{\mu} = \overline{y}$ and $1/\hat{\alpha} = \overline{(1/y)} - 1/\overline{y}$, when expressed in these parameters.
(c) Derive the information matrix for the canonical θ and use for example the Reparameterization lemma to show that the information matrix for (μ, α) is diagonal, i.e. that these parameters are information orthogonal.
(d) Use the theory of mixed parameterizations to conclude the result in (c) without calculations. △

Exercise 3.26 ***Moments for density estimate***
Let $f_0(y)$ be some estimate of some density, based on a sample $y_1, \ldots, y_n$. This estimate need not have the same mean and variance as the sample, but by ML estimation in the tilted exponential family with density

$$f(y;\, \theta_1, \theta_2) \propto f_0(y)\, e^{\theta_1 y + \theta_2 y^2},$$

the estimate f_0 can be refined to a density with the same first and second moments as the sample (Efron and Tibshirani, 1996), cf. Example 2.12. Derive the likelihood equations and show the moment property. △

3.7 Completeness and Basu's Theorem

Completeness of a (minimal sufficient) statistic $\boldsymbol{t}$, say, essentially means that the parametric model (parameter space) for $\boldsymbol{t}$ is rich enough to match the dimension of $\boldsymbol{t}$. By abuse of language, the statistic is often said to be complete, even though completeness is a demand on the corresponding family of distributions. An assumption of completeness is typically used in statistical theory for demonstration of uniqueness properties. The concept is of a relatively technical character, and will not be needed outside this section, but for completeness (sic!), a precise definition is given here:

Definition 3.22 ***Completeness***
A statistic $\boldsymbol{t}$ is *complete*, or more properly expressed, the family of distributions of $\boldsymbol{t}$ (parameter space $\boldsymbol{\Theta}$, say) is complete, if the property

$$E_\theta\{h(\boldsymbol{t})\} = 0 \text{ for all } \boldsymbol{\theta} \in \boldsymbol{\Theta}, \tag{3.43}$$

for a function $h(\boldsymbol{t})$, requires that the function $h(\boldsymbol{t})$ is the zero function (being zero with probability 1).

The definition is not special to exponential families, but canonical statistics of full exponential families of distributions provide most of the important examples. A slightly weaker criterion is *boundedly complete*, which only requires (3.43) for bounded functions $h(\boldsymbol{t})$. This is actually enough for Basu's theorem (Proposition 3.24), but full exponential families possess the stronger property.

Proposition 3.23 ***Exponential family completeness***
If $\boldsymbol{t}$ *is the canonical statistic of a full exponential family, the model for* $\boldsymbol{t}$ *is complete.*

Proof This can be proved by using the uniqueness theorem for Laplace transforms, noting that $E_\theta\{h(\boldsymbol{t})\}$ is the Laplace transform of $h(\boldsymbol{t})g(\boldsymbol{t})$, where $g(\boldsymbol{t})$ is the structure function. The reader is referred to Barndorff-Nielsen (1978, Lemma 8.2) or Lehmann and Romano (2005, Sec. 4.3) for details. □

One use of the completeness property is in Basu's theorem, that comes next. Another use (but not here) is for establishing uniqueness in the Rao–Blackwell theorem, see for example Pawitan (2001, Sec. 8.6).

Proposition 3.24 ***Basu's theorem***
If $\boldsymbol{t}$ *is a complete sufficient statistic, typically the canonical statistic* $\boldsymbol{t}$ *in a*

full exponential family, and $\boldsymbol{u}$ *is another statistic, then* $\boldsymbol{u}$ *and* $\boldsymbol{t}$ *are mutually independent precisely when the distribution of* $\boldsymbol{u}$ *is parameter-free.*

Proof First we note that $\boldsymbol{t}$ sufficient means that the conditional distribution of $\boldsymbol{u}$, given $\boldsymbol{t}$, is parameter-free. Basu's theorem, however, concerns the *marginal* distribution of $\boldsymbol{u}$.

The simplest direction of proof is when we start by assuming $\boldsymbol{u}$ and $\boldsymbol{t}$ independent. Then, the marginal distribution of $\boldsymbol{u}$ is identical with the conditional distribution of $\boldsymbol{u}$ given $\boldsymbol{t}$, for any $\boldsymbol{t}$. Hence, the marginal distribution of $\boldsymbol{u}$ is also parameter-free, which was to be shown.

For the converse part, completeness comes in. Starting with a parameter-free distribution for $\boldsymbol{u}$, $E\{s(\boldsymbol{u})\}$ must be parameter-free for any statistic $s(\boldsymbol{u})$. Since $\boldsymbol{t}$ is sufficient, the conditional expected value, $E\{s(\boldsymbol{u})|\boldsymbol{t}\}$, will also be parameter-free. Next, $E\{s(\boldsymbol{u})|\boldsymbol{t}\}$ has expected value $E\{s(\boldsymbol{u})\}$, so

$$h(\boldsymbol{t}) = E\{s(\boldsymbol{u})|\boldsymbol{t}\} - E\{s(\boldsymbol{u})\}$$

is a (parameter-free) function of $\boldsymbol{t}$ with $E_{\theta}\{h(\boldsymbol{t})\} = 0$ for all $\boldsymbol{\theta}$. Since $\boldsymbol{t}$ is complete, it follows that the function h must be the zero function, or in other words, $E\{s(\boldsymbol{u})|\boldsymbol{t}\}$ is a constant, independent of $\boldsymbol{t}$. This is enough for the conclusion that $\boldsymbol{u}$ is independent of $\boldsymbol{t}$; for example we can let $s(\boldsymbol{u})$ be the indicator function for a set of $\boldsymbol{u}$-values. The expected value of an indicator function is the probability of the corresponding set. □

Most examples for exponential families of the use of Basu's theorem are related to the normal distribution. The following simple properties for a sample from a normal distribution with sample mean $\overline{y}$ and variance s^2 are typical, but easily extended to, for example, linear models:

- The residuals $y_i - \overline{y}$ are independent of $\overline{y}$ (regard σ^2 as already specified)
- $\overline{y}$ and s^2 are mutually independent (regard σ^2 as already specified)
- The normalized residuals $(y_i - \overline{y})/s$ are independent of $\overline{y}$ and s^2.

In Section 5.1 we will repeatedly refer to Basu's theorem.

Exercise 3.27 ***Correlation coefficient***
Suppose (y_1, y_2) follows a bivariate normal distribution, but a special one with $\rho = 0$. Given a sample from this distribution, show that the sample correlation coefficient r is statistically independent of the two sample means and the two sample variances. This property is useful in Exercise 5.2. △

Exercise 3.28 ***First-order ancillarity***
Show that if t is the canonical statistic of a full exponential family, and

u is another statistic, then u and t are uncorrelated if and only if $E_\theta(u)$ is parameter-free. *Hint:* The derivative of a constant is zero.

That $E_\theta(u)$ is parameter-free is a weaker requirement than that the distribution of u is parameter-free, of course, and likewise uncorrelated is weaker than independent. When $E_\theta(u)$ is parameter-free, u is called *first-order ancillary* by Lehmann and Casella (1998, Sec. 1.6). △

3.8 Mean Value Parameter and Cramér–Rao (In)equality

Exponential families in mean value parameterization have a remarkable role in the Cramér–Rao inequality (also called the information inequality), which gives a lower bound for the variance of an estimator, see Proposition 3.25, or alternatively for example Pawitan (2001, Sec. 8.5) for a one-parameter version of it, and Bickel and Doksum (2001, Sec. 3.4) for the multiparameter case. Equality can only be attained if the parametric family is of exponential type and the estimator used is its canonical statistic (or an affine function of it). For unbiased estimators, the bound is the inverse of the Fisher information, and an unbiased estimator is efficient (or minimum variance) in the sense of equality in the Cramér–Rao inequality, precisely when it is the canonical statistic in an exponential family, which is used to estimate its mean value, that is, the mean value parameter $\boldsymbol{\mu}_t$.

The Cramér–Rao inequality is here of only peripheral interest. However, because it connects with the mean value parameterization of an exponential family, this chapter is the natural place to discuss it.

For the inequality, suppose data are described by some model family satisfying suitable regularity conditions (details omitted) and parameterized by a parameter vector $\boldsymbol{\eta}$.

Proposition 3.25 ***Equality in the Cramér–Rao inequality***
Any unbiased estimator $\hat{\boldsymbol{\eta}} = \hat{\boldsymbol{\eta}}(\boldsymbol{y})$ of $\boldsymbol{\eta}$ satisfies

$$\mathrm{var}(\hat{\boldsymbol{\eta}}; \boldsymbol{\eta}) \geq I(\boldsymbol{\eta})^{-1}, \tag{3.44}$$

with equality if and only if the distribution family is an exponential family, with $\hat{\boldsymbol{\eta}}(\boldsymbol{y})$ as its canonical statistic and $\boldsymbol{\eta}$ as its mean value parameter, i.e. $\boldsymbol{\eta} = \boldsymbol{\mu}_t$ and $\hat{\boldsymbol{\eta}}(\boldsymbol{y}) = \boldsymbol{t}(\boldsymbol{y}) = \hat{\boldsymbol{\mu}}_t$.

An inequality written $\boldsymbol{A} \geq \boldsymbol{B}$ as in (3.44), referring to positive definite matrices $\boldsymbol{A}$ and $\boldsymbol{B}$, means that $\boldsymbol{A}-\boldsymbol{B}$ is positive semidefinite, or equivalently that $\boldsymbol{B}^{-1}\boldsymbol{A}$ (and $\boldsymbol{A}\boldsymbol{B}^{-1}$ and $\boldsymbol{B}^{-1/2}\boldsymbol{A}\boldsymbol{B}^{-1/2}$) has all its eigenvalues ≥ 1.

Proof We will use the knowledge that the score plays an important role, and express $\hat{\eta}$ by means of its theoretical linear regression on the score function U. That is, we write

$$\hat{\eta}(y) = \eta + \mathrm{cov}\{\hat{\eta}, U(\eta)\}\,\mathrm{var}(U(\eta))^{-1}\,U(\eta; y) + Z(y; \eta), \tag{3.45}$$

where the rest (residual) $Z(y; \eta)$ necessarily has expected value 0. Formula (3.45) simplifies, because $\mathrm{cov}(\hat{\eta}, U(\eta))$ equals the identity matrix $\mathrm{I}_{\dim \eta}$ and can therefore be omitted. This is seen as follows, by a change of order of differentiation and integration/summation (notationally, let D_η generate row vectors):

$$\begin{aligned}
\mathrm{cov}\{\hat{\eta}, U(\eta)\} &= [\text{since } E(U) = 0] = E\{\hat{\eta}\, U(\eta)^T\} \\
&= [\text{since } U(\eta)^T = D_\eta \log L(\eta)] = E\{\hat{\eta}\, D_\eta L(\eta)/L(\eta)\} \\
&= [\text{since } L(\eta; y) = f(y; \eta)] = \int \hat{\eta}\, D_\eta f(y; \eta)\, \mathrm{d}y \\
&= D_\eta E(\hat{\eta}) = D_\eta \eta = \mathrm{I}_{\dim \eta}\,.
\end{aligned}$$

For the second term of (3.45), finally note that $\mathrm{var}(U(\eta)) = I(\eta)$. Concerning the third term, Z, we need only observe that this error term is uncorrelated with U, since residuals are uncorrelated with regressors. This implies that variances add, that is,

$$\begin{aligned}
\mathrm{var}(\hat{\eta}(y)) &= \mathrm{var}(I(\eta)^{-1}\, U) + \mathrm{var}(Z) \geq \\
&\geq I(\eta)^{-1}\,\mathrm{var}(U)\, I(\eta)^{-1} = I(\eta)^{-1}\,.
\end{aligned}$$

Primarily, this shows the inequality part of the proposition. If $Z = 0$, equality is attained, but it is not obvious that this is possible in any model, since the resulting right-hand side looks like a function of both data and the parameter η. However, if there is a solution, the estimator can necessarily be written

$$\hat{\eta}(y) = \eta + I(\eta)^{-1}\, U(\eta;\, y)\,.$$

This is equivalent to

$$U(\eta;\, y) = I(\eta)\,(\hat{\eta}(y) - \eta)\,.$$

Consequently, if there is a primitive function (the log-likelihood) with this U as gradient, it must have the form

$$\log L(\eta;\, y) = a(\eta)^T\,\hat{\eta}(y) + b(\eta) + c(y)\,,$$

for some primitive functions $a(\eta)$ and $b(\eta)$ of $I(\eta)$ and $-I(\eta)\eta$, respectively, and some integration constant $c(y)$. However, this is precisely the form of an exponential family likelihood with $\hat{\eta}(y)$ as canonical statistic, so exponential type is necessary. Conversely, exponential type is sufficient, because if we have an exponential family, then $\boldsymbol{t}$ estimates is expected value $\boldsymbol{\mu}_t$ with the desired variance $\mathrm{var}(\boldsymbol{t}) = I(\boldsymbol{\eta})^{-1}$. □

Remark 3.26 ***Extension to not necessarily unbiased estimators.*** The Cramér–Rao inequality is usually formulated without the restriction to unbiased estimators. If $\hat{\boldsymbol{\eta}}$ is not unbiased, $\mathrm{cov}(\hat{\boldsymbol{\eta}}, U(\boldsymbol{\eta}))$ appearing in the proof will not be the identity matrix I, but equal to the gradient

$$D_\eta E(\hat{\boldsymbol{\eta}}) = \mathrm{I} + D_\eta \,\mathrm{Bias}(\hat{\boldsymbol{\eta}}),$$

with elements $\partial E(\hat{\eta}_i)/\partial \eta_j$. This yields a corresponding factor on each side of $I(\boldsymbol{\eta})^{-1}$ in the Cramér–Rao inequality, more precisely

$$\mathrm{var}(\hat{\boldsymbol{\eta}}; \boldsymbol{\eta}) \geq D_\eta E(\hat{\boldsymbol{\eta}})^T I(\boldsymbol{\eta})^{-1} D_\eta E(\hat{\boldsymbol{\eta}}). \tag{3.46}$$

△

Remark 3.27 ***Alternative multivariate proof.*** An alternative proof of (3.44) goes via linear forms, considering unbiased estimators of the scalar $c^T \eta$ and showing that their minimum variance is $c^T I(\eta)^{-1} c$, for any c. △

Exercise 3.29 ***Cramér–Rao***
If $E(\hat{\boldsymbol{\eta}}) = g(\boldsymbol{\eta}) \neq \boldsymbol{\eta}$, we could either use the inequality with $\hat{\boldsymbol{\eta}}$ regarded as an unbiased estimator of $g(\boldsymbol{\eta})$, or use the inequality with $\hat{\boldsymbol{\eta}}$ as a biased estimator of $\boldsymbol{\eta}$, (3.46). Show that these two approaches yield the same lower bound for the variance. △

4

Asymptotic Properties of the MLE

The exponential families allow more precise and far-reaching results than parametric families in general. In this chapter we derive some large sample properties for the MLE, including $\sqrt{n}$-consistency and asymptotic normality. We also derive the saddlepoint approximation for the distribution of the MLE, providing a refinement of the standard approximate normality, and in many cases extraordinarily good for quite small samples.

As usual in statistics, the basic results here are formulated and proved for iid observations, assuming an increasing sample size n. However, in concrete applications the data set is always of a fixed and limited size, and rarely a simple iid sample, but the applicability of the results goes far beyond iid situations.

4.1 Large Sample Asymptotics

For simplicity of notation we suppose the exponential family is regular, so $\boldsymbol{\Theta}$ is open, but the results hold for any full family if $\boldsymbol{\Theta}$ is only restricted to its interior. Proposition 3.10 showed that the log likelihood function is strictly concave and therefore has a unique maximum, corresponding to a unique root of the likelihood equations, provided there is a (finite) maximal value. We have seen examples in Chapter 3 of likelihoods without a finite maximum, for example in the binomial and Poisson families. We shall here first show that if we have a sample of size n from a regular exponential family, the risk for such an event tends to zero with increasing n, and the MLE $\hat{\boldsymbol{\theta}}$ approaches the true $\boldsymbol{\theta}$, i.e. $\hat{\boldsymbol{\theta}}$ is a consistent estimator. The next step, to an asymptotic Gaussian distribution of $\hat{\boldsymbol{\theta}}$, is not big.

We shall use the following notation. Let $\boldsymbol{t}(\boldsymbol{y})$ be the canonical statistic for a single observation, and $\boldsymbol{t}_n = \sum_i \boldsymbol{t}(\boldsymbol{y}_i)$ the corresponding canonical statistic for the whole sample. Let $\boldsymbol{\mu}_t(\boldsymbol{\theta}) = E_\theta\{\boldsymbol{t}(\boldsymbol{y})\} = E_\theta\{\boldsymbol{t}_n/n\}$, i.e. the mean value per observational unit. The one-to-one canonical and mean-value parameterizations by $\boldsymbol{\theta}$ and $\boldsymbol{\mu}$ are related by $\hat{\boldsymbol{\theta}}(\boldsymbol{t}) = \boldsymbol{\mu}^{-1}(\boldsymbol{t})$, and for a regular family both $\boldsymbol{\Theta}$ and $\boldsymbol{\mu}(\boldsymbol{\Theta})$ are open sets in $\mathbb{R}^k$, where $k = \dim \boldsymbol{\theta}$.

The existence and consistency of the MLE is essentially only a simple application of the law of large numbers under the mean value parameterization (when the MLE is $\hat{\boldsymbol{\mu}}_t = \boldsymbol{t}_n/n$), followed by a reparameterization. In an analogous way, asymptotic normality of the MLE follows from the central limit theorem applied on $\hat{\boldsymbol{\mu}}_t$. However, we will also indicate stronger versions of these results, utilizing more of the exponential family structure.

Proposition 4.1 ***Existence and consistency of the MLE of*** θ
For a sample of size n from a regular exponential family, and for any $\boldsymbol{\theta} \in \boldsymbol{\Theta}$,

$$\Pr\{\hat{\boldsymbol{\theta}}(\boldsymbol{t}_n/n) \text{ exists}; \boldsymbol{\theta}\} \to 1 \text{ as } n \to \infty,$$

and furthermore,

$$\hat{\boldsymbol{\theta}}(\boldsymbol{t}_n/n) \to \boldsymbol{\theta} \text{ in probability as } n \to \infty.$$

These convergences are uniform on compact subsets of $\boldsymbol{\Theta}$.

Proof By the (weak) law of large numbers (Khinchine version), $\boldsymbol{t}_n/n \to \boldsymbol{\mu}_t$ $(= \boldsymbol{\mu}_t(\boldsymbol{\theta}))$ in probability as $n \to \infty$. More precisely expressed, for any fixed $\delta > 0$,

$$\Pr\{|\boldsymbol{t}_n/n - \boldsymbol{\mu}_t| < \delta\} \to 1\,. \tag{4.1}$$

Note that as soon as $\delta > 0$ is small enough, the δ-neighbourhood (4.1) of $\boldsymbol{\mu}_t$ is wholly contained in the open set $\boldsymbol{\mu}(\boldsymbol{\Theta})$, and that $\boldsymbol{t}_n/n$ is identical with the MLE $\hat{\boldsymbol{\mu}}_t$. This is the existence and consistency result for the MLE of the mean value parameter, $\hat{\boldsymbol{\mu}}_t$. Next we transform this to a result for $\hat{\boldsymbol{\theta}}$ in $\boldsymbol{\Theta}$.

Consider an open δ'-neighbourhood of $\boldsymbol{\theta}$. For $\delta' > 0$ small enough, it is wholly within the open set $\boldsymbol{\Theta}$. There the image function $\boldsymbol{\mu}_t(\boldsymbol{\theta})$ is well-defined, and the image of the δ'-neighbourhood of $\boldsymbol{\theta}$ is an open neighbourhood of $\boldsymbol{\mu}(\boldsymbol{\theta})$ in $\boldsymbol{\mu}(\boldsymbol{\Theta})$ (since $\hat{\boldsymbol{\theta}} = \boldsymbol{\mu}_t^{-1}$ is a continuous function). Inside that open neighbourhood we can always find (for some $\delta > 0$) an open δ-neighbourhood of type (4.1), whose probability goes to 1. Thus, the probability for $\hat{\boldsymbol{\theta}}(\boldsymbol{t}_n/n)$ to be in the δ'-neighbourhood of $\boldsymbol{\theta}$ also goes to 1. This shows the asymptotic existence and consistency of $\hat{\boldsymbol{\theta}}$ for any fixed $\boldsymbol{\theta}$ in $\boldsymbol{\Theta}$.

Finally, this can be strengthened (Martin-Löf, 1970) to uniform convergence on compact subsets of the parameter space $\boldsymbol{\Theta}$. Specifically, Chebyshev's inequality can be used to give a more explicit upper bound of order $1/n$ to the complementary probability, $\Pr\{|\boldsymbol{t}_n/n - \boldsymbol{\mu}_t(\boldsymbol{\theta})| \geq \delta\}$, proportional to $\mathrm{var}_t(\boldsymbol{\theta})$, that is bounded on compact subsets of $\boldsymbol{\Theta}$, since it is a continuous function of $\boldsymbol{\theta}$. Details are omitted. □

Remark 4.2 ***Other parameterizations.*** In analogy with how the law of large numbers was carried over from the mean value parameter to the canonical parameter in the proof of Proposition 4.1, we can reparameterize by any other parameter, ψ say. The next proposition will be expressed in terms of a general parameter, or a lower-dimensional function of it. △

The central limit theorem applied on $\hat{\mu}_t = t_n/n$ immediately yields a Gaussian large sample distribution for the MLE in the mean value parameterization. This is easily extended from μ_t to an arbitrary parameter by a Taylor expansion of the reparameterization function (assumed smooth). Let $\psi = \psi(\mu_t)$ be the new parameter. The MLE of ψ is

$$\hat{\psi} = \psi(\hat{\mu}_t) = \psi(t_n/n) = \psi + \left(\frac{\partial \psi}{\partial \mu_t}\right)^T (t_n/n - \mu_t) + o(|t_n/n - \mu_t|).$$

The asymptotic distribution for $\hat{\psi}$ is immediate from this:

Proposition 4.3 ***Asymptotic distribution of the MLE***
For a sample of size n from a regular exponential family, and for any parameter function $\psi = \psi(\mu_t)$, the MLE $\hat{\psi}$ is $\sqrt{n}$-consistent and asymptotically Gaussian as $n \to \infty$, such that $\sqrt{n}(\hat{\psi}-\psi)$ is asymptotically distributed as

$$\mathrm{N}_k\left(\mathbf{0}, \left(\frac{\partial \psi}{\partial \mu_t}\right)^T \mathrm{var}(t) \left(\frac{\partial \psi}{\partial \mu_t}\right)\right). \tag{4.2}$$

Here $\mathrm{var}(t) = I(\mu_t)^{-1}$ *is the inverse Fisher information for μ_t per observation, that is here of course more naturally expressed in terms of ψ, and if $\psi(\mu_t)$ is one-to-one, (4.2) can be written* $\mathrm{N}_k\left(\mathbf{0}, I(\psi)^{-1}\right)$.

Remark 4.4 ***Uniform convergence on compacts.*** In the same way as for Proposition 4.1, the present result can be strengthened to imply uniform convergence towards normality on compact subsets of the parameter space (Martin-Löf, 1970). △

In comparison with typical large sample local results in a wider generality than for exponential families, Propositions 4.1 and 4.3 go further in several respects:

- The full exponential family implies that the MLE is unique and is a global maximum, and not only a possibly local one.
- The full exponential family implies that local regularity conditions are automatically satisfied (if only ψ is an everywhere differentiable function of μ, such that $\left(\frac{\partial \psi}{\partial \mu}\right)$ has full rank).

- For an exponential family in any parameterization, its inverse Fisher information, needed for the MLE variance, can be expressed in terms of $\mathrm{var}(\boldsymbol{t})$, going via its mean value parameterization.
- In general, when estimating the large sample variance of the MLE, we must choose between expected and observed information in the MLE point $\hat{\boldsymbol{\psi}}$. In a full exponential family, these information statistics are equal in the MLE point, for any parameterization, see Exercise 3.7.

General results tells that under a set of suitable regularity conditions, the MLE is correspondingly asymptotically normally distributed as $n \to \infty$, but with the asymptotic variance expressed as being the inverse of the Fisher information. First note that when we are in a regular exponential family, we require no additional regularity conditions. Secondly, the variance $\mathrm{var}(\boldsymbol{t}_n/n) = \mathrm{var}(\boldsymbol{t})/n$ for $\hat{\boldsymbol{\mu}}_t$ is precisely the inverse of the Fisher information matrix for $\boldsymbol{\mu}_t$ (Proposition 3.15), and unless $\boldsymbol{\psi}(\boldsymbol{\mu}_t)$ is of lower dimension than $\boldsymbol{\mu}_t$ itself, the variance formula in Proposition 4.3 can alternatively be obtained from $\mathrm{var}(\boldsymbol{t})$ by use of the Reparameterization lemma (Proposition 3.14). When $\boldsymbol{\psi}$ and $\boldsymbol{\mu}_t$ are not one-to-one, but $\dim \boldsymbol{\psi}(\boldsymbol{\mu}_t) < \dim \boldsymbol{\mu}_t$, e.g. $\boldsymbol{\psi}$ a subvector of $\boldsymbol{\mu}_t$, the theory of profile likelihoods (Section 3.3.4) can be useful, yielding correct formulas expressed in Fisher information terms. In a concrete situation we are of course free to calculate the estimator variance directly from the explicit form of $\hat{\boldsymbol{\psi}}$. In any case, a conditional inference approach, eliminating nuisance parameters (Section 3.5), should also be considered. Asymptotically, however, we should not expect the conditional and unconditional results to differ, unless the nuisance parameters are incidental and increase in number with n, as in Exercise 3.16.

Note the special case of the canonical parameterization, with $\boldsymbol{\theta}$ as canonical parameter for the distribution of the single observation ($n = 1$). For this parameter, the asymptotic variance of the MLE is $\mathrm{var}(\boldsymbol{t})^{-1}/n$, where $\boldsymbol{t}$ is the single observation statistic. This is the special case of Proposition 4.3 for which

$$\left(\frac{\partial \boldsymbol{\psi}}{\partial \boldsymbol{\mu}_t}\right) = \left(\frac{\partial \boldsymbol{\theta}}{\partial \boldsymbol{\mu}_t}\right) = \left(\frac{\partial \boldsymbol{\mu}_t}{\partial \boldsymbol{\theta}}\right)^{-1} = \mathrm{var}_{\theta}(\boldsymbol{t})^{-1}. \tag{4.3}$$

Like in the general large sample statistical theory, we may use the result of Proposition 4.3 to construct asymptotically correct confidence regions. For example, expressed for the canonical parameter $\boldsymbol{\theta}$, it follows from Proposition 4.3 that the quadratic form in $\hat{\boldsymbol{\theta}}$,

$$Q(\boldsymbol{\theta}) = n(\hat{\boldsymbol{\theta}} - \boldsymbol{\theta})^T \mathrm{var}_{\theta}(\boldsymbol{t})\,(\hat{\boldsymbol{\theta}} - \boldsymbol{\theta}), \tag{4.4}$$

is asymptotically χ^2 distributed with $\dim \boldsymbol{\theta}$ degrees of freedom. This implies that the region

$$\{\boldsymbol{\theta} : \; Q(\boldsymbol{\theta}) \leq \chi^2_{1-\alpha}(\dim \boldsymbol{\theta})\}$$

has the approximate confidence level $1 - \alpha$. If furthermore $\boldsymbol{\theta}$ is approximated by $\hat{\boldsymbol{\theta}}$ in $\mathrm{var}_{\theta}(\boldsymbol{t})$ in (4.4), the region becomes a confidence ellipsoid centred at $\hat{\boldsymbol{\theta}}$.

Here are a few illustrations of Propositions 4.1 and 4.3.

Example 4.1 ***Bernoulli trials, continued from Example 3.4***
The parameter of interest is the mean value parameter, here called π_0, and the canonical parameter is $\theta = \mathrm{logit}(\pi_0)$. The probability for nonexistence of $\hat{\theta} = \mathrm{logit}(\bar{y})$, or equivalently, for $\hat{\pi}_0 = \bar{y}$ falling outside the open unit interval (0, 1), is

$${\pi_0}^n + (1 - \pi_0)^n.$$

This probability tends to zero as $n \to \infty$ for any π_0 in the open interval (0, 1), uniformly on compact subsets, e.g. for $\pi_0(1 - \pi_0) \geq \delta > 0$, but not uniformly on the whole open interval (0, 1).

The asymptotic distribution

$$\sqrt{n}\,(\bar{y} - \pi_0) \sim \mathrm{N}\,(0,\, \pi_0\,(1 - \pi_0))$$

holds for any π_0 in the open (0,1). For θ the corresponding result is

$$\sqrt{n}\,(\hat{\theta} - \theta) \sim \mathrm{N}\Big(0,\; (\pi_0(1 - \pi_0))^{-1}\Big),$$

compare with the Fisher informations for π_0 and θ from Example 3.4. △

Example 4.2 ***Time to n successes, continued from Example 2.6***
The number of Bernoulli trials up to and including the first success has $t(y) = y$ (Example 2.6). Repeating this n times, thus until n successes, yields a sample $\boldsymbol{y} = \{y_1, \ldots, y_n\}$ with $t_n = t(\boldsymbol{y}) = \sum_1^n y_i$ and probability

$$f(\boldsymbol{y};\, \pi_0) = \prod_{i=1}^{n} \Big\{(1 - \pi_0)^{y_i - 1}\, \pi_0\Big\} = (1 - \pi_0)^{t_n - n}\, \pi_0^n\,.$$

By counting all realizations with a fixed t_n ($t_n \geq n$) we obtain the classic result, e.g. Feller (1968, p. 36 or p. 155), that t_n is negative binomial,

$$f(t_n;\, \pi_0) = \binom{t_n - 1}{t_n - n} (1 - \pi_0)^{t_n - n}\, \pi_0^n\,. \tag{4.5}$$

The first factor in (4.5), distinguishing $f(t_n;\, \pi_0)$ from $f(\boldsymbol{y};\, \pi_0)$, is the structure function, $g(t_n)$. The canonical parameter is $\theta = \log(1 - \pi_0)$. For example by differentiation of $\log C(\theta) = \log\{e^\theta/(1-e^\theta)\}$ we obtain

$$E(t) = \frac{1}{1-e^\theta} = \frac{1}{\pi_0},$$
$$\mathrm{var}(t) = \frac{e^\theta}{(1-e^\theta)^2} = \frac{1-\pi_0}{\pi_0^2}.$$

Provided $\pi_0 > 0$, t_n is finite with probability 1, and the probability for an MLE of π_0 in the open unit interval is

$$\Pr(t_n > n) = 1 - \pi_0^n,$$

which goes fast to 1 as $n \to \infty$ (uniformly in intervals $\pi_0 \leq \pi_1 < 1$).

Finally, the mean value parameter is $\mu = 1/\pi_0$, and its MLE is $\hat{\mu} = t/n$ with $\mathrm{var}(t_n/n) = (1-\pi_0)/(\pi_0^2 n)$ (exactly). Now, the error propagation implicit in Proposition 4.3 yields the large sample (large n) distribution, with the result that $\hat{\pi}_0$ is approximately $\mathrm{N}(\pi_0, \pi_0^2(1-\pi_0)/n)$.
See also Exercise 4.2 and Example 4.4. △

Example 4.3 ***A sample from the normal, continued from Example 3.10***
For nonexistence of the MLE we must have $s = 0$. This outcome has probability 0, provided $n > 1$ and $\sigma > 0$ (i.e. $\boldsymbol{\theta} \in \boldsymbol{\Theta}$). △

Exercise 4.1 ***Asymptotics for the Poisson model***
A sample of size n is taken from a Poisson, $\mathrm{Po}(\lambda)$, $\lambda > 0$.
(a) Derive the probability that the MLE of θ does not exist, and verify that it tends to zero with n.
(b) Apply Proposition 4.3 to this situation, with λ as parameter. Check that the result is as expected.
(c) In the different situation when there is only one observation y from $\mathrm{Po}(\lambda)$, but λ is large, $\lambda \to \infty$, we may use Proposition 4.3 in an artificial way by assuming we had a sample $z_1, ..., z_n$ of n observations from $\mathrm{Po}(a)$, with a fixed and $\lambda = n\,a$. Then $\sum z_i = y$ is the canonical statistic. Carry this out in detail, to establish convergence to normality of y and consistency of the MLE of a, and to find the asymptotic distribution of this MLE. Finally carry this over to the MLE of λ. △

Exercise 4.2 ***Negative binomial variances***
Check the two variances in Example 4.2. △

Exercise 4.3 ***Normal distribution sample variance***
Derive the large sample distribution for the unconditional and conditional sample variances in a normal distribution. Use for example results from Example 3.10. Explain the results from the well-known exact results. △

4.2 Small Sample Refinement: Saddlepoint Approximations

Saddlepoint approximation is a technique for refined approximation of densities and distribution functions when the normal (or χ^2) approximation is too crude. The method is typically used on distributions for

- Sums of random variables.
- ML estimators.
- Likelihood ratio test statistics,

and for approximating structure functions. These approximations often have surprisingly small errors, and in some remarkable cases they yield exact results. We confine ourselves to the basics, and consider only density approximation for canonical statistics $\boldsymbol{t}$ (which involves the structure function $g(\boldsymbol{t})$), and for the MLE in any parameterization of the exponential family.

The saddlepoint approximation for probability densities (continuous distribution case) is due to Daniels (1954). Approximations for distribution functions, being of particular interest for test statistics (for example the so-called Lugannani–Rice formula for the signed log-likelihood ratio), is a more complicated topic, because it will require approximation of integrals of densities instead of just densities, see for example Jensen (1995).

Suppose we have a regular exponential family, with canonical statistic $\boldsymbol{t}$ (no index n here) and canonical parameter $\boldsymbol{\theta}$. If the family is not regular, we restrict $\boldsymbol{\theta}$ to the interior of $\boldsymbol{\Theta}$. Usually $\boldsymbol{t} = t(\boldsymbol{y})$ has some sort of a sum form, even though the number of terms may be small. We shall derive an approximation for the density $f(\boldsymbol{t};\, \boldsymbol{\theta}_0)$ of $\boldsymbol{t}$, for some specified $\boldsymbol{\theta} = \boldsymbol{\theta}_0$. One reason for wanting such an approximation in a given exponential family is the presence of the structure function $g(\boldsymbol{t})$ in $f(\boldsymbol{t};\, \boldsymbol{\theta}_0)$, which is the factor, often complicated, by which $f(\boldsymbol{t};\, \boldsymbol{\theta}_0)$ differs from $f(\boldsymbol{y}; \boldsymbol{\theta}_0)$. Another reason is that we might want the distribution of the ML estimator, and $\boldsymbol{t} = \hat{\boldsymbol{\mu}}_t$ is one such MLE.

We will use the central limit theorem for densities, but not applied directly on $f(\boldsymbol{t};\, \boldsymbol{\theta}_0)$ itself, but on $f(\boldsymbol{t};\, \boldsymbol{\theta})$ for a suitable different $\boldsymbol{\theta}$, that will be chosen depending on $\boldsymbol{t}$ such that we can expect as good an approximation as possible. This suitable $\boldsymbol{\theta}$ turns out to be the MLE regarded as a function of $\boldsymbol{t}$, $\boldsymbol{\theta} = \hat{\boldsymbol{\theta}}(\boldsymbol{t}) = \boldsymbol{\mu}_t^{-1}(\boldsymbol{t})$. The derivation goes as follows. For arbitrary $\boldsymbol{\theta}$ we can jointly write

$$f(\boldsymbol{t};\, \boldsymbol{\theta}) \begin{cases} = f(\boldsymbol{t}; \boldsymbol{\theta}_0)\, \dfrac{C(\boldsymbol{\theta}_0)}{C(\boldsymbol{\theta})}\, e^{(\boldsymbol{\theta} - \boldsymbol{\theta}_0)^T\, \boldsymbol{t}} \\ \approx \dfrac{1}{(2\pi)^{k/2}\, \sqrt{\det V_t(\boldsymbol{\theta})}}\, e^{-\frac{1}{2}(\boldsymbol{t} - \boldsymbol{\mu}_t(\boldsymbol{\theta}))^T\, V_t(\boldsymbol{\theta})^{-1}\, (\boldsymbol{t} - \boldsymbol{\mu}_t(\boldsymbol{\theta}))} \end{cases}, \tag{4.6}$$

where we have used a multivariate normal density approximation at $\boldsymbol{t}$ when the parameter is $\boldsymbol{\theta}$, and $k = \dim \boldsymbol{t} = \dim \boldsymbol{\theta}$. We can solve for $f(t;\, \boldsymbol{\theta}_0)$ in (4.6) and get an approximation for it, whatever choice of $\boldsymbol{\theta}$ we have made. The normal approximation is typically best in the centre of the distribution, so we choose $\boldsymbol{\theta}$ such that the fixed $\boldsymbol{t}$ is at the centre of $f(\boldsymbol{t};\, \boldsymbol{\theta})$. More precisely, we let $\boldsymbol{\theta}$ be the MLE $\hat{\boldsymbol{\theta}}(\boldsymbol{t})$. Not only the exponent in the normal density but also the dominating (lowest order) approximation error term will then vanish, and the result is the so-called saddlepoint approximation:

Proposition 4.5 ***Saddlepoint approximation***
The saddlepoint approximation of a density $f(\boldsymbol{t};\, \boldsymbol{\theta}_0)$ in an exponential family is

$$f(\boldsymbol{t};\, \boldsymbol{\theta}_0) \approx \frac{1}{(2\pi)^{k/2}\sqrt{\det V_t(\hat{\boldsymbol{\theta}}(\boldsymbol{t}))}} \frac{C(\hat{\boldsymbol{\theta}}(\boldsymbol{t}))}{C(\boldsymbol{\theta}_0)}\, e^{(\boldsymbol{\theta}_0 - \hat{\boldsymbol{\theta}}(t))^T \boldsymbol{t}}, \tag{4.7}$$

and the corresponding approximation of the structure function $g(\boldsymbol{t})$ is

$$g(\boldsymbol{t}) \approx \frac{1}{(2\pi)^{k/2}\sqrt{\det V_t(\hat{\boldsymbol{\theta}}(\boldsymbol{t}))}}\, C(\hat{\boldsymbol{\theta}}(\boldsymbol{t}))\, e^{-\hat{\boldsymbol{\theta}}\boldsymbol{t}^T \boldsymbol{t}}. \tag{4.8}$$

Here $\hat{\boldsymbol{\theta}}(\boldsymbol{t}) = \mu^{-1}(\boldsymbol{t})$ is the MLE as a function of the variate $\boldsymbol{t}$, and $V_t(\hat{\boldsymbol{\theta}}(\boldsymbol{t}))$ is the matrix of second order derivatives of $\log C(\boldsymbol{\theta})$ *at $\boldsymbol{\theta} = \hat{\boldsymbol{\theta}}(\boldsymbol{t})$.*

Remark 4.6 ***Extensions and further refinements.***

- The same technique can at least in principle be used outside exponential families to approximate a density $f(\boldsymbol{t})$. The idea is to embed $f(\boldsymbol{t})$ in an exponential family by exponential tilting, see Example 2.12 with $f(\boldsymbol{t})$ in the role of its $f_0(y)$, For practical use, f must have an explicit and manageable Laplace transform, since the norming constant $C(\boldsymbol{\theta})$ will be the Laplace transform of the density f, cf. (3.9).
- For iid samples of size n one might guess the error is expressed by a factor of magnitude $1+O(1/\sqrt{n})$. It is in fact a magnitude better, $1+O(1/n)$, and this holds uniformly on compact subsets of $\boldsymbol{\Theta}$. The error factor can be further refined by using a more elaborate Edgeworth expansion instead of the simple normal density in the second line of (4.6). An Edgeworth expansion is typically excellent in the centre of the distribution, where it would be used here (for $\boldsymbol{\theta} = \hat{\theta}(t)$). On the other hand, it can be terribly bad in the tails and is generally not advisable for a direct approximation of $f(t; \boldsymbol{\theta}_0)$. See for example Jensen (1995, Ch. 2) for details and proofs. Such a refinement of the saddlepoint approximation was used by Martin-Löf to prove Boltzmann's theorem, see Chapter 6.

- Further refinement is possible, at least in principle, through multiplication by a norming factor such that the density approximation becomes a proper probability density. Note, this factor will typically depend on θ_0.
- Even if the relative approximation error is often remarkably small, it generally depends on the argument t. For the distribution of the sample mean it is constant only for three univariate densities of continuous distributions: the normal density, the gamma density, and the inverse Gaussian density (Exercise 4.6). The representation is exact for the normal, see Exercise 4.4, and the inverse Gaussian. For the gamma density the error factor is close to 1, representing the Stirling's formula approximation of $n!$ (Section B.2.2). Correction by a norming factor, as mentioned in the previous item, would of course eliminate such a constant relative error, but be pointless in this particular case. △

Before we give a concrete example of a saddlepoint approximation, we apply the theoretical result to derive a saddlepoint approximation for the distribution of the ML estimator. Despite the risk for confusion, we use the letter f also for this density and let the argument of the function show to which density we refer.

Note that t is the ML estimator of μ_t, so (4.7) is already the saddlepoint approximation for the MLE of a particular parameterization of the family. In order to get a more generally valid formula, note first that V_t in the denominator is the inverse of the Fisher information for μ_t, and second that the two last factors of (4.7) may be expressed as the likelihood ratio $L(\theta_0)/L(\hat{\theta})$, and the likelihood ratio value is the same whatever parameterization we use. Rewriting (4.7) in this way turns out to yield a formula valid for any (smooth) parameterization of the family, say $\psi = \psi(\theta)$ or $\psi(\mu_t)$:

Proposition 4.7 ***Saddlepoint approximation for the MLE***
The saddlepoint approximation for the density $f(\hat{\psi}; \psi_0)$ of the ML estimator $\hat{\psi} = \hat{\psi}(t)$ in any smooth parameterization of a regular exponential family is

$$f(\hat{\psi}; \psi_0) \approx \frac{\sqrt{\det I(\hat{\psi})}}{(2\pi)^{k/2}} \frac{L(\psi_0)}{L(\hat{\psi})}. \tag{4.9}$$

Here $I(\hat{\psi})$ is the Fisher information for the chosen parameter ψ, with inserted $\hat{\psi}$.

Proof We transform from the density for $t = \hat{\mu}_t$ to the density for $\hat{\psi}(t)$. To achieve this we only have to see the effect of a variable transformation on the density. When we replace the variable $\hat{\mu}_t$ by $\hat{\psi}$, we must also multiply by the Jacobian determinant for the transformation that has Jacobian matrix $\left(\frac{\partial \hat{\mu}_t}{\partial \hat{\psi}}\right)$. However, this is also precisely what is needed to change the square root of $\det I(\mu_t) = \det V_t^{-1} = 1/\det V_t$ to take account of the reparameterization, see the Reparameterization lemma, Proposition 3.14. Hence the form of formula (4.9) is invariant under reparameterizations. □

Proposition 4.7 works even much more generally, as shown by Barndorff-Nielsen and others, and the formula is usually named *Barndorff-Nielsen's formula*, or the p^* ('p-star') *formula*, see Reid (1988) and Barndorff-Nielsen and Cox (1994). Pawitan (2001, Sec. 9.8) refers to it as the 'magical formula'. The formula holds also after conditioning on distribution-constant (= ancillary) statistics (see Proposition 7.5 and Remark 7.6), and it holds exactly in transformation models, given the ancillary configuration statistic. It is more complicated to derive outside the exponential families, however, because we cannot then start from (4.7) with the canonical parameter θ as the parameter of interest.

By analogy with the third item of Remark 4.6 about approximation (4.7), we may here refine (4.9) by replacing the approximate normalization constant $1/\sqrt{2\pi}$ by the exact one, that makes the density integrate to 1.

Example 4.4 ***Time to n successes, continued from Example 4.2***
When Bernoulli trials are repeated until n successes, we found in Example 4.2 that the large sample (large n) distribution of $\hat{\pi} = n/t_n$ is $\mathrm{N}(\pi_0, \pi_0^2(1-\pi_0)/n)$. When π_0 is small, n must be quite large for this to be a good approximation, and t_n will then be of an even larger magnitude. On the other hand, $\hat{\pi}$ is a known one-to-one function of t_n and in this particular example the exact distribution for t_n is known, the negative binomial probability function (4.5). The exact form, with its factorials, is not computationally attractive for large n and large t_n, however. An excellent compromise between the exact expression and the crude normal approximation is provided by the saddlepoint approximation. We could either apply it to make an approximation of the structure function $g(t_n) = \binom{t_n-1}{t_n-n}$ (Proposition 4.5) or of the whole probability function for $\hat{\pi}$ (Proposition 4.7). We choose the latter here, for illustration. Both will yield the same result, and they will also be equivalent to the use of Stirling's formula on the factorials in the structure function.

This is an approximation of the discrete probability function for $\hat{\pi}$ by a continuous distribution density. The discrete character is not a complication when, as here, the probability function for t_n is on the set of integers, and not on a more sparse or more dense set.

To apply formula (4.9), we need the Fisher information for π_0 in $\hat{\pi}$, which we have from the large sample variance of $\hat{\pi}$ in Example 4.2 as $I(\hat{\pi}) = n/\hat{\pi}^2(1-\hat{\pi})$. We also need the likelihood ratio

$$\frac{L(\pi_0)}{L(\hat{\pi})} = \frac{(1-\pi_0)^{t_n-n}\,\pi_0^n}{(1-\hat{\pi})^{t_n-n}\,\hat{\pi}^n} = \frac{(1-\pi_0)^{n/\hat{\pi}-n}\,\pi_0^n}{(1-\hat{\pi})^{n/\hat{\pi}-n}\,\hat{\pi}^n}.$$

Combination according to formula (4.9) yields the result desired:

$$f_{\hat{\pi}}(\hat{\pi};\pi_0) \approx \frac{\sqrt{n}}{\sqrt{2\pi}}\,\frac{(1-\pi_0)^{n/\hat{\pi}-n}\,\pi_0^n}{(1-\hat{\pi})^{n/\hat{\pi}-n+1/2}\,\hat{\pi}^{n+1}}.$$

△

Exercise 4.4 ***Saddlepoint approximation for the normal sample mean***
Show that the saddlepoint approximation for the density of $\overline{y}$, when a sample of size n is taken from a $N(\mu, \sigma^2)$, is exactly the $N(\mu, \sigma^2/n)$ density. △

Exercise 4.5 ***A saddlepoint approximation for an MLE***
Suppose we have a sample $y_1, \ldots, y_n$ from a distribution on the unit interval, with density

$$f(y;\alpha) = \alpha\, y^{\alpha-1}, \quad 0 < y < 1.$$

Find the MLE $\hat{\alpha}$ and calculate the observed or expected information for α (why are they the same?)
Derive the p^* formula for the density of $\hat{\alpha}$, and demonstrate that it corresponds to a gamma density for $1/\hat{\alpha}$, except that the gamma function has been replaced by its Stirling formula approximation. Why does the gamma density appear? △

Exercise 4.6 ***The p^* formula for the inverse Gaussian (IG)***
For a sample from the inverse Gaussian distribution, the MLE and the (diagonal) information matrix for the mixed parameterization by (μ, α) were derived in Exercise 3.25. Use these results in the p^* formula for the density of the MLE, to show that $\hat{\mu}$ and $\hat{\alpha}$ are mutually independent, $\hat{\mu}$ is $IG(\mu, n\alpha)$ distributed and $1/\hat{\alpha}$ is gamma distributed, more precisely $n\alpha/\hat{\alpha}$ is $\chi^2(n-1)$.
Remark: The approximations are actually exact, except for a Stirling's formula type approximation of $\Gamma((n-1)/2)$. △

5

Testing Model-Reducing Hypotheses

Suppose we trust, as a satisfactory description of data $\boldsymbol{y}$, a certain model type, a regular exponential family of order r with canonical statistic $\boldsymbol{t}$ and canonical parameter $\boldsymbol{\theta}$. Do we need all r dimensions? Is perhaps a lower-dimensional submodel sufficient, of order $p < r$ and with a certain p-dimensional canonical statistic $\boldsymbol{u}$? If $\boldsymbol{t} = (\boldsymbol{u}, \boldsymbol{v})$ and correspondingly $\boldsymbol{\theta} = (\boldsymbol{\theta}_u, \boldsymbol{\theta}_v)$, or $(\boldsymbol{\lambda}, \boldsymbol{\psi})$, we want to test the q-dimensional hypothesis $(q = r - p)$ of a specified value of $\boldsymbol{\psi}$. Without loss of generality we may assume that this $\boldsymbol{\psi}$-value is zero, so the hypothesis can be formulated as

$$\mathrm{H}_0 : \quad \boldsymbol{\psi} = \mathbf{0}, \quad \text{versus } \boldsymbol{\psi} \neq \mathbf{0}, \text{ with } \dim \boldsymbol{\psi} = \dim \boldsymbol{v} = q > 0, \tag{5.1}$$

A simple example is that we have a multiple regression model and want to delete a certain set of its regressors to reduce the dimension of the model. A variation is that one of its components represents a treatment effect, of primary concern. The hypothesis to be tested, which in this case we might want to reject, is then the absence of a treatment effect.

Alternatively, the primary model might be the smaller p-dimensional model with canonical statistic $\boldsymbol{u}$. We want to test if this model fits data, by embedding it in a wider model. This aspect is further discussed in Section 5.3.

Even more generally, any hypothesis restricting $\boldsymbol{\theta}$ to a linear (affine) subspace $\boldsymbol{\Theta}_0$ of $\boldsymbol{\Theta}$ can be written in the form $\boldsymbol{\psi} = \mathbf{0}$ after a suitable linear (affine) transformation of the canonical statistic $\boldsymbol{t}$. If the linear subspace of $\boldsymbol{\Theta}$ is spanned by $\boldsymbol{\theta} = A\,\boldsymbol{\psi}$ for some $r \times p$ matrix A and parameter vector $\boldsymbol{\psi}$ with dimension $p = \dim \boldsymbol{\psi}$, this is equivalent to a model with p-dimensional canonical statistic $\boldsymbol{u} = A^T \boldsymbol{t}$ and canonical parameter $\boldsymbol{\psi}$.

In this chapter we consider significance tests for such *model-reducing* hypotheses. This means we choose an adequate test statistic, whose distribution under H_0 is under control and used to form critical regions and to calculate p-values. This does not say that model reduction as a step in a model building or model selection process should necessarily be based on such formal significance tests. Depending on the intended use of the model, criteria such as the AIC or BIC may be more motivated for model reduction. In other situations we really want to test a hypothesis about $\boldsymbol{\theta}$, but the hypothesis is not linear in $\boldsymbol{\theta}$ and does not reduce $\dim \boldsymbol{t}$ as much as $\dim \boldsymbol{\theta}$. The resulting model is then of type *curved* exponential family, see Chapter 7. Note also that we do not consider here hypotheses of interval type, such as H_0: $\psi \leq \psi_0$, since they do not reduce $\dim \boldsymbol{t}$, nor $\dim \boldsymbol{\theta}$.

We will concentrate on hypothesis testing, but the reader should be aware of the close relation with confidence statements. A *confidence region* can be defined as consisting of those hypothesis values which are not rejected in a corresponding hypothesis test, and vice versa. More precisely, in terms of confidence functions, which extend the fixed confidence degree for one region to confidence values for a nested sequence of regions, and with restriction to one-sided situations (for simplicity of formulation):

Given a certain test procedure for testing H_0: $\psi = \psi_0$ against ψ-values $> \psi_0$, the degree of confidence of an interval $(-\infty, \psi_0]$ is the same as the p-value of the test of H_0.

This motivates our seemingly unfair treatment of confidence aspects. Another motivation is the publication of the book by Schweder and Hjort (2016), who treat the confidence concept in detail and depth, including a chapter on confidence properties for exponential families. Also it should not be forgotten that the construction of the confidence region can sometimes be made simple by finding an adequate *pivot*, that is, a function of the data and the parameter of interest which has a parameter-free distribution. Probability intervals for the pivot can be inverted to confidence intervals for the parameter.

We first derive the so-called 'exact test', with a terminology originating from Fisher's exact test for independence in a contingency table. We will also look at other tests and test criteria, and in particular note some optimality properties satisfied by exponential families. In the last section we will consider large sample tests, their relation to the exact test, and their asymptotic properties.

5.1 Exact Tests

We consider a model reduction of the canonical statistic from $\boldsymbol{t}$ to $\boldsymbol{u}$,

$$\boldsymbol{t} = (\boldsymbol{u}, \boldsymbol{v}) \longrightarrow \boldsymbol{u}\,,$$

and the corresponding parametric formulation (5.1). It will be simplest to work and argue under a mixed parameterization, so as nuisance parameter

$\boldsymbol{\lambda}$, when specified, we take the mean value parameter $\boldsymbol{\mu}_u$ for $\boldsymbol{u}$. Notation and dimensions will be $\dim \boldsymbol{t} = \dim \boldsymbol{\theta} = r = p + q$, $\dim \boldsymbol{u} = \dim \boldsymbol{\mu}_u$ (or $\dim \boldsymbol{\lambda}$) $= p$, and $\dim \boldsymbol{v} = \dim \boldsymbol{\psi} = q$.

Remember the conditionality principle of Section 3.5, that statistical inference about $\boldsymbol{\psi}$ should be conditional on $\boldsymbol{u}$. The argument was that $\boldsymbol{u}$ provides no information about $\boldsymbol{\psi}$, as expressed in the likelihood factorization

$$L(\boldsymbol{\mu}_u, \boldsymbol{\psi}; \boldsymbol{u}, \boldsymbol{v}) = L_1(\boldsymbol{\psi}; \boldsymbol{v} \,|\, \boldsymbol{u})\, L_2(\boldsymbol{\mu}_u, \boldsymbol{\psi}; \boldsymbol{u}), \tag{5.2}$$

valid whether H_0 is true or not. All information provided by the statistic $\boldsymbol{u}$ is consumed in estimating its own mean value $\boldsymbol{\mu}_u$. Since $\boldsymbol{\mu}_u$ and $\boldsymbol{\psi}$ are variation independent, no type of knowledge about $\boldsymbol{\mu}_u$ can yield any information about $\boldsymbol{\psi}$.

From the proof of Proposition 3.16 we get the conditional distribution for $\boldsymbol{v}$ given $\boldsymbol{u}$ in (3.22) and in L_1 of (5.2). Inserting the hypothesis value $\boldsymbol{\psi} = 0$ in (3.22) simplifies the conditional density to a ratio of two structure functions,

$$f_0(\boldsymbol{v} \,|\, \boldsymbol{u}) = \frac{g(\boldsymbol{u}, \boldsymbol{v})}{g_0(\boldsymbol{u})}, \tag{5.3}$$

where index 0 indicates distributions under H_0, and in particular g_0 is the structure function in the marginal exponential family for $\boldsymbol{u}$ under H_0. Note the analogy with model checking procedures, making use of the parameter-free conditional distribution of data $\boldsymbol{y}$, given the sufficient statistic $\boldsymbol{t}$.

The following formulation of the 'exact' test follows along the lines of R.A. Fisher (1935) and P. Martin-Löf (see Section 6.3).

Proposition 5.1 ***The principle of the 'exact test' of*** $\mathrm{H}_0 : \boldsymbol{\psi} = \mathbf{0}$

(1) Use $\boldsymbol{v}$ *as test statistic, with null distribution density* $f_0(\boldsymbol{v} \,|\, \boldsymbol{u})$ *from (5.3);*
(2) Reject H_0 *if this density is too small, and calculate the p-value as*

$$Pr\{f_0(\boldsymbol{v} \,|\, \boldsymbol{u}_{obs}) \leq f_0(\boldsymbol{v}_{obs} \,|\, \boldsymbol{u}_{obs})\} = \int_{f_0(\boldsymbol{v} \,|\, \boldsymbol{u}_{obs}) \leq f_0(\boldsymbol{v}_{obs} \,|\, \boldsymbol{u}_{obs})} f_0(\boldsymbol{v} \,|\, \boldsymbol{u}_{obs})\, \mathrm{d}\boldsymbol{v}\,. \tag{5.4}$$

The second part of the principle is controversial, but this is how Fisher originally constructed his exact test. Since densities of continuous distributions change form under nonlinear variable transformations, it is important for the principle that the density considered is the (conditional) density for the canonical $\boldsymbol{v}$, when the size α rejection region and the p-value are specified as in (5.4). Usually in practice $f_0(\boldsymbol{v} \,|\, \boldsymbol{u})$ is unimodal, and sometimes even symmetric (implying that the p-value of the test is $Pr\{|\boldsymbol{v}| \geq \boldsymbol{v}_{obs})$, and then there is little to argue about. Multimodality occurs, however, at least in pathological cases, and then the principle may be regarded as counter-intuitive. One such case (with symmetry) is the test for Spearman rank correlation in models for rank data, see Section 12.7.

Figure 5.1 gives a schematic picture in the one-dimensional case ($\dim \boldsymbol{v} = 1$) of how the p-value is calculated according to the second part of the principle, formula (5.4). The density is $f_0(\boldsymbol{v} \,|\, \boldsymbol{u}_{obs})$.

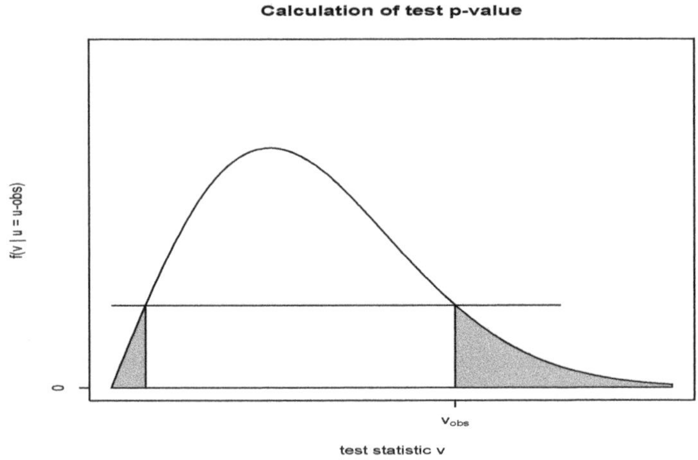

Figure 5.1 Calculation of p-value according to (5.4)

Without symmetry, but with $\dim \boldsymbol{v} = 1$ and unimodality, the split between the two tails when constructing a rejection region is not obvious; think of an F-distribution for comparing two sample variances. One alternative principle of p-value calculation regards a two-sided test as represented by two one-sided tests, see Section 5.3. When $\dim \boldsymbol{v} > 1$, it is less obvious how an alternative p-value to (5.4) should at all be defined. See also Proposition 5.2 for a large sample justification of the p-value formula.

In a specific situation it is a more or less difficult distributional problem to find the conditional distribution for a given observed $\boldsymbol{u}_{obs}$. This goes back on the problem to find an explicit form for the structure function of the family. In many classical situations, this leads to well-known tests, see the next examples. In other cases, the structure function is not explicit, but the distribution might be numerically approximated, for example by simulation.

Example 5.1 ***Variance test for a normal distribution***
The argument for constructing the exact test of H_0: $\sigma = \sigma_0$ for a sample of size n from $N(\mu, \sigma^2)$ was essentially given already in Example 3.3. The hypothesis can be expressed in terms of the canonical parameter $-\frac{1}{2}/\sigma^2$ of $\sum y_i^2$, and the conditional distribution of $\sum y_i^2$ given $\sum y_i$ may be replaced by the distribution of s^2, which is independent of $\sum y_i$ (Proposition 3.24). Under H_0, the test statistic $(n-1)s^2/\sigma_0^2$ is $\chi^2(n-1)$-distributed. Calculation of a p-value according to the second part of the exact test principle will be somewhat complicated. It is much easier to use the principle that a two-sided test is equivalent to two one-sided tests. Thus, let a computer package compute a one-sided tail probability, and double it. △

Example 5.2 ***Mean value test for a normal distribution***
We consider testing H_0: $\mu = 0$ for a sample of size n from $N(\mu, \sigma^2)$. Note that if the hypothesis value were some $\mu_0 \neq 0$ we could simply subtract μ_0 from all y-values to achieve the desired H_0. With $v = \sum y_i$, or $v = \bar{y}$, the hypothesis is equivalent to $\theta_v = 0$. The exact test should be conditional, given $u = \sum y_i^2$. Thus we are free to eliminate the scale parameter by considering $v/\sqrt{u}$ instead of v. The advantage is that under H_0, $v/\sqrt{u}$ is free of both parameters, and then Basu's theorem (Proposition 3.24) guarantees that $v/\sqrt{u}$ is in fact independent of u. Thus, we can forget about the conditioning on u, i.e., we may restrict the attention to the marginal distribution of $v/\sqrt{u}$. The last step is to show that this test is equivalent to the t-test. Let us rewrite $v/\sqrt{u}$ as

$$\frac{\bar{y}}{\sqrt{\sum y_i^2}} = \frac{\bar{y}}{\sqrt{(n-1)s^2 + n\bar{y}^2}} = \frac{\bar{y}/s}{\sqrt{(n-1) + n(\bar{y}/s)^2}},$$

The right-hand side is seen to be an odd and monotone function of the t-test statistic $\sqrt{n}\,\bar{y}/s$. Thus, a single tail or a symmetric pair of tails in $\bar{y}/\sqrt{\sum y_i^2}$ is equivalent to a single or symmetric pair of tails in the usual t-test.

An alternative derivation would use the explicit form of the structure function for the Gaussian sample.

As noted in Example 3.3, a complete conditional inference about μ is not possible. In particular, μ is not linear in the canonical parameters, and we require an estimate of σ to express the precision of $\hat{\mu}$. However, it was possible to derive the exact test as a conditional test, because the hypothesis H_0: $\mu = \mu_0$ is a model-reducing hypothesis for each fixed μ_0. △

Exercise 5.1 ***Two-sample test for variance equality***
The two-sample tests for mean and variance are technically more compli-

cated to derive than the single sample tests. Given two normally distributed samples, of sizes n_1 and n_2, construct a test statistic for an exact test of H_0: Equal variances. Discuss the calculation of p-value.
Hints: First condition on the sample means to make it possible to restrict consideration to the sample variances, which are independent of the sample means. Next condition on the pooled variance, and express the test statistic in terms of the ratio of the sample variances. Establish suitable independence and conclude that the test is an F-test. You need not know the form of the χ^2 or F-distributions, only what they represent. △

Exercise 5.2 ***Correlation test***
Given a sample of size $n > 2$ from a bivariate normal distribution, use the same type of procedure as in Example 5.2 to derive an exact test of the hypothesis that the two variates are uncorrelated. That is: specify u and v, find a function of them that is parameter-free under H_0, conclude independence (see Exercise 3.27), and go over to the marginal distribution. Finally transform to a test statistic of known distribution (use hint).
Hint: $\sqrt{n-2}\, r \,/\, \sqrt{1-r^2}$ is exactly $t(n-2)$-distributed under H_0. △

Exercise 5.3 ***A tool for scan tests in disease surveillance***
The following model and test are the basic tools for the prospective scan statistic proposed by Kulldorff (2001). Question: Has a disease outbreak occurred in a certain region? This region is compared with another region, acting as a baseline. The numbers observed, y_i, $i = 1, 2$, are assumed to be $\text{Po}(\lambda_i A_i)$, where A_i is known and represents the population size in the region. We want to test H_0: $\lambda_1 = \lambda_2$ (against $\lambda_1 > \lambda_2$). Construct the (one-sided) exact test under H_0 (cf. Section 3.5, in particular Exercise 3.13). △

5.2 Fisher's Exact Test for Independence (or Homogeneity or Multiplicativity) in 2 × 2 Tables

This is likely to be a well-known test from previous courses, but perhaps not in all its disguises and all aspects, as derived from different sampling schemes. Suppose n items or individuals are classified with respect to two criteria of two categories each, y_{ij} being the observed number within category combination (i, j), $i = 1, 2$, $j = 1, 2$. For example, suppose a random sample of individuals is taken from a large cohort (of controlled age) and is classified as male/female and smoker/nonsmoker, and tabulated in a so-called *contingency table*. The number of individuals in the different category combinations are then assumed to be multinomially distributed, with probabilities corresponding to the four cohort fraction sizes. In a stratified sampling, on the other hand, the number of males and females are determined before sampling. We then have two binomial distributions instead of one multinomial. Or, in a third type of situation, suppose all individuals coming to a clinic (a certain year, say) and diagnosed as having some (not very common) disease are classified by, say, sex and smoker/nonsmoker. If we think of these patients as appearing randomly in time, according to a Poisson process for each category combination, the number of individuals is Poisson distributed in each category combination and in subtotals and total, y_{ij} being Po(λ_{ij}) and the total $y_{..}$ being Po($\sum_{i,j} \lambda_{ij}$). In the following table we introduce some more notation.

Table 5.1 *Notation for 2 × 2 tables of counts.*

Counts	$j = 0$	$j = 1$	Row sums
$i = 0$	y_{00}	y_{01}	$r_0 = y_{0.}$
$i = 1$	y_{10}	y_{11}	$r_1 = y_{1.}$
Column sums	$s_0 = y_{.0}$	$s_1 = y_{.1}$	$n = y_{..}$

We summarize the three sampling schemes as follows.

Poisson: All y_{ij} are mutually independent and Po(λ_{ij}).
Multinomial: The total n is fixed and $\{y_{ij}\}$ is multinomially distributed with probabilities $\{\pi_{ij}\}$, $\sum_{ij} \pi_{ij} = 1$.
Two binomials: The row sums r_0 and r_1 are fixed, and y_{00} and y_{10} are mutually independent and Bin(r_0, π_0) and Bin(r_1, π_1), respectively.

(A fourth, less-used sampling scheme is when both row and column sums are fixed, for example in the 'tea-tasting lady' experiment (Fisher), when a person tries to classify two tastes of tea and is told how many cups there are of each kind.)

In each of these situations there is a natural first hypothesis to be tested, to simplify the model by specifying in some way or other that the classifications operate mutually independently. These are named and specified as follows.

Poisson: Test for *multiplicativity*, H_0: $\lambda_{ij} = \alpha_i \beta_j$, or $\psi = 0$, with $\psi = \log\{(\lambda_{00}\lambda_{11})/(\lambda_{01}\lambda_{10})\}$.
Multinomial: Test for *independence* between classifications, H_0: Log odds ratio = 0, that is, $(\pi_{00}\pi_{11})/(\pi_{10}\pi_{01}) = 1$, or $\pi_{ij} = \alpha_i \beta_j$, with the suitable normalization $\sum_i \alpha_i = \sum_j \beta_j = 1$.
Two binomials: Test for *homogeneity*, H_0: $\pi_0 = \pi_1$, or $\psi = 0$, where $\psi = \text{logit}(\pi_0) - \text{logit}(\pi_1)$

Before we arrive at Fisher's exact test for these three test situations, we will see successively that they can all be reduced by conditioning, to one and the same case. We start in the Poisson model. Multiplicativity in the mean value λ is equivalent to additivity on canonical log-scale. Under the multiplicative Poisson model, row sums $r_i = y_{i.}$ and column sums $s_j = y_{.j}$ form the canonical statistic:

$$f(\{y_{ij}\}; \{\lambda_{ij}\}) = \prod_{i,j} \lambda_{ij}^{y_{ij}} e^{-\lambda_{ij}} / y_{ij}! \propto \prod_i \alpha_i^{r_i} \prod_j \beta_j^{s_j} h(\mathbf{y})$$
$$= e^{\sum_i \log(\alpha_i) r_i + \sum_j \log(\beta_j) s_j} h(\mathbf{y}) . \qquad (5.5)$$

However, the order of the family is only 3, because the four subtotals r_0, r_1, s_0, s_1 are linearly restricted by the relation $r_0 + r_1 = s_0 + s_1 = n$. Thus we can as well or better use r_0, s_0 and n as components of the canonical statistic. This is the u on which we condition in the exact test. Let us do the conditioning in stages. If we first condition on the total n we arrive in the multinomial model. To check this, first note that conditioning on the total of a number of independent Poisson variables transforms the distribution to a multinomial. In this case:

$$f(\mathbf{y} \,|\, y_{..} = n; \boldsymbol{\lambda}) = \frac{\prod_{i,j} \lambda_{ij}^{y_{ij}} e^{-\lambda_{ij}} / y_{ij}!}{(\sum_{ij} \lambda_{ij})^n e^{-\sum_{ij} \lambda_{ij}} / n!}$$
$$= \prod_{i,j} \left(\frac{\lambda_{ij}}{\sum_{ij} \lambda_{ij}} \right)^{y_{ij}} \frac{n!}{\prod_{ij} y_{ij}!} .$$

We see here that given the total n, the table $\mathbf{y}$ is multinomially distributed with probabilities $p_{ij} = \lambda_{ij}/(\sum \lambda_{ij})$. Furthermore, when $\lambda_{ij} = \alpha_i \beta_j$ we get

$$p_{ij} = \frac{\alpha_i}{\alpha_0 + \alpha_1} \frac{\beta_j}{\beta_0 + \beta_1} ,$$

so the Poisson multiplicativity hypothesis has become the contingency table independence hypothesis. (Or, if you prefer, check that the odds ratio equals 1.) Thus we can draw the important conclusion that the two situations must have the same exact test.

After conditioning on the total n, we are in the contingency table situation under its independence hypothesis, with remaining canonical statistic r_0 and s_0. We continue by conditioning on r_0 (and $r_1 = n - r_0$). But then we arrive in the two binomials situation, and the independence hypothesis goes over in the homogeneity hypothesis $\pi_0 = \pi_1$. We prove the latter fact here:

$$\pi_0 = \frac{\pi_{00}}{\pi_{00} + \pi_{01}} = \frac{\alpha_0\beta_0}{\alpha_0\beta_0 + \alpha_0\beta_1} = \frac{\beta_0}{\beta_0 + \beta_1},$$

and obviously the same result is obtained for π_1. Thus, the homogeneity test for two binomials must also have the same exact test.

It only remains to express this common test. To represent v, given all marginal subtotals, for example expressed by $\boldsymbol{u} = \{r_0, s_0, n\}$, we could use $v = y_{00}$ or any other individual y_{ij}, or the 'interaction' effect $y_{11} - y_{10} - y_{01} + y_{00}$, or anything else that allows us to reconstruct the whole table. To find the conditional probabilities $f_0(v \mid \boldsymbol{u})$ may be regarded as primarily a combinatorial problem, but it is perhaps more easily done by a statistical argument in the two binomials version of the model, that is, when considering the row sums r_0 and r_1 as given (with sum n). Under H_0, the binomials have the same π, denoted π_0 say, and this implies that $s_0 = y_{00} + y_{10}$ is $\text{Bin}(n; \pi_0)$. The conditional probability $f_0(v \mid \boldsymbol{u})$ is parameter-free under H_0, so it will not depend on π_0. The probability of observing a specific table, with y_{00} and y_{10} in the first column, given s_0, can then be expressed as

$$\begin{aligned} f_0(y_{00}, y_{10} \mid s_0, r_0, r_1; \pi_0) &= \frac{f(y_{00}, s_0 - y_{00} \mid r_0, r_1; \pi_0)}{f(s_0 \mid r_0, r_1; \pi_0)} \\ &= \frac{\binom{r_0}{y_{00}} \pi_0^{y_{00}} (1-\pi_0)^{r_0 - y_{00}} \binom{r_1}{y_{10}} \pi_0^{y_{10}} (1-\pi_0)^{r_1 - y_{10}}}{\binom{n}{s_0} \pi_0^{s_0} (1-\pi_0)^{n - s_0}} \\ &= \frac{\binom{r_0}{y_{00}} \binom{r_1}{s_0 - y_{00}}}{\binom{n}{s_0}}. \end{aligned}$$

This is the *hypergeometric probability distribution* for y_{00}, which is the basis for Fisher's exact test of independence (or multiplicativity or homogeneity).

Note the different nature of the conditioning in the exact test and in the modelling of data. The conditioning requested by the exact test is a technical type of conditioning, whereas the characterization of a realistic model for a set of data as being Poisson, multinomial or two binomials depends on the conditioning used in the design of the study. This distinction is stressed by Cox (2006, p. 54). The technical conditioning can also be used in the opposite direction, and often a model with independent Poisson data is easier to handle than a contingency table model with dependencies generated by the fixed total. Whether realistic or not, we can use the technical trick to regard n as generated by some Poisson distribution, and then all counts y_{ij}

become independent and Poisson distributed, see Example 5.3 (last part) and Section 5.6.

The example of the 2×2–table can easily be extended to larger tables, see Example 2.5 and Exercise 5.5. Much of this example was actually expressed in a more general form, with sums and products in its notations to facilitate an extension to larger two-dimensional tables. We leave part of this as Exercise 5.5. The new difficulty appears in the last stage, where v in $f_0(v \mid \boldsymbol{u})$ is no longer one-dimensional, more precisely of dimension $(k-1)(l-1)$ in a $k \times l$ table. This makes the test more difficult, at least in practice, when the simpler hypergeometric is no longer enough. An alternative is to restrict attention to a univariate test statistic, and use the deviance test or some other approximately χ^2 test, which can be justified asymptotically, see Section 5.4.

The theory can also be extended to higher-dimensional tables (> 2 classification factors). It turns out that we have to consider and distinguish total independence and conditional independence between classification factors in a contingency table, see further Section 10.3.

Exercise 5.4 ***Model with odds ratio parameter***
In a saturated multinomial model for the 2×2 contingency table, show that the canonical $\boldsymbol{t}$ may be represented by $\{y_{00}, r_0, s_0\}$, and that the canonical parameter component for y_{00} is then the log odds ratio. △

Exercise 5.5 ***Larger tables***
In a $k \times l$ table, $k \geq 2$, $l \geq 2$, formulate models for data under the three different sampling situations, extend the hypotheses of multiplicativity, independence and homogeneity, and show that the exact test is necessarily the same in all these three situations. However, you need not try to derive an explicit expression for $f_0(\boldsymbol{v} \mid \boldsymbol{u})$ (now a multivariate density). △

5.3 Further Remarks on Statistical Tests

A test p-value is calculated under the H_0 model and is valid as such. So far we have assumed that the question leading to a test is formulated as a matter of possible simplification from the bigger, 'alternative' model, with canonical statistic $\boldsymbol{t} = (\boldsymbol{u}, \boldsymbol{v})$, to the smaller model with canonical statistic $\boldsymbol{u}$. However, other types of situation are also allowed. We might have begun with the more specific H_0 model and ask if it fits the data reasonably. Either we do then have a specific, wider model in mind, that we believe in as an alternative, or we are afraid of departures from H_0 in some direction, without believing in a specific wider model as 'true' model. A typical example is a trend test, when we check the H_0 model by trying it against an alternative with a linear time trend, well aware that a possible trend need not be linear but perhaps monotone, thus containing a linear component.

Thus the theory of Section 5.1 allows the test to be a *pure significance test* (Cox & Hinkley, 1974, ch. 3), the concept meaning that only a null model is properly specified (a family with canonical statistic $\boldsymbol{u}$). If we want to test that the statistic $\boldsymbol{v}$ has no role in the summarization of data, we may embed the null model in a wider exponential family with canonical statistic $\boldsymbol{t} = (\boldsymbol{u}, \boldsymbol{v})$ by exponential tilting (see Example 2.12), in order to get an alternative model of exponential type within which we may construct the exact test (or one of its large sample approximations, see Section 5.4). An alternative view is described in Section 6.3.

A statistical test in a model with nuisance parameters is called *similar* if it is defined by a critical region with a *size* (probability under H_0 for rejection of H_0 – type I error probability) that is independent of the nuisance parameter (in this case $\boldsymbol{\lambda}$, or $\boldsymbol{\mu}_u$). Clearly, this is a highly desirable property. It is satisfied by the exact test derived in Section 5.1, since if the test is size α for each value of the conditioning statistic, it is also marginally size α.

Note that under H_0, when $\boldsymbol{\psi} = \boldsymbol{0}$ ($\boldsymbol{\theta}_v = \boldsymbol{0}$), $\boldsymbol{u}$ is a minimal sufficient statistic for the nuisance parameter $\boldsymbol{\theta}_u$, and full exponential families are (boundedly) complete, see Section 3.7. A classic theorem by Lehmann and Scheffé tells that if, under H_0, there is a sufficient statistic whose distribution family is (boundedly) complete, a similar test of size α must be conditionally size α, given the sufficient statistic (i.e. $\boldsymbol{u}$). Thus, in order to achieve similarity of a test for $\boldsymbol{\psi} = \boldsymbol{0}$, there is no alternative for elimination of the influence of the nuisance parameter, than to condition on $\boldsymbol{u}$, that is, to use $f_0(\boldsymbol{v} \mid \boldsymbol{u})$. Such a conditionally valid critical region is sometimes said to have *Neyman structure*, see for example Cox and Hinkley (1974, Sec. 5.2).

In retrospect, this result could be used for example in an implicit derivation of the t-test statistic of Example 5.2. Since we know (already from introductory courses) that the t-test statistic has the similarity property, it must have Neyman structure and thus be the conditional test, derived with more effort in Example 5.2.

If possible, a test should be chosen such that it is more powerful than any other test against all alternatives to H_0. However, such tests exist only in the very restrictive case $\dim \boldsymbol{v} = 1$ and $\dim \boldsymbol{u} = 0$ (no nuisance parameters), and under restriction to one-sided alternatives. Such a test is then obtained by application of the *Neyman–Pearson Lemma*, considering the likelihood ratio against some alternative. The test will be independent of the (one-sided) alternative chosen, because exponential families have so-called monotone likelihood ratio. This will be a special case both for its one-sidedness and for its severe dimension constraints. However, in the presence of nuisance parameters, it is natural to first eliminate these by restricting consideration to similar tests, before considering optimality considerations.

The optimality considerations for $\dim \boldsymbol{v} = 1$ carry over to confidence regions, see e.g. Schweder and Hjort (2016, Th. 5.11 in particular).

As mentioned, only one-sided tests can be most powerful against a specific alternative. Two-sided tests are preferably regarded as representing two one-sided tests, by the following procedure:

- Calculate the two p-values
- Select the smallest of them (of course)
- Compensate exactly for this selection by doubling the p-value

This corresponds to critical regions with equal probability in the two tails. A modification is needed when the test statistic has a discrete distribution. For further details, see Cox and Hinkley (1974, Sec. 33.4(iv)) or Cox (2006, Sec. 3.3).

5.4 Large Sample Approximation of the Exact Test

The general theory of statistics provides several large sample statistics for testing hypotheses of type H_0: $\boldsymbol{\psi} = \mathbf{0}$, asymptotically equivalent as the amount of information (e.g. sample size) increases, and asymptotically $\chi^2(q)$ distributed, see Section 5.5. They are primarily the log-likelihood ratio (or deviance) test, the score test (Rao's test), and the MLE test (or Wald's test). We here first note, for later reference, the form of the score test in an exponential family setting with the hypothesis (5.1) to be tested. Next we will demonstrate that the exact test introduced in Section 5.1 is asymptotically equivalent to the score test, and thus also to the other large sample tests, to be more generally discussed in Section 5.5.

As a start, the score test statistic W_u can generally be written

$$W_u = U(\hat{\boldsymbol{\theta}}_0)^T \, I(\hat{\boldsymbol{\theta}}_0)^{-1} \, U(\hat{\boldsymbol{\theta}}_0), \qquad (5.6)$$

where the score $\boldsymbol{U}$ is calculated in the MLE $\hat{\boldsymbol{\theta}}_0$ of the H_0 model. This form holds not only in exponential families, and it holds for any model-reducing hypothesis, whatever linear restrictions it imposes on $\boldsymbol{\theta}$. In exponential families with $\boldsymbol{\theta}$ canonical, $\boldsymbol{U}(\boldsymbol{\theta}) = \boldsymbol{t} - \boldsymbol{\mu}_t(\boldsymbol{\theta})$. Parenthetically, then, since $I(\hat{\boldsymbol{\theta}}) = \mathrm{var}(\boldsymbol{t})$, W_u in (5.6) is a squared Mahalanobis type distance from $\boldsymbol{t}$ to the hypothesis model. Under H_0: $\boldsymbol{\theta}_v = \mathbf{0}$ (or $\boldsymbol{\psi} = \mathbf{0}$), note that the $\boldsymbol{u}$-part of $\boldsymbol{t}$ is sufficient, and in the MLE $\hat{\boldsymbol{\theta}}_0$ ($= \hat{\boldsymbol{\theta}}_0(\boldsymbol{u})$), the $\boldsymbol{u}$-part of U is zero:

$$U(\hat{\boldsymbol{\theta}}_0) = \boldsymbol{t} - \boldsymbol{\mu}_t(\hat{\boldsymbol{\theta}}_0) = \begin{pmatrix} \mathbf{0} \\ \boldsymbol{v} - \boldsymbol{\mu}_v(\hat{\boldsymbol{\theta}}_0) \end{pmatrix}$$

Here $\boldsymbol{\mu}_v$ denotes the v-component of $\boldsymbol{\mu}_t$. Inserting this expression for U in (5.6) and noting that $I(\theta) = V_t(\theta)$ yields

$$W_u = (\boldsymbol{v} - \boldsymbol{\mu}_v(\hat{\boldsymbol{\theta}}_0))^T \, V_t^{vv}(\hat{\boldsymbol{\theta}}_0) \, (\boldsymbol{v} - \boldsymbol{\mu}_v(\hat{\boldsymbol{\theta}}_0)) \tag{5.7}$$

where V_t^{vv} denotes the submatrix of the inverse, $(V_t^{-1})_{vv}$. Although notationally expressed in $\hat{\boldsymbol{\theta}}_0$, this form allows any choice of parameterization, since we only need the first two moments of $\boldsymbol{t}$ estimated under H_0. We may also note that W_u can be regarded as a function of the exact test statistic, because in the latter we may subtract from v any statistic only depending on u, and likewise normalize by such a statistic, and $\hat{\boldsymbol{\theta}}_0$ depends only on $\boldsymbol{u}$. What remains as a justification of W_u is to show that the conditional distribution of $\boldsymbol{v}$ given $\boldsymbol{u}$ tends to the normal distribution underlying the asymptotic χ^2 distribution of W_u. We prove the following result (Martin-Löf, 1970):

Proposition 5.2 ***Large sample approximation of the exact test***
The exact test, rejecting for small values of the conditional density $f_0(\boldsymbol{v}\,|\,\boldsymbol{u})$, is asymptotically equivalent to the $\chi^2(\dim \boldsymbol{v})$-distributed score test (5.7) and the other large sample tests equivalent to the score test.

Proof The exact test statistic is the ratio of two structure functions,

$$f_0(\boldsymbol{v}\,|\,\boldsymbol{u}) = \frac{g(\boldsymbol{u}, \boldsymbol{v})}{g_0(\boldsymbol{u})}.$$

We will use the saddlepoint approximation method on each of them. The denominator is in fact a constant in the conditional density for $\boldsymbol{v}$, but it may be instructive to see factors cancel between the approximated numerator and denominator. Thus, we first approximate the denominator within the marginal model for $\boldsymbol{u}$ when $\boldsymbol{\psi} = \mathbf{0}$, using (4.8) in Proposition 4.5:

$$g_0(\boldsymbol{u}) \approx \frac{C(\hat{\boldsymbol{\theta}}_0)}{(2\pi)^{p/2}\sqrt{\det V_u(\hat{\boldsymbol{\theta}}_0)}} \, e^{-\hat{\boldsymbol{\theta}}_u^T \boldsymbol{u}}, \tag{5.8}$$

where $\hat{\boldsymbol{\theta}}_u$ is the $\boldsymbol{u}$-part of $\hat{\boldsymbol{\theta}}_0 = (\hat{\boldsymbol{\theta}}_u, \mathbf{0})$. If we next approximated the numerator in a standard way, too, by using (4.8) in the point $\boldsymbol{t} = (\boldsymbol{u}, \boldsymbol{v})$ for the full model, we would get $\boldsymbol{v}$ involved in an intricate way. Instead, we go back to the derivation of the saddlepoint approximation with the combination of the two equations (4.6), but in the choice of $\boldsymbol{\theta}$ now select $\boldsymbol{\theta} = \hat{\boldsymbol{\theta}}_0$ instead of the standard $\boldsymbol{\theta} = \hat{\boldsymbol{\theta}}(\boldsymbol{t})$. The procedure yields the following result:

$$g(\boldsymbol{u}, \boldsymbol{v}) \approx \frac{C(\hat{\boldsymbol{\theta}}_0)}{(2\pi)^{r/2}\sqrt{\det V_t(\hat{\boldsymbol{\theta}}_0)}} e^{-\hat{\boldsymbol{\theta}}_u^T \boldsymbol{u}} e^{-\frac{1}{2}(\boldsymbol{t} - \boldsymbol{\mu}_t(\hat{\boldsymbol{\theta}}_0))^T V_t^{-1}(\hat{\boldsymbol{\theta}}_0)(\boldsymbol{t} - \boldsymbol{\mu}_t(\hat{\boldsymbol{\theta}}_0))} \tag{5.9}$$

The statistic $\boldsymbol{v}$ (subvector of $\boldsymbol{t}$) is only found in the exponent, where we recognize the score statistic (5.6). We can see that the approximation of the density $f_0(\boldsymbol{v} \,|\, \boldsymbol{u})$ for $\boldsymbol{v}$ is a monotone decreasing function of the score test statistic. This completes the proof. □

It might be elucidating to study the approximation a bit closer. Rewriting the score test statistic in the exponent of (5.9) as in (5.7), we see that we have a multivariate normal distribution for $\boldsymbol{v}$. It follows that the conditional distribution $f_0(\boldsymbol{v} \,|\, \boldsymbol{u})$ is asymptotically a dim $\boldsymbol{v}$-dimensional multivariate normal distribution,

$$f_0(\boldsymbol{v} \,|\, \boldsymbol{u}) \approx \frac{1}{(2\pi)^{q/2}\sqrt{\det(V_t^{vv}(\hat{\boldsymbol{\theta}}_0))^{-1}}} e^{-\frac{1}{2}(\boldsymbol{v} - \boldsymbol{\mu}_v(\hat{\boldsymbol{\theta}}_0))^T V_t^{vv}(\hat{\boldsymbol{\theta}}_0)(\boldsymbol{v} - \boldsymbol{\mu}_v(\hat{\boldsymbol{\theta}}_u, \boldsymbol{0}))}, \tag{5.10}$$

where V_t^{vv} denotes the vv-block of the inverse of V_t. Since we are approximating the conditional distribution of $\boldsymbol{v}$, it follows that the large sample test can be given a conditional interpretation:

If a p-value is calculated as the probability over all points $\boldsymbol{v}$ for which $f_0(\boldsymbol{v} \,|\, \boldsymbol{u}_{obs}) \leq f_0(\boldsymbol{v}_{obs} \,|\, \boldsymbol{u}_{obs})$, it can be approximated by a χ^2 probability, or, if a critical value is specified for the score test statistic, the critical value for $f_0(\boldsymbol{v} \,|\, \boldsymbol{u})$ can be approximated by using the corresponding χ^2 quantile in the exponent.

Example 5.3 ***The score test for homogeneity or independence or multiplicativity in*** 2×2***-tables, continuation from Section 5.2***
We here apply Proposition 5.2 and derive the score test approximating the exact test. We also look at the well-known Pearson χ^2 test, which will turn out to be identical with the score test.

We use the knowledge from Section 5.2 to conclude that we need only pursue the derivation for one of the three models, but for illumination we do it in two of them. As a first choice, in order to deal with covariance matrices of small dimension, we derive an expression for the score test in the 'two binomials' set-up (with row sums fixed and given). Under homogeneity, $u = s_0$ is sufficient for the single nuisance parameter, the probability π_0 say, with MLE $\hat{\pi}_0 = s_0/n$ under H_0. We let $v = y_{00}$. Under H_0, $\mu_v(\hat{\boldsymbol{\theta}}_0) =$

$r_0\,\hat{\pi}_0 = r_0 s_0/n$ and $\text{var}(v) = r_0\,\pi_0(1-\pi_0)$. The whole variance-covariance matrix for $\boldsymbol{t} = (u, v)$ is

$$V_t(\boldsymbol{\theta}_0) = \begin{pmatrix} n\,\pi_0(1-\pi_0) & r_0\,\pi_0(1-\pi_0) \\ r_0\,\pi_0(1-\pi_0) & r_0\,\pi_0(1-\pi_0) \end{pmatrix}$$

We invert this matrix and extract the vv corner element:

$$V_t^{vv}(\boldsymbol{\theta}_0) = \frac{1}{\pi_0(1-\pi_0)\,r_0(n-r_0)} \begin{pmatrix} r_0 & -r_0 \\ -r_0 & n \end{pmatrix}_{vv} = \frac{n}{\pi_0(1-\pi_0)\,r_0 r_1}$$

The MLE of $\pi_0(1-\pi_0)$ is $\hat{\pi}_0(1-\hat{\pi}_0) = s_0 s_1/n^2$. This yields the test statistic

$$W_u = \frac{n^3\,(y_{00} - r_0 s_0/n)^2}{r_0 r_1\,s_0 s_1} = \frac{n\,(y_{00}y_{11} - y_{10}y_{01})^2}{r_0 r_1\,s_0 s_1} \tag{5.11}$$

The last expression of (5.11) is obtained from the preceding expression through multiplication by n inside and simplification of the resulting expression. Note that $W_u = 0$ when the observed odds ratio is 1. However, (5.11) does not at all look like the conventional Pearson's χ^2 test statistic. It can in fact be rewritten on that form, or vice versa, but we choose here another path, that will yield Pearson's χ^2 statistic more immediately.

We now start from version (5.6), instead of version (5.7), and we make the calculations in the Poisson model. In the full model, with one parameter λ_{ij} per cell and $\boldsymbol{t}$ represented by the full 2×2-table, the score $U(\boldsymbol{\theta})$ has components $y_{ij} - \lambda_{ij}$ ($\boldsymbol{\lambda} = \exp(\boldsymbol{\theta})$), to be calculated and estimated under the multiplicativity hypothesis, when the MLE is $\hat{\lambda}_{ij}^{(0)} = r_i s_j/n$. The corresponding variance–covariance matrix is extremely simple, being diagonal with elements $\text{var}(y_{ij}) = \lambda_{ij}$. To obtain the inverse, we only invert the diagonal elements. This yields the score test statistic

$$W_u = \sum_{i,j} \frac{(y_{ij} - \hat{\lambda}_{ij}^{(0)})^2}{\hat{\lambda}_{ij}^{(0)}} = \sum_{i,j} \frac{(y_{ij} - r_i s_j/n)^2}{r_i s_j/n}. \tag{5.12}$$

This statistic can be recognized as the standard Pearson test statistic, approximately χ^2 distributed with 1 degree of freedom (dimension reduction from 4 to 3). A historical anecdote is that Karl Pearson proposed this test statistic under the belief that it had 3 degrees of freedom. Fisher (1922a) derived the correct number, $df = 1$. This was hard for Pearson to accept.

△

Exercise 5.6 ***Determinants***

Use Proposition B.1 in Section B.1 to check that the determinant in (5.10) is consistent with the ratio of the determinants in (5.8) and (5.9). △

Exercise 5.7 ***Explicit proof of identity***
Show that expressions (5.11) and (5.12) are in fact identical, by verifying that

- $(y_{ij} - r_i s_j/n)^2$ is the same for all (i, j);
- $1/(r_i s_j)$ sum over (i, j) to $n^2/(r_0 r_1 s_0 s_1)$. △

5.5 Asymptotically Equivalent Large Sample Tests, Generally

In Section 5.4, the generally valid asymptotic equivalence of several test statistics was mentioned. For completeness we motivate and formulate this important property in two propositions. Without reference to exponential families, suppose we have a model whose parameter $\boldsymbol{\theta}$ is partitioned into a parameter of interest $\boldsymbol{\psi}$ and a nuisance parameter $\boldsymbol{\lambda}$, so $\boldsymbol{\theta} = (\boldsymbol{\lambda}, \boldsymbol{\psi})$ (with abuse of notation, since $\boldsymbol{\theta}$, $\boldsymbol{\psi}$ and $\boldsymbol{\lambda}$ are all treated as column vectors). We consider testing the hypothesis H_0: $\boldsymbol{\psi} = \boldsymbol{\psi}_0$, reducing the parameter dimension from $\dim \boldsymbol{\lambda} + \dim \boldsymbol{\psi}$ to $\dim \boldsymbol{\lambda}$.

The model must satisfy suitable regularity assumptions. These could be based on those of Cramér (1946, Sec. 33.3), in his proof of the large sample existence of a unique local MLE, consistent and asymptotically normally distributed, but then in a slightly stronger version, implying that the assumptions are satisfied uniformly for $\boldsymbol{\theta}$ in a neighbourhood of the true parameter point $\boldsymbol{\theta}_0$ under H_0, say $\boldsymbol{\theta}_0 = (\boldsymbol{\lambda}_0, \boldsymbol{\psi}_0)$. We assume n is a sample size, even though in applications we often do not have iid data.

Following the notation of Cox and Hinkley (1974), the deviance, or twice the log likelihood ratio in the MLEs, will here be denoted W, that is,

$$W = 2 \log \frac{L(\hat{\boldsymbol{\theta}})}{L(\hat{\boldsymbol{\lambda}}_0)}, \tag{5.13}$$

where we write $\hat{\boldsymbol{\theta}} = (\hat{\boldsymbol{\lambda}}, \hat{\boldsymbol{\psi}})$ and $\hat{\boldsymbol{\theta}}_0 = (\hat{\boldsymbol{\theta}}_0, \boldsymbol{\psi}_0)$, in order to distinguish the two estimates of $\boldsymbol{\lambda}$.

For the first proposition we assume that H_0 holds. Taylor expansion of W in $\hat{\boldsymbol{\theta}} = \hat{\boldsymbol{\theta}}_0$ around the point $\hat{\boldsymbol{\theta}}$ yields

$$\begin{aligned} W &= -2\{\log L(\hat{\boldsymbol{\theta}}_0) - \log L(\hat{\boldsymbol{\theta}})\} \\ &= -2U(\hat{\boldsymbol{\theta}})^T(\hat{\boldsymbol{\theta}}_0 - \hat{\boldsymbol{\theta}}) + (\hat{\boldsymbol{\theta}}_0 - \hat{\boldsymbol{\theta}})^T J(\hat{\boldsymbol{\theta}})(\hat{\boldsymbol{\theta}}_0 - \hat{\boldsymbol{\theta}}) + o(n\,|\hat{\boldsymbol{\theta}}_0 - \hat{\boldsymbol{\theta}}|^2). \end{aligned} \tag{5.14}$$

The first term on the right-hand side is identically zero and the second term dominates as $n \to \infty$, being of magnitude $O(1/\sqrt{n})O(n)O(1/\sqrt{n}) = O(1)$. Here we may asymptotically change $J(\hat{\boldsymbol{\theta}})$ to its expected value $I(\hat{\boldsymbol{\theta}})$, and these matrices may be calculated in any of the points $\hat{\theta}$ or $\hat{\theta}_0$.j The first test equivalence has now been established: The quadratic form test statistic

$$W_e^* = (\hat{\theta}_0 - \hat{\theta})^T I(\hat{\theta}_0)(\hat{\theta}_0 - \hat{\theta}) \tag{5.15}$$

(or with $I(\hat{\theta})$, or with J instead of I) is asymptotically equivalent to the statistic W.

Note: For a simple hypothesis ($\theta = \psi$; no need for λ), this test statistic is the same as the Wald statistic, but for composite hypotheses, the two test statistics differ.

Next we derive the score test statistic by utilizing a linearization of U:

$$\mathbf{0} = U(\hat{\theta}) = U(\hat{\theta}_0) - J(\hat{\theta}_0)(\hat{\theta} - \hat{\theta}_0) + o(n\,|\hat{\theta} - \hat{\theta}_0|). \tag{5.16}$$

Here the left-hand side is zero because $\hat{\theta}$ is a root to the likelihood equations. Inserting

$$\hat{\theta} - \hat{\theta}_0 = J(\hat{\theta}_0)^{-1} U(\hat{\theta}_0) \tag{5.17}$$

in W_e^*, and changing from J to I, we obtain a first expression for the *score test statistic* W_u,

$$W_u = U(\hat{\theta}_0)^T\, I(\hat{\theta}_0)^{-1}\, U(\hat{\theta}_0). \tag{5.18}$$

As before, we may use J instead of I. However, it is not natural now to replace $J(\hat{\theta}_0)$ by $J(\hat{\theta})$, or analogously for I, because then we would lose the great advantage with the score test statistic, that it only requires $\hat{\theta}_0$, not the full-dimensional $\hat{\theta}$.

The score test statistic is also called Rao's test statistic, after C.R. Rao, who introduced it for composite hypotheses in 1948.

The statistic W_u is a quadratic form in the dim θ-dimensional vector U, for which we might expect a $\chi^2(\dim\theta)$ distribution. That is an illusion, however, because it is really only dim ψ-dimensional. Since $\hat{\theta}_0 = (\hat{\lambda}_0, \psi_0)$ is the MLE under H_0, the λ-part subvector U_λ of U, being the score vector in the H_0-model, is zero in $\hat{\theta}_0$. Thus, inserting

$$U(\hat{\theta}_0) = \begin{pmatrix} \mathbf{0} \\ U_\psi(\hat{\theta}_0) \end{pmatrix} \tag{5.19}$$

in the expression for W_u, it is seen that we only need the lower right corner submatrix $(I^{-1})_{\psi\psi}$ of I^{-1} corresponding to the nonzero part of U. We use the notation $(I^{-1})_{\psi\psi} = I^{\psi\psi}$ and can then write

$$W_u = U_\psi(\hat{\theta}_0)^T\, I^{\psi\psi}(\hat{\theta}_0)\, U_\psi(\hat{\theta}_0). \tag{5.20}$$

Thus, W_u is only of underlying dimension dim ψ. Note, however, that $I^{\psi\psi}$

is not the inverse of the variance matrix of U_ψ, which is $I_{\psi\psi}^{-1}$. Instead, $I^{\psi\psi}$ represents the inverse of the conditional variance of U_ψ, given U_λ, see Section B.1. Thus, to get the right test statistic we must utilize that we know $U_\lambda(\hat{\boldsymbol{\theta}}_0)$, and this is done by conditioning on it. Because $U_\lambda(\hat{\boldsymbol{\theta}}_0) = \mathbf{0}$, its influence is seen only in the norming variance matrix based on I. More clarity may be gained when we now finally derive the Wald test statistic from W_u.

In formula (5.17) for $\hat{\boldsymbol{\theta}} - \hat{\boldsymbol{\theta}}_0$, let us restrict attention only to the subvector $\hat{\boldsymbol{\psi}} - \boldsymbol{\psi}_0$, which should be its informative part. Remembering from (5.19) that the $\boldsymbol{\lambda}$-part of U is zero, we obtain the approximate expression

$$\hat{\boldsymbol{\psi}} - \boldsymbol{\psi}_0 = I^{\psi\psi}(\hat{\boldsymbol{\theta}}_0)\, U_\psi(\hat{\boldsymbol{\theta}}_0),$$

where we can multiply from the left by the inverse of $I^{\psi\psi}(\hat{\boldsymbol{\theta}}_0)$ to obtain instead an equivalent formula for U_ψ in terms of $\hat{\boldsymbol{\psi}} - \boldsymbol{\psi}_0$. We insert this formula in expression (5.20) for the score test statistic and obtain the Wald test statistic W_e,

$$W_e = (\hat{\boldsymbol{\psi}} - \boldsymbol{\psi}_0)^T\, (I^{\psi\psi}(\hat{\boldsymbol{\theta}}))^{-1}\, (\hat{\boldsymbol{\psi}} - \boldsymbol{\psi}_0) \tag{5.21}$$

This statistic, also called the ML statistic, was derived by Wald (1943). Here we use the information in $\hat{\boldsymbol{\theta}}$, rather than $\hat{\boldsymbol{\theta}}_0$, of course, since we otherwise need both estimates (unless H_0 is simple). If both $\hat{\boldsymbol{\theta}}$ and $\hat{\boldsymbol{\theta}}_0$ are available, the deviance is recommended. Note that the middle matrix is the inverse of the asymptotic variance–covariance matrix of $\hat{\boldsymbol{\psi}}$. More specifically, we know from before that $\hat{\boldsymbol{\theta}}$ is asymptotically $\mathrm{N}(\boldsymbol{\theta}_0, I(\hat{\boldsymbol{\theta}}_0)^{-1})$. Thus, the subvector $\hat{\boldsymbol{\psi}}$ is $\mathrm{N}(\boldsymbol{\psi}_0, (I^{-1})_{\psi\psi}(\hat{\boldsymbol{\theta}}_0) = I^{\psi\psi}(\hat{\boldsymbol{\theta}}_0))$. This also shows that W_e is χ^2 distributed with $\dim \boldsymbol{\psi}$ degrees of freedom. The derivation starting from the deviance, shows that all four test statistics have the same distribution. Thus we have (essentially) derived the following result:

Proposition 5.3 ***Four asymptotically equivalent large sample tests for composite hypotheses***
As $n \to \infty$, the following four large sample tests are asymptotically equivalent under $\mathrm{H}_0 : \boldsymbol{\psi} = \boldsymbol{\psi}_0$, *and their common distribution is the* $\chi^2(\dim \boldsymbol{\psi})$ *distribution:*

- *The twice log likelihood ratio statistic W (the deviance), see (5.13)*
- *The score or Rao test statistic W_u, see (5.18) and (5.20)*
- *The Wald (or maximum likelihood) test statistic W_e, see (5.21)*
- *The double maximum likelihood test statistic W_e^*, see (5.15)*

Proposition 5.4 ***Test equivalence under local alternatives to*** H_0

The four large sample tests of Proposition 5.3 are asymtotically equivalent also under alternatives approaching H_0 *at rate* $O(1/\sqrt{n})$ *as* $n \to \infty$. *The common asymptotic distribution in this case is a noncentral* χ^2 *distribution with noncentrality parameter*

$$(\psi - \psi_0)^T \,(I^{\psi\psi}(\theta_0))^{-1}\,(\psi - \psi_0)$$

Proof For the latter proposition it only remains to comment on the equivalence under alternatives to H_0. For a fixed alternative, all test statistics go to infinity with n and the power of all tests approach 1, so this is not a fruitful situation for comparisons. To keep the power away from 1, we must consider alternatives approaching H_0. If θ approaches θ_0 at rate $O(1/\sqrt{n})$ or faster, the remainder terms in the basic approximations (5.14) and (5.16) will still be of a smaller magnitude than the corresponding main terms, Reconsidering the approximation steps of the derivations, it can be seen that this is sufficient for the asymptotic equivalence of all four test statistics. The noncentral χ^2 distribution follows from the form of W_e and the definition of that distribution (details omitted). □

Remark 5.5 ***Practical aspects of the choice of test statistic.*** The fact that test statistics are asymptotically equivalent does of course not imply that they are equally useful. There are several aspects to consider:

- W and W_u are parameterization invariant, which is an important advantage, because the χ^2 approximation for W_e (or W_e^*) can be much worse than for W and W_u if the parameterization is not well chosen (see Figure 5.2).
- W_u only requires $\hat{\theta}_0$, which is an advantage if $\hat{\theta}$ is difficult to find.
- W_e only requires $\hat{\theta}$, which is an advantage if $\hat{\theta}_0$ is difficult to find.
- W_e is most directly connected with MLE standard error and the corresponding confidence intervals, but note that in contrast to deviance-based confidence intervals, the MLE-based intervals are sensitive to the choice of parameterization, see Figure 5.2 for an illustration of how the deviance and Wald test intervals differ depending on parameter choice.
- If the information matrix is large-dimensional and difficult to invert, W (and W_e^*) have the advantage of not requiring matrix inversion.
- Note that W, W_e^* and the general version (5.18) of W_u can be particularly handy to use when the hypothesis is not formulated as explicitly as $\psi = \psi_0$, but more indirectly in terms of subspaces or flats (affine subspaces) of the parameter space Θ for θ, cf. Exercise 5.11.

△

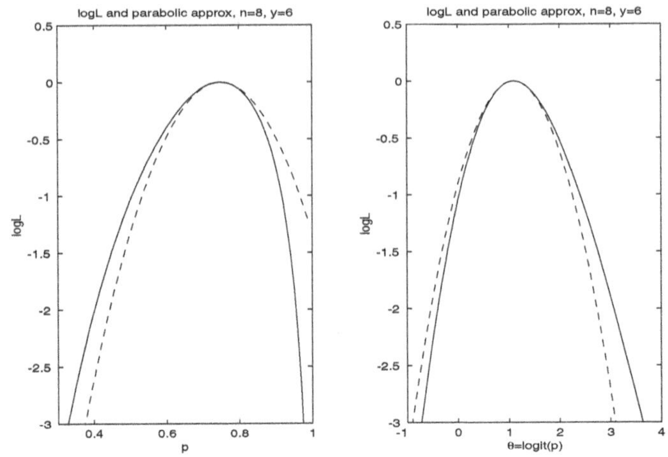

Figure 5.2 Illustration under Bin(n; p) with $n = 8$, $y = 6$ of $\log L$ approximations, as function of the binomial probability p (to the left) and of the canonical $\theta = \text{logit}(p)$ (to the right).
The likelihood has been normalized to have $\log L$ maximum zero. Note form and size of likelihood-based (invariant) and Wald confidence intervals, obtained when cutting $\log L$ and its approximation by a horizontal line, e.g. at level -2 for $\approx 95\%$.
— log L
– – parabolic approximation (Wald)

In Section 4.2 it was pointed out that refinement of p-values, or of test statistics (to fit χ^2) is a more difficult topic than refinement of the distribution of the MLE, and we will not enter this complex area here. Instead, Barndorff-Nielsen and Cox (1994), Jensen (1995) and Skovgaard (2001) are proposed as a choice of references to the interested reader.

5.6 A Poisson Trick for Deriving Large Sample Test Statistics in Multinomial and Related Cases

If we try to extend the results of Example 5.3 from 2×2 to a larger $k\times l$ table, the derivation of an explicit form in the Poisson model remains simple, because its variance matrix remains diagonal. Formula (5.12) still holds for a $k\times l$ table. For fixed sample size n and multinomial sampling, on the other hand, the restriction in the sample space implies that we need invert a non-diagonal variance matrix, and this is not trivial. The homogeneity test in the general case is not simpler, since it involves a number of multinomials with fixed totals. However, by the following result, going back to Fisher (1922a), we are allowed to find the form of the score test statistic in these and similar cases by pretending we are in the Poisson situation, even if this is not the case. In this way, for the large sample tests, we have a correspondence to the equivalence results in Section 5.2 for the exact test.

Proposition 5.6 ***A Poisson trick***
Suppose we have a sample of size n from a full family with canonical statistic $\boldsymbol{t}$ and canonical parameter $\boldsymbol{\theta}$, and that we want to test a model-reducing hypothesis $\boldsymbol{t} \to \boldsymbol{u}$ (or $\boldsymbol{\psi} = \boldsymbol{\psi}_0$ for a subvector of $\boldsymbol{\theta}$ in suitable parameterization). Then the standard large sample test statistics (LR, Score, Wald) are identically the same as in the joint model for $(\boldsymbol{t}, n)$, where n is regarded as Poisson distributed with mean λ and the original sample as conditional, given n, and when the MLE $\hat{\lambda} = n$ is inserted for λ.

Proof From the construction, the joint model for $(\boldsymbol{t}, n)$ has likelihood

$$L(\boldsymbol{\theta}, \lambda;\, \boldsymbol{t}, n) = L_1(\boldsymbol{\theta};\, \boldsymbol{t}\,|\,n)\, L_2(\lambda;\, n), \tag{5.22}$$

where L_1 is identical with the original likelihood, and L_2 is the Poisson likelihood. This factorization is an example of a so-called cut, and n is S-ancillary for $\boldsymbol{\theta}$. More precisely, the joint distribution for $(\boldsymbol{t}, n)$ is seen to be

$$f(\boldsymbol{t}, n;\, \boldsymbol{\theta}, \lambda) = e^{\boldsymbol{\theta}^T \boldsymbol{t}}\, (\lambda / C(\boldsymbol{\theta}))^n\, g_n(\boldsymbol{t})\, e^{-\lambda} / n!\,.$$

Clearly, this is an exponential family with canonical statistic $(\boldsymbol{t}, n)$. The S-ancillarity of n is connected with the fact that $(\boldsymbol{\theta}, \lambda)$ is a mixed type parameterization, λ (= μ_u in notation from Section 3.5) being the mean value of n, and used instead of the quite complicated canonical parameter $\log\lambda - \log C(\boldsymbol{\theta})$ for n in the joint family, cf. Section 3.5.

It is evident from (5.22) that

(1) the MLE of λ is $\hat{\lambda} = n$ both in the full model for $\boldsymbol{\theta}$ and in the hypothesis model.
(2) the MLE of $\boldsymbol{\theta}$ is the same in the joint model as in the original model, and this holds both in the full model and in the hypothesis model.

It follows that the likelihood ratio test statistic W is the same, whether n is regarded as random or nonrandom, since L_2 cancels from the ratio and the L_1 ratio is the same in the joint as in the original model.

The score function in the joint model with $\hat{\lambda}$ for λ is

$$U(\boldsymbol{\theta}, \hat{\lambda};\, \boldsymbol{t}, n) = \begin{pmatrix} U_1(\boldsymbol{\theta};\, \boldsymbol{t}\,|\,n) \\ U_2(\hat{\lambda};\, n) \end{pmatrix} = \begin{pmatrix} U_1(\boldsymbol{\theta};\, \boldsymbol{t}\,|\,n) \\ 0 \end{pmatrix}$$

Thus, for the score test statistic we only also need the upper left $\boldsymbol{\theta}$-corner. In combination with the fact that the observed and expected information matrices for $(\boldsymbol{\theta}, \lambda)$ are block diagonal, it is seen that the score test statistic is the same in the joint as in the original model:

$$W_u(\boldsymbol{\theta}, \hat{\lambda}) = \begin{pmatrix} U_1(\boldsymbol{\theta}) \\ 0 \end{pmatrix}^T \begin{pmatrix} I_{11}(\boldsymbol{\theta}) & \mathbf{0} \\ \mathbf{0}^T & I_{22}(\hat{\lambda}) \end{pmatrix}^{-1} \begin{pmatrix} U_1(\boldsymbol{\theta}) \\ 0 \end{pmatrix}$$
$$= U_1(\boldsymbol{\theta})^T I_{11}(\boldsymbol{\theta})^{-1} U_1(\boldsymbol{\theta}) = W_u(\boldsymbol{\theta})$$

More specifically, $U_1(\boldsymbol{\theta}) = \boldsymbol{t} - \boldsymbol{\mu}_t(\boldsymbol{\theta})$, $I_{11}(\boldsymbol{\theta}) = J_{11}(\boldsymbol{\theta}) = V_t(\boldsymbol{\theta})$, and $I_{22}(\lambda) = \mathrm{var}(n/\lambda) = 1/\lambda$, but these formulas are not needed in the proof.

Finally, the Wald test W_e is based on $\boldsymbol{\psi} - \hat{\boldsymbol{\psi}}$, which is the same in the original and joint models. The normalization is by the $\psi\psi$ part of the inverted information matrix, and by the block structure for mixed parameters again, this is the same in the original and joint models. □

To see how the trick Proposition 5.6 is used, we consider testing for independence or homogeneity in a contingency table. Other situations where the trick is an efficient tool, is in deriving the form of the usual χ^2 goodness of fit test, and in a goodness of fit test for the more complex conditional Rasch model, see Proposition 12.1. It is also practical for estimation in multinomial models, where the Poisson trick can allow the use of generalized linear models software with Poisson link, see Chapter 9, in particular Example 9.2. Another potential use in multinomial models is when data are incomplete, and we want to prepare for the use of the EM algorithm by computing its expected rate of convergence, see Remark 8.6 in Section 8.3.

Example 5.4 ***Test for independence or homogeneity***
We consider a general $k{\times}l$ table with fixed total n. The underlying canonical statistic $\boldsymbol{t}$ is the whole table, except for the linear restriction $\sum y_{ij} = n$ that makes all entries mutually dependent. However, when regarding n as an outcome of $\mathrm{Po}(\lambda)$, all entries y_{ij} in the table become Poisson and mutually independent. As an equivalent representation of $(\boldsymbol{t}, n)$ we may as well take the whole Poisson table $\{y_{ij}\}$. Thus the score test quadratic form takes the form $\sum\sum(y_{ij} - \lambda\pi_{ij})^2/(\lambda\pi_{ij})$, where we replace λ by $\hat{\lambda} = n$ and π_{ij} by $\hat{\pi}_{ij} = \hat{\alpha}_i \hat{\beta}_j$ to obtain W_u, cf. the special case (5.12).

To obtain the same quadratic form for the homogeneity test, with all row sums fixed, say, it is natural to make a trivial extension of Proposition 5.6 to allow several fixed sample sizes (the row sums) to be regarded as Poisson outcomes. Equivalently we may assume the overall total n to be Poisson and the row sums to be multinomial, given n. △

Exercise 5.8 ***Can several Bernoulli sequences be combined?***
Suppose we have $k > 1$ Bernoulli sequences of lengths $n_1, \ldots, n_k$, binary primary data $\{y_{ij}\}$. Formulate a model for data and consider the hypothesis that these sequences can be combined in one long series, that is, that their success probabilities are the same (π_0, say).
(a) Show that the exact test statistic is

$$\binom{n_1}{t_1}\binom{n_2}{t_2}\cdots\binom{n_k}{t_k}\Big/\binom{n}{u}$$

where t_i is the number of successes in sequence i, u is the total number of successes, and $n = \sum n_i$.
(b) Use Proposition 5.6 to show that the score test statistic is

$$W_u = \sum_{i=1}^{k} \frac{(t_i - n_i\,\hat{\pi}_0)^2}{n_i\,\hat{\pi}_0\,(1 - \hat{\pi}_0)}$$

where $\hat{\pi}_0 = u/n$ is the ML-estimated success probability under H_0. Tell the degrees of freedom of the score test.
Remark: W_u is a monotone function of the so-called Lexis ratio. △

Exercise 5.9 ***Test for common binomial***
Suppose we have data on the number of boys and girls in a sample of n families among those having k children. We want to know if the data fit a common binomial distribution for the number of boys, Bin(k; λ) for some probability λ. In the (large) population of families with k children, some proportion π_j has j boys, $j = 0, \ldots, k$. This implies that the sample is from a multinomial distribution with the unknown probabilities π_j. The hypothesis is that these probabilities fit a binomial. The data to consider are the numbers y_j of families with j boys, $j = 0, \ldots, k$, with $\sum y_j = n$.
(a) In the hypothesis model, the canonical statistic u is one-dimensional. What is u, and what is its mean value?
(b) Show that the exact test statistic is

$$\frac{n!\binom{k}{0}^{y_0}\binom{k}{1}^{y_1}\cdots\binom{k}{k}^{y_k}}{y_0!\,y_1!\cdots y_k!\binom{nk}{u}}$$

(c) Use Proposition 5.6 to show that the score test statistic is

$$W_u = n \sum_{j=0}^{k} \frac{\left(y_j/n - \pi_j(\hat{\lambda})\right)^2}{\pi_j(\hat{\lambda})}$$

where $\pi_j(\lambda) = \binom{k}{j} \lambda^j (1 - \lambda)^{k-j}$, and $\hat{\lambda}$ is the MLE of the probability λ for 'boy' in the hypothesis model. What is $\hat{\lambda}$, and what are the degrees of freedom? △

Exercise 5.10 ***The same Poisson?***
Suppose y_1 and y_2 are independent and Poisson distributed observations, $y_i \sim Po(\lambda_1)$, $y_2 \sim Po(\lambda_2)$. The question is if they in fact come from the same Poisson, and the task is to construct statistical tests of this hypothesis.

(a) First construct the exact test, including its test statistic distribution.
(b) Construct an approximate test, based on a normal approximation.
(c) Construct another approximate test based on the normal approximation.
(d) Discuss briefly under what additional conditions the normal approximation in (b) and (c) might be adequate.
(e) Find out how the previous test constructions in (a)-(c) should be changed when we have a sample of m observations, $y_{11}, ..., y_{1m}$ from $\text{Po}(\lambda_1)$ and a sample of n observations, $y_{21}, ..., y_{2m}$ from $\text{Po}(\lambda_2)$. △

Exercise 5.11 ***Test for specified multinomial distribution***
Given a sample from a multinomial distribution, construct various large sample tests for the hypothesis of a particular such distribution (e.g. with equal cell probabilities). △

Exercise 5.12 ***Test for specified coefficient of variation***
Given a sample from a normal distribution, construct at least one large sample test for the hypothesis of a particular coefficient of variation, H_0: $\sigma/\mu = c$ (=specified).
Remark: The H_0 model here is a curved family, discussed in Example 7.1. △

Exercise 5.13 ***Does a gamma simplify to an exponential?***
Construct a suitable large sample test for testing the hypothesis that a sample from a gamma distribution is in fact from an exponential distribution. The test should not involve the MLE under the gamma model, for which we do not have an explicit form. Useful mathematical properties are that the values of $d \log \Gamma(\psi)/d\psi$ and $d^2 \log \Gamma(\psi)/d\psi^2$ in the point $\psi = 1$ are known; they equal Euler's constant $C = -0.5772$, and $\pi^2/6$, respectively. △

Exercise 5.14 ***Weighted multiplicative Poisson, continued***
We return to the weighted multiplicative Poisson model, Example 3.21, with Poisson means $\lambda_{ij} = N_{ij}\,\alpha_i\beta_j$ for known N_{ij}-values. Consider testing the hypothesis H_0: No row effects.
(a) Solve the likelihood equations under H_0.
(b) Determine the score-based vector that is a part of the score test statistic for testing H_0. You need not calculate the information matrix part of the test statistic.
(c) Tell what large sample distribution the score test statistic should have (under H_0), and make an intelligent guess of the form of this statistic. This means that you need not derive the statistic, except for what you did in (a) and (b). However, try to argue for your specific suggestion. △

6

Boltzmann's Law in Statistics

Here we present Per Martin-Löf's philosophy and theory of statistical modelling. This was inspired by Boltzmann's law in statistical physics, that derives a distribution of exponential type for the physical macrostate from the so-called microcanonical distribution for the state at molecular level. Its statistical version postulates that data should follow an exponential family model, provided some basic 'microcanonical' assumptions are satisfied. This postulate can be mathematically proved in simple situations, including the 'iid case', and such a derivation of (a class of) exponential families will be presented here. The most comprehensive account of Per Martin-Löf's work and ideas on this topic is given in Martin-Löf (1970, unpublished), but they are also presented in Martin-Löf (1974a,b, 1975).

6.1 Microcanonical Distributions

As before we let $\boldsymbol{y}$ denote the outcome of the data collection, the primary data. A fundamental idea in statistical inference is to reduce the high-dimensional primary data $\boldsymbol{y}$ to some relatively low-dimensional statistic $t = t(\boldsymbol{y})$, while retaining all information in the data relevant for the inference about the model via its parameters. In statistics this idea goes back to Fisher (1922b) and his introduction of the concept of sufficiency, cf. Proposition 3.4. In statistical physics it is even older. Inspired by the terminology there, the statistic $t(\boldsymbol{y})$ will be called the *canonical statistic*, in contrast to the *microcanonical state* $\boldsymbol{y}$, and t will of course later be seen to appear as the canonical statistic of an exponential family. We will primarily consider three examples in this chapter.

Example 6.1 ***Bernoulli trials (again)***
The outcome of n Bernoulli trials is a sequence $\boldsymbol{y} = (y_1, \ldots, y_n)$ of zeros and ones. These primary data are reduced to the number of 'successes', that is, to the statistic $t(\boldsymbol{y}) = \sum y_i$. △

Example 6.2 ***Cross-classification of n items by two binary criteria***
The primary data in this case are $\boldsymbol{y} = (\boldsymbol{y}_1, \ldots, \boldsymbol{y}_n)$, where each $\boldsymbol{y}_i$ is a vector of two binary components. One reduction of the data regards the n items or individuals as representing some population, and $\boldsymbol{t}$ is formed as the well-known 2×2-table of frequencies. The more restrictive model that assumes independence between the two binary criteria corresponds to a further reduction of this table to its row and column sums only, cf. Section 5.2. △

Example 6.3 ***Ideal gas model***
For an ideal gas in a container the full data would tell the position and velocity vector for each of its n molecules. These microstate data are not even observable. The macroscopic state of the gas is described by the total kinetic energy for the collection of molecules, which is related to the temperature of the gas. This example will be used to point out some analogies with statistical physics. △

Now let $\boldsymbol{t}(\boldsymbol{y}) = \boldsymbol{t}$ be observed, and consider the set Y_t of possible outcomes which could have generated $\boldsymbol{t}$,

$$Y_t = \{\boldsymbol{y}; \boldsymbol{t}(\boldsymbol{y}) = \boldsymbol{t}\}.$$

Fisher (1922b) called this an *isostatistical region*. Martin-Löf (1974a) states:

> To say that an outcome $\boldsymbol{y}$ is described by a statistic $\boldsymbol{t}(\boldsymbol{y})$ is tantamount to saying that $\boldsymbol{y}$ can be considered as drawn at random from the set Y_t.

This is the *microcanonical distribution* of $\boldsymbol{y}$ on Y_t. Hence our reduction of data goes from $\boldsymbol{y}$ to $\boldsymbol{t}$ under the assumption or belief that $\boldsymbol{y}$ can be assumed randomly selected from Y_t. If the outcome space Y is discrete, the random selection is easily specified as a simple random draw. If the number of outcomes in Y_t is $g(\boldsymbol{t})$, assumed finite (for any $\boldsymbol{t}$), each of these outcomes $\boldsymbol{y}$ in Y_t should have the probability

$$p_t(\boldsymbol{y}) = 1/g(\boldsymbol{t})$$

to be the realized outcome. In the continuous case this characterization is somewhat more complicated mathematically, because we have to pay attention to the fact that the distance between close isostatistical regions, Y_t and Y_{t+dt} say, may vary in space. By the uniform distribution on Y_t in this case we mean the restriction of the Lebesgue measure in Y to the infinitesimally thin shell between the surfaces Y_t and Y_{t+dt}, normalized to be a probability measure. The result is based on the local partitioning

$$d\boldsymbol{y} = \frac{\mathrm{d}\sigma_t \, \mathrm{d}\boldsymbol{t}}{\sqrt{\det(J\,J^T)}},$$

where $\mathrm{d}\sigma_t$ represents the Lebesgue surface area measure on $\mathcal{Y}_t$ and $J = J(y)$ is the $\dim(\boldsymbol{t}) \times \dim(\boldsymbol{y})$ Jacobian matrix for the function $\boldsymbol{t} = \boldsymbol{t}(\boldsymbol{y})$ (assumed continuous and of full rank). It follows that the uniform probability measure on $\mathcal{Y}_t$ is proportional to $\mathrm{d}\sigma_t / \sqrt{\det(J\,J^T)}$. The normalizing denominator is the structure function, determined by integration over $\mathcal{Y}_t$ as

$$g(\boldsymbol{t}) = \int_{\mathcal{Y}_t} \frac{1}{\sqrt{\det(J\,J^T)}}\,\mathrm{d}\sigma_t\,.$$

In analogy with the discrete case, a regularity condition needed but also typically satisfied is that $\mathcal{Y}_t$ is compact, so the integral over $\mathcal{Y}_t$ exists.

In this chapter we will mostly assume that the outcome space is discrete, for simplicity only, but with occasional remarks about the continuous case.

Example 6.4 ***Bernoulli trials, continued from Example 6.1***
There are $g(t) = \binom{n}{t}$ outcomes in $\mathcal{Y}_t$, all equally probable, so $p_t(\boldsymbol{y}) = 1/\binom{n}{t}$. △

Example 6.5 ***Binary cross-classifications, continued from Example 6.2***
When $\boldsymbol{t}$ represents the whole 2×2 table we have $p_t(\boldsymbol{y}) = n_{11}!\,n_{12}!\,n_{21}!\,n_{22}!\,/n!$. The independence model, with $\boldsymbol{t} = (r, s)$ reduced to one row and one column sum, has $p_t(\boldsymbol{y}) = 1/\binom{n}{r}\binom{n}{s}$. △

Example 6.6 ***Ideal gas, continued from Example 6.3***
In statistical physics the 'uniform' distribution on $\mathcal{Y}_t$ is called the *microcanonical distribution* (Gibbs), and we will use the same term in the statistical context. △

6.2 Boltzmann's Law

We will here introduce Boltzmann's law, that relates the microcanonical set-up with exponential families, and we will prove the law under some more or less crucial assumptions. First we must prescribe a *repetitive structure* underlying the data. In the mathematical derivation that follows we assume data represent independent replicates of the same trial (i.e. in usual statistical terminology we have an iid sample). This is not a crucial assumption for Boltzmann's law *per se*, but for a reasonably simple proof of it. A more important assumption is that the canonical statistic $\boldsymbol{t}(\boldsymbol{y})$ is a sum type of function, e.g. $\sum y_i$, $\sum x_i y_i$ (regression) or $\sum_j y_{ij}$ (cross-classifications). For exponential families, sum type is implied by the iid assumption (Proposition 1.2), but note that here we are not (yet) assuming any distribution over different values of $\boldsymbol{t}$. If the canonical statistic is not of sum type, the following theory will not be applicable. For example, if the maximum observation is the adequate statistic for describing data, the exponential family will be replaced by another type of distribution family.

To be able to derive the exponential family we make one more fundamental assumption about the repetitive structure. We assume that the data available can be regarded as part of a much larger, typically hypothetical population of data of the same type. The outcome $\boldsymbol{y}$ and statistic value $\boldsymbol{t} = \boldsymbol{t}(\boldsymbol{y})$ of the part realized must then be notationally distinguished from the corresponding characteristics of the large data set (the population), which will be denoted $\boldsymbol{Y}$ and $\boldsymbol{T} = \boldsymbol{T}(\boldsymbol{Y})$. If the population is described by the statistic $\boldsymbol{T}$, this implicitly means that $\boldsymbol{Y}$ follows the uniform (microcanonical) distribution for $\boldsymbol{Y}$ on $\mathcal{Y}_T$. The aim of the inference is to make statements about the population characteristics.

Example 6.7 ***Bernoulli trials, continued from Example 6.4***
A sequence of n Bernoulli trials may be regarded as (the first) part of a much longer, hypothetical sequence of N such trials, still under the same conditions. If we are interested in the probability for success, it can be interpreted as the long run frequency of successes (by letting $N \to \infty$). △

Example 6.8 ***Cross-classifications, continued from Example 6.5***
Here we may think of the n individuals as a random sample from a large population of size N. We are presumably not interested in just characterizing a moderately large existing population of individuals, but rather think of the large population as an arbitrarily large hypothetical 'superpopulation'. △

Example 6.9 ***Ideal gas, continued from Example 6.6***
In statistical physics the gas quantity under consideration is thought of as connected with (i.e. being a part of) a very much larger heat container. △

Let us now start from the microcanonical (i.e. uniform) distribution for $\boldsymbol{Y}$ and consider the implications for the relatively small part $\boldsymbol{y}$. We first do this in the setting of Example 6.1.

Example 6.10 ***Bernoulli trials, continued from Example 6.7***
When $\sum_1^N y_i = T$,

$$p_T(\boldsymbol{Y}) = 1 \Big/ \binom{N}{T}$$

Now, for fixed N and T, if we have observed $\boldsymbol{y} = (y_1, \ldots, y_n)$, with $\sum_1^n y_i = t$, the sequence $y_{n+1}, \ldots, y_N$ can be arbitrary, provided it sums to $T - t$. The number of such $(N-n)$-long sequences is $\binom{N-n}{T-t}$. Hence the probability for observing a particular sequence $\boldsymbol{y}$ with $t(\boldsymbol{y}) = t$ in the smaller data set, as induced by the uniform distribution for $\boldsymbol{Y}$ over $\mathcal{Y}_T$ in the larger data set, is

$$\frac{\binom{N-n}{T-t}}{\binom{N}{T}} = \frac{g_{N-n}(T-t)}{g_N(T)}, \tag{6.1}$$

where g_{N-n} and g_N are the corresponding structure functions. When $N >> n$, $T >> t$, and $N - n >> T - t$, (6.1) is well approximated by (check!)

$$\left(\frac{T}{N}\right)^t \left(1 - \frac{T}{N}\right)^{n-t}$$

We recognize the standard exponential family probability for an outcome sequence with t successes and $n - t$ failures, when the probability for success is taken to be the relative frequency of successes in the larger trial, T/N. If we let $N \to \infty$ such that $T/N \to \pi_0$, $0 < \pi_0 < 1$, then the probability tends to $\pi_0^t(1-\pi_0)^{n-t}$, and π_0 can be interpreted as the (infinitely) long run relative frequency of successes. △

What we have done in this special example has been to demonstrate the validity of the more generally applicable Boltzmann's law:

Proposition 6.1 ***Boltzmann's law***
Consider a given trial with outcome $\boldsymbol{y}$ *and statistic* $\boldsymbol{t}(\boldsymbol{y}) = \boldsymbol{t}$ *(of sum type) as part of a much larger trial. The uniform distribution for the outcome* $\boldsymbol{Y}$ *over the isostatistical region* $\mathcal{Y}_T$ *in the large trial induces a distribution for* $\boldsymbol{y}$ *in the given trial that can be approximated by the exponential family*

$$f(\boldsymbol{y};\boldsymbol{\theta}) = \frac{1}{C(\boldsymbol{\theta})}\, e^{\boldsymbol{\theta}^T \boldsymbol{t}(\boldsymbol{y})}, \tag{6.2}$$

where $\boldsymbol{\theta}$ *is a function of the value of the statistic* $\boldsymbol{T}(\boldsymbol{Y})$ *in the large trial, and* $C(\boldsymbol{\theta})$ *is a norming constant.*

The principle was formulated and proved in Martin-Löf (1970). Note that (6.2) differs from the general definition of exponential families given in Chapter 1 by not having a factor $h(\boldsymbol{y})$. This is of course because we started from a uniform microcanonical distribution, which is a requirement according to Boltzmann's law. Note that almost all examples in Chapter 2 have a constant function h, e.g. the multinomial, geometric, exponential, normal, gamma and von Mises distributions, and many of them have very important roles in model building. The Poisson distribution is also included, by the standard Bernoulli approximation argument (the count is the statistic representing many trials, each with small success probability).

Proof We will justify the approximation under these assumptions:

- Data $\boldsymbol{y}$ represent a subsample of n out of N mutually independent repetitions of a single trial.
- The canonical statistic is formed as the sum over the n or N trials of the canonical statistic for the single trial.
- The space of possible $\boldsymbol{t}$-values is (a subset of) the space $\mathbb{Z}^k$ of integers.

Afterwards, comments are given on how the proof is extended to the continuous case, and additionally to a table of random counts under multiplicativity (a noniid situation). Stringent proofs are found in Martin-Löf (1970).

The proof has two parts. First we derive an exact expression for the induced distribution, and next we approximate this distribution by use of the saddlepoint approximation technique, letting $N \to \infty$. For the normal density approximation involved in the discrete case we assume the function $\boldsymbol{t}(\boldsymbol{y})$ has been chosen such that the range of possible $\boldsymbol{t}$-values is not part of a proper subgroup of $\mathbb{Z}^k$ (e.g. not the set of even numbers), nor a coset of such a subgroup (e.g. the odd numbers). This demand can always be achieved by a suitable affine transformation.

Since the outcome $\boldsymbol{y}$ in the smaller trial is considered fixed, the probability induced by the equal probability density on Y_T is found by counting the possible outcomes of the complementary part, with its $N - n$ observations and statistic $\boldsymbol{T} - \boldsymbol{t}$, and comparing with the corresponding count with N and $\boldsymbol{T}$. The counts are given by the corresponding structure functions, $g_{N-n}(\boldsymbol{T} - \boldsymbol{t})$ and $g_N(\boldsymbol{T})$, respectively. This yields for the induced probability of $\boldsymbol{y}$ the expression

$$f_T(\boldsymbol{y}) = \frac{g_{N-n}(\boldsymbol{T} - \boldsymbol{t})}{g_N(\boldsymbol{T})}, \tag{6.3}$$

cf. the special case (6.1), in which the same argumentation was used.

When we now want to use saddlepoint approximations of the structure function, the situation is analogous to when the exact test statistic was approximated in Proposition 5.2. In both cases a ratio of structure functions is approximated, and one of them can be handled simpler than the other. In the present case we want an approximation in the point $\boldsymbol{T}/N$. This fits perfectly with the denominator. We use (4.8) from Proposition 4.5 for the distribution family constructed by exponential tilting, that is, by embedding the structure function in the corresponding exponential family for the canonical statistic $\boldsymbol{T}$ (cf. Example 2.12):

$$g_N(\boldsymbol{T}) \approx \frac{1}{(2\pi)^{k/2}\sqrt{\det V_N(\hat{\boldsymbol{\theta}})}}\, C_N(\hat{\boldsymbol{\theta}})\, e^{-\hat{\boldsymbol{\theta}}^T \boldsymbol{T}} \tag{6.4}$$

$$= \frac{1}{(2\pi N)^{k/2}\sqrt{\det V_1(\hat{\boldsymbol{\theta}})}}\, C_1(\hat{\boldsymbol{\theta}})^N\, e^{-\hat{\boldsymbol{\theta}}^T \boldsymbol{T}} \tag{6.5}$$

Here $C_1(\boldsymbol{\theta})$ is the norming function for a single trial and $V_1(\boldsymbol{\theta})$ is the corresponding matrix of second order derivatives of $\log C(\boldsymbol{\theta})$. All are calculated in the MLE $\hat{\boldsymbol{\theta}} = \hat{\boldsymbol{\theta}}(\boldsymbol{T}) = \boldsymbol{\mu}^{-1}(\boldsymbol{T}/N)$. where $\boldsymbol{\mu}$ is the single trial mean value vector.

The numerator $g_{N-n}(\boldsymbol{T} - \boldsymbol{t})$ will be approximated in the same point $\hat{\boldsymbol{\theta}} = \hat{\boldsymbol{\theta}}(\boldsymbol{T})$, to make factors in the ratio cancel, but this point does not exactly correspond to the MLE when $\boldsymbol{T} - \boldsymbol{t}$ is observed in a set of $N - n$ trials. Let us first neglect that difference. Then we get in the same way as in (6.4)

$$g_{N-n}(\boldsymbol{T} - \boldsymbol{t}) \approx \frac{C_{N-n}(\hat{\boldsymbol{\theta}})}{(2\pi)^{k/2}\sqrt{\det V_{N-n}(\hat{\boldsymbol{\theta}})}}\; e^{-\hat{\boldsymbol{\theta}}^T (\boldsymbol{T}-\boldsymbol{t})}$$

$$= \frac{C_1(\hat{\boldsymbol{\theta}})^{(N-n)}}{(2\pi(N-n))^{k/2}\sqrt{\det V_1(\hat{\boldsymbol{\theta}})}}\; e^{-\hat{\boldsymbol{\theta}}^T (\boldsymbol{T}-\boldsymbol{t})} \tag{6.6}$$

When forming the ratio (6.3), several factors vanish exactly or approximately (because $(N-n)/N \approx 1$ for large N). If we write $\boldsymbol{\theta}$ for $\hat{\boldsymbol{\theta}} = \boldsymbol{\mu}^{-1}(\boldsymbol{T}/N)$, that characterizes the population of size N, the result takes the form

$$f_T(y) \approx \frac{g_{N-n}(\boldsymbol{T} - \boldsymbol{t})}{g_N(\boldsymbol{T})} \to C_1(\boldsymbol{\theta})^{-n}\, e^{\boldsymbol{\theta}^T \boldsymbol{t}}. \tag{6.7}$$

This is the same as (6.2) when the smaller trial is an iid sample of size n, so we have got the desired result.

It remains to handle the complication that the saddlepoint approximation is not precisely in the centre of the approximating normal density, but this time we do not want it to matter. Similarly to (5.9), we obtain the Gaussian type adjustment factor

$$e^{-\frac{1}{2}\{\boldsymbol{T} - \boldsymbol{t} - \boldsymbol{\mu}_{N-n}(\hat{\boldsymbol{\theta}})\}^T V_{N-n}^{-1} \{\boldsymbol{T} - \boldsymbol{t} - \boldsymbol{\mu}_{T-t}(\hat{\boldsymbol{\theta}})\}}$$
$$= e^{-\frac{1}{2}\{\boldsymbol{t} - n\boldsymbol{\mu}(\hat{\boldsymbol{\theta}})\}^T V(\hat{\boldsymbol{\theta}})^{-1} \{\boldsymbol{t} - n\boldsymbol{\mu}(\hat{\boldsymbol{\theta}})\}/(N-n)}.$$

From the first to the second line, not only have $\boldsymbol{\mu}_{N-n}$ and V_{N-n} been re-expressed in the notations of the single trial, but we have also used that

$\boldsymbol{T} - N\boldsymbol{\mu}(\hat{\boldsymbol{\theta}}) = 0$. Since $\boldsymbol{t}$ is fixed, and both $\boldsymbol{t} - n\boldsymbol{\mu}(\hat{\boldsymbol{\theta}})$ and $V(\hat{\boldsymbol{\theta}}(T))^{-1}$ as functions of $\hat{\boldsymbol{\theta}}$ are bounded on compact sets of the argument, the factor $1/(N-n)$ in the exponent makes the whole factor converge to 1 with increasing N. For a strict proof of the validity of the approximations we must have the approximation errors under better control, but this would require more technicalities. Martin-Löf (1970) shows in his more stringent proof that the relative approximation error is uniformly $O(1/N)$ for $\boldsymbol{t}$ bounded and $\boldsymbol{\theta}$ within a compact subset of $\boldsymbol{\Theta}$. □

Remark 6.2 ***The continuous case.*** The first part of the proof, showing that the density can be written as in (6.3) is technically somewhat more complicated. The reason was explained earlier, that we have to work with volumes of thin shells between close isostatistical surfaces, instead of just counting points. The rest of the proof works very much as in the discrete case. One additional regularity assumption need be satisfied. The characteristic function for the distribution in which $\boldsymbol{T}$ is embedded must be integrable for N large enough. In particular this guarantees that for such N the density for $\boldsymbol{T}$ cannot be unbounded. △

Remark 6.3 ***Two-way tables of counts.*** The validity of Boltzmann's law is not restricted to the iid case, but the law is difficult to prove in more generality. A special case in which there exists such a proof is that of the two-way tables of counts. This should be compared with the iid case of two-way classifications, with fixed table format and fixed total n (= sample size) for the smaller table, and a large increasing population size N, when Proposition 6.1 applies. Now let instead the large population correspond to a large number of rows and columns. It is not enough that the larger table total N is large, but its row and column sums must be large, too. The following result is shown in Martin-Löf (1970):

Proposition 6.4 ***Boltzmann's law for two-way tables of counts***
Consider a fixed format two-factor table of counts, $\boldsymbol{y} = \{y_{ij}\}$ *(with* $\boldsymbol{t}$ *formed by its row and columns sums), embedded in another, larger table with an increasing number of rows and columns. Let the larger table row and column sums,* R_i *and* S_j*, for the same row and column indices as in the smaller table, all increase with the table size in such a way that* $R_i S_j/N \to \alpha_i \beta_j > 0$ *for all* (i, j)*, where* N *is the large table total. Then the distribution for the smaller (fixed size) table, as induced by the microcanonical distribution for the larger table, will converge to the multiplicative Poisson distribution with cell means* $\alpha_i \beta_j$*.*

△

Exercise 6.1 ***Some examples***
Check that the following models are consistent with the application of Boltzmann's law: The Gaussian linear model, the Ising model, and the geometric, gamma, beta and von Mises distributions (Examples 2.9, 2.16, 2.6, 2.7, 2.4, 2.14, respectively). △

6.3 Hypothesis Tests in a Microcanonical Setting

Model-reducing hypotheses and the exact test, derived in a parametric setting in Chapter 5, can be introduced already in the more primitive microcanonical setting. This is done in Martin-Löf (1970, 1974a,b, 1975).

We consider the (reductive) hypothesis that a description of data with the statistic $\boldsymbol{t} = (\boldsymbol{u}, \boldsymbol{v})$ can be reduced to a description with the lower-dimensional statistic $\boldsymbol{u}$. The number of outcomes with the observed $\boldsymbol{t} = (\boldsymbol{u}, \boldsymbol{v})$ is $g(\boldsymbol{u}, \boldsymbol{v})$, but when only $\boldsymbol{u}$ is specified, the corresponding number is $g(\boldsymbol{u}) = \sum_{\boldsymbol{v}'} g(\boldsymbol{u}, \boldsymbol{v}')$, where the sum is over all possible $\boldsymbol{v}'$ outcomes. We assume $g(\boldsymbol{u})$ is finite. Martin-Löf formulates as a *fundamental principle*:

> Relative to $g(\boldsymbol{u})$, the smaller the number $g(\boldsymbol{u}, \boldsymbol{v})$ of outcomes $\boldsymbol{y}$ with the particular observed $\boldsymbol{v}$, the more does our observation contradict the hypothesis that the statistic can be reduced from $(\boldsymbol{u}, \boldsymbol{v})$ to $\boldsymbol{u}$.

A fundamental principle means that it does not seem possible to reduce it to any other more basic or convincing principles.

This principle immediately leads to a test having the p-value (or critical level)

$$\epsilon(\boldsymbol{t}(y)) = \frac{\sum_{\boldsymbol{v}'} g(\boldsymbol{u}, \boldsymbol{v}')}{g(\boldsymbol{u})}, \tag{6.8}$$

where the summation is now over all $\boldsymbol{v}'$ such that $g(\boldsymbol{u}, \boldsymbol{v}') \leq g(\boldsymbol{u}, \boldsymbol{v})$ for the observed $(\boldsymbol{u}, \boldsymbol{v})$. Martin-Löf calls this the *exact test*, since it generalizes what Fisher called the exact test. Note that this test, albeit both model and argument are now different (more basic and primitive), is formally identical with the exact test in exponential families derived in Section 5.1. As noted also in Section 5.1, there are pathological examples when the p-value calculation of the test is counterintuitive. See Section 12.7 for such an example, with t a function of Spearman's correlation coefficient.

The approach can be extended to *prediction*, and along this line also to *confidence regions* (Martin-Löf, 1970, 1974a). Assume the statistic $\boldsymbol{u}$ would have been satisfactory to describe data, but that $\boldsymbol{u}$ is not available.

Instead we want to make a statement about $\boldsymbol{u}$ by help of an observed statistic $\boldsymbol{v}$, jointly represented by $\boldsymbol{t} = (\boldsymbol{u}, \boldsymbol{v})$. Consider the set of $\boldsymbol{u}$ outcomes such that the p-value (6.8) is at least as large as a prescribed critical level ϵ_0. This is the set of $\boldsymbol{u}$-values for which the observed $\boldsymbol{v}$ can be said to have a $(1 - \epsilon_0)$-*plausible* value. In particular we can use this construction in the same set-up as in the proof of Boltzmann's law. Let $\boldsymbol{u} = \boldsymbol{T}$ and $\boldsymbol{v} = \boldsymbol{t}$, corresponding to all N trials and the first n trials, respectively. This yields a prediction region for $\boldsymbol{T}$, or $\boldsymbol{T}/N$ if we prefer. Now, according to Boltzmann's law, $\boldsymbol{T}/N$ has an approximate interpretation as the single trial mean value parameter of the exponential family (6.2). This means the prediction region for $\boldsymbol{T}/N$ turns into a $(1 - \epsilon_0)$ confidence region for $\boldsymbol{\mu}(\boldsymbol{\theta})$, that may equivalently be transformed into a region for $\boldsymbol{\theta}$ itself.

The resulting confidence procedure satisfies the correspondence between confidence regions and 'acceptance' regions for significance tests, with some further demands on the test to be used. The procedure can be extended to large sample approximations of the exact test, and to allow additional components of $\boldsymbol{t}$ (nuisance parameters).

6.4 Statistical Redundancy: An Information-Theoretic Measure of Model Fit

A dilemma in statistical significance testing is that, since no model or sampling scheme is perfect, the p-value of the test will be small enough to reject the model if only the amount of data is large enough. Many ways to deal with this problem have been proposed, typically as various measures of fit designed for special classes of situations. Martin-Löf (1970, 1974b) proposed a quite different solution, namely to supplement the p-value with an information-theoretic measure of fit, the *statistical redundancy*. We give a brief description here and refer to his 1974 paper for a more comprehensive account. We first introduce a microcanonical version of the redundancy and next a more useful canonical (or parametric) redundancy, that approximates the microcanonical, in analogy with the distribution approximation provided by Boltzmann's law.

6.4.1 Microcanonical Redundancy

Let $\epsilon(\boldsymbol{t}(\boldsymbol{y}))$ be the p-value of an exact (i.e. microcanonical) test of a reductive hypothesis H_0, specifying that the statistic $\boldsymbol{t} = (\boldsymbol{u}, \boldsymbol{v})$ can be reduced to $\boldsymbol{u}$. For the procedure of the fundamental principle formulated in Section 6.3, the p-value is given by formula (6.8), but the procedure can be extended to any test procedure having the natural nestedness property that the outcomes can be ordered according to some measure of deviation from H_0. In each case $\epsilon(\boldsymbol{t}(\boldsymbol{y}))$ is given by a formula of type (6.8), only with differently defined sets of $\boldsymbol{v}'$ to sum over. The p-value allows an information-theoretic interpretation, quoting from Martin-Löf (1974b):

> $-\log_2 \epsilon(\boldsymbol{t}(\boldsymbol{y}))$ is the absolute decrease in the number of binary units needed to specify the outcome $\boldsymbol{y}$ when we take into account the regularities that we detect by means of the exact test.

This is seen in the following way. Given $\boldsymbol{u}(\boldsymbol{y})$, let all possible outcomes $\boldsymbol{y}$ be (weakly) binary ordered according to their p-values $\epsilon(\boldsymbol{t}(\boldsymbol{y}))$, i.e. successively 1, 10, 11, 100, 101, 110, 111, 1000, etc. The number of binary digits needed to code all outcomes, whose number is $g(\boldsymbol{u})$, is roughly $\log_2 g(\boldsymbol{u})$. The length of the binary code required to arrive at the observed $\boldsymbol{y}$ by the test ordering is roughly $\log_2 \sum_{\boldsymbol{v}'} g(\boldsymbol{u}, \boldsymbol{v}')$, summed over the $\boldsymbol{v}'$ preceding (and including) the observed $\boldsymbol{v}$. Hence the absolute decrease in length achieved by using the test outcome in addition to $\boldsymbol{u}$ is the difference $\log_2 g(\boldsymbol{u}) - \log_2 \sum_{\boldsymbol{v}'} g(\boldsymbol{u}, \boldsymbol{v}') = -\log_2 \epsilon(\boldsymbol{t}(\boldsymbol{y}))$. The corresponding relative decrease in length will be called the *microcanonical redundancy*, $R(\boldsymbol{t})$. Thus, we define:

Definition 6.5 ***Microcanonical redundancy***
The *microcanonical redundancy* is the relative decrease in the number of binary units needed to specify the outcome $\boldsymbol{y}$ when we take into account the regularities identified by the exact test, and it is given by

$$R(\boldsymbol{t}) = \frac{\log_2 g(\boldsymbol{u}) - \log_2 \sum_{\boldsymbol{v}'} g(\boldsymbol{u}, \boldsymbol{v}')}{\log_2 g(\boldsymbol{u})} = -\frac{\log \epsilon(\boldsymbol{t})}{\log g(\boldsymbol{u})}, \tag{6.9}$$

with fixed $\boldsymbol{u}(\boldsymbol{t}) = \boldsymbol{u}$, and summation over all $\boldsymbol{v}'$ more extreme than $\boldsymbol{v}$.

It is clear from its ratio form that $R(t)$ is in fact independent of the choice of base of the logarithm, for which reason $\log_2$ is replaced by log in the rightmost part of (6.9). In the sequel we assume log is $\log_e$. Also note that $0 \leq R(\boldsymbol{t}) \leq 1$. Finally, we note that the denominator in (6.9) coincides with the *microcanonical entropy* of the uniform probability distribution under the null hypothesis. More generally, the entropy H of a distribution with density (probability function) $p(\boldsymbol{y})$ is defined as

$$H = E[-\log p(\boldsymbol{y})],$$

and the microcanonical entropy is the special case when all probabilities are the same, $1/g(\boldsymbol{u})$, given $\boldsymbol{u}$. The corresponding entropy for the exponential family (6.2) depends on $\boldsymbol{\theta}$ and will be referred to as the *canonical entropy*, see Section 6.4.2.

Example 6.11 ***Sex of newborns as Bernoulli trials***
Of 88 273 children born in Sweden in 1935, 45 682 were boys, that is, 51.75%, see Cramér (1946, Table 30.6.1). This is a highly significant deviation from the hypothesis of equal probability (i.e. we regard u as empty and t is for example the number of boys). What is the redundancy? The exact entropy for the denominator is simple, $g(u) = 2^n$, with $\log g(u) = n \log 2$. It is too much to demand the exact p-value, but that is perhaps not needed. We may conventionally mean- and variance-standardize the test statistic and use a normal approximation (trusted or not) to obtain the p-value $2.5 \cdot 10^{-25}$. This yields $R \approx 0.0009$, which is pretty low. We will return to the interpretation aspect in Section 6.4.3, but first we give a more general approach to this tail approximation problem. △

6.4.2 Canonical Redundancy

The approximation needed will be expressed in terms of canonical entropies. The approximation is strongly related to the saddlepoint approximation of the structure function, to Boltzmann's law, and to the large sample approximations of the exact test derived in Section 5.4. In particular we make the same assumptions as in the proof of Boltzmann's law, including the assumption of data being discrete, and that the situation is of iid type, a sample of size n, where we will also let $n \to \infty$. We distinguish sample from single trial by an index n when we refer to the whole sample.

The canonical entropy for the exponential family (6.2), with n replicates and statistic $\boldsymbol{t}_n$, is

$$H_n(\boldsymbol{\theta}) = -\boldsymbol{\theta}^T E_{\theta}(\boldsymbol{t}_n) + \log C_n(\boldsymbol{\theta}) = -n\,\{\boldsymbol{\theta}^T \boldsymbol{\mu}(\boldsymbol{\theta}) - \log C(\boldsymbol{\theta})\}, \qquad (6.10)$$

where $H_n(\boldsymbol{\theta})$ is proportional to the sample size and $\boldsymbol{\mu}$ and C represent the single trial. Also, since the MLE $\hat{\boldsymbol{\theta}}$ satisfies $\boldsymbol{\mu}(\boldsymbol{\theta}) = \boldsymbol{t}_n/n$ (when it exists),

$$H_n(\hat{\boldsymbol{\theta}}) = -\hat{\boldsymbol{\theta}}^T \boldsymbol{t}_n + \log C_n(\hat{\boldsymbol{\theta}})) = -\log p(\boldsymbol{y}; \hat{\boldsymbol{\theta}}), \qquad (6.11)$$

so in the point $\boldsymbol{\theta} = \hat{\boldsymbol{\theta}}$ the entropy coincides with the maximal log-likelihood.

We have two models to distinguish, each with its own MLE, and as in Section 5.4 we use notation $\hat{\boldsymbol{\theta}}_0$ under H_0. From the saddlepoint approximation of the structure function, see Proposition 4.5 formula (4.8), applied on $g_n(\boldsymbol{t}_n)$ in the full model or $g_n(\boldsymbol{u}_n)$ in the hypothesis model, it is seen that the asymptotically dominating term in $\log g_n(\boldsymbol{t}_n)$ is $n\,(\log C(\hat{\boldsymbol{\theta}}) - \hat{\boldsymbol{\theta}}(\boldsymbol{t}_n/n)) =$

$n\,H(\boldsymbol{t}_n/n)$, and analogously for $\log g_n(\boldsymbol{u}_n)$ under H_0. This holds provided $\hat{\theta}(\boldsymbol{t}_n/n)$ stays within a compact subset of $\boldsymbol{\Theta}$. Thus a ratio of microcanonical entropies such as $\log g(\boldsymbol{t})/\log g(\boldsymbol{u})$ is asymptotically the same as the corresponding ratio of canonical entropies with MLEs for the parameters.

After this preparation we are ready for the *canonical redundancy*:

Definition 6.6 ***Canonical redundancy***
The *canonical redundancy* is defined as the ratio

$$R(\hat{\boldsymbol{\theta}}(\boldsymbol{t})) = \frac{H_n(\hat{\boldsymbol{\theta}}_0) - H_n(\hat{\boldsymbol{\theta}})}{H_n(\hat{\boldsymbol{\theta}}_0)} = \frac{H(\hat{\boldsymbol{\theta}}_0) - H(\hat{\boldsymbol{\theta}})}{H(\hat{\boldsymbol{\theta}}_0)} \tag{6.12}$$

One more argument is needed to see that (6.12) approximates the primary concept (6.9), with its information-theoretic interpretation. We show that $H(\hat{\boldsymbol{\theta}})$ approximates the microcanonical $\log \sum_{\boldsymbol{v}'} g(\boldsymbol{u},\, \boldsymbol{v}')/n$ for the terms up to and including the observed $\boldsymbol{v}$, with $g(\boldsymbol{u}, \boldsymbol{v}') \leq g(\boldsymbol{u}, \boldsymbol{v})$. The truth of this statement might be surprising, but the result is due to the increasing steepness of the saddlepoint approximation at the observed $\boldsymbol{t} = (\boldsymbol{u}, \boldsymbol{v})$ as n increases, which in turn is due to the value of the parameter $\boldsymbol{\theta}$ chosen such that it maximizes the likelihood with the observed $\boldsymbol{t}$, namely as $\hat{\boldsymbol{\theta}}$. In fact the very crude inequalities

$$g(\boldsymbol{u},\, \boldsymbol{v}) \leq \sum_{\boldsymbol{v}'} g(\boldsymbol{u},\, \boldsymbol{v}') \leq k_n(\boldsymbol{u})\, g(\boldsymbol{u},\, \boldsymbol{v}) \tag{6.13}$$

are sufficient, where $k_n(\boldsymbol{u})$ is the number of distinct values of $\boldsymbol{t}$ (or $\boldsymbol{v}$) possible for the given $\boldsymbol{u}$. Martin-Löf demonstrates that $k_n(\boldsymbol{u}) = O(n^{\dim \boldsymbol{v}})$, and this factor is asymptotically negligible in comparison with the exponential increase of $g(\boldsymbol{u},\, \boldsymbol{v})$ with n.

Example 6.12 ***Sex of newborns, continued from Example 6.11***
We return to the $n = 88273$ children, 51.75% of whom were boys. The canonical entropy per observation with estimated probability parameter $\hat{\pi}_0$ is $H(\hat{\pi}_0) = -\{\hat{\pi}_0 \log \hat{\pi}_0 + (1 - \hat{\pi}_0)\log(1 - \hat{\pi}_0)\} = 0.69253.$ and correspondingly $H(1/2) = 0.69315$. This yields the redundancy $R(\hat{\pi}_0) = 0.0009$, so we get the same first significant digit as before. In this case the crude bound $k_n(u)$ in (6.13) equals $n + 1$. This implies that the upper bound factor $k_n(u)$ contributes to an upper bound for the redundancy by not more than 0.0002. Thus the magnitude of the redundancy is not affected. △

Remark 6.7 ***Shannon redundancy.*** In the binomial case and its extension to the multinomial, the canonical (but not the microcanonical) redundancy for the test of equal probabilities coincides with the classic Shannon measure of redundancy in information theory. △

There are alternative expressions and approximations for the numerator of the canonical redundancy. Formula (6.11) shows that the entropy estimates can be written $H_n(\hat{\theta}) = -\log L(\hat{\theta}; y)$, from which it follows that the numerator can be expressed in terms of the log-likelihood ratio test statistic:

$$R(\hat{\theta}(t)) = \frac{\log L(\hat{\theta}; y) - \log L(\hat{\theta}_0; y)}{H_n(\hat{\theta}_0)} = \frac{W}{2\, H_n(\hat{\theta}_0)} \tag{6.14}$$

In this formula W can be replaced by the score test statistic W_u or any other of the asymptotically equivalent test statistics introduced in Section 5.5.

6.4.3 A Redundancy Scale

The information-theoretic interpretation of the redundancy measure provides an absolute scale for redundancy. That is, the interpretation is the same whatever model and reductive hypothesis we consider. On the other hand, the interpretation of the values on this scale is somewhat arbitrary and subjective, like for degrees of coldness on a thermometer scale, or the statistical significance of a p-value. Martin-Löf (1970, 1974b) has proposed the following scale, where some redundancy values and their qualitative interpretation is supplemented with how the corresponding probabilities for 'success' in Bernoulli trials deviate from the ideal coin tossing value 1/2:

Table 6.1 *Redundancy scale.*

Redundancy	Model fit	Bernoulli prob.	
1	Worst possible	0.000	1.000
0.1	Very bad	0.316	0.684
0.01	Bad	0.441	0.559
0.001	Good	0.482	0.518
0.0001	Very good	0.494	0.506

Note that the value for the sex distribution in Example 6.12, $R = 0.0009$ for the relative frequency 0.5175, falls close to the mark *Good*. The Bernoulli probability scale is valuable as a reference, because it is more difficult to judge model fit in more complicated models and for hypotheses corresponding to more degrees of freedom. Here is therefore one more of the illustrations provided by Martin-Löf, again with data taken from Cramér (1946).

Example 6.13 ***Independence between income and number of children*** Cramér (1946, Table 30.5.2) discusses the association between two classification criteria in a contingency table. Data for 25 263 families, taken from a Swedish census, represent family annual income in four classes and family number of children in five classes.

Table 6.2 *Married couples by annual income and number of children.*

Children	Income				Total
	group 1	group 2	group 3	group 4	number
0	2161	3577	2184	1636	9558
1	2755	5081	2222	1052	11110
2	936	1753	640	306	3635
3	225	419	96	38	778
≥ 4	39	98	31	14	182
Total	6116	10928	5173	3046	25263

The test for independence has $\chi^2 = 568.5$ on 12 degrees of freedom, which is 'extremely significant'. The corresponding redundancy value is $R = 0.005$, which is not extreme, but too close to *Bad* to be a satisfactory fit, in the scale of Table 6.1. △

6.5 A Modelling Exercise in the Light of Boltzmann's Law

Exercise 6.2 ***Modelling of pairwise comparisons data***
Imagine data of either of the following two types.

- In a tournament with m participants, all meet pairwise once. All meetings result in a win (1) for one of the competitors and a loss (0) for the other.
- A judge is asked to evaluate the achievements of m competitors by making a complete set of pairwise comparisons, deciding who 'won' each comparison.

In both cases the outcome of the trial (the primary data) may be tabulated as a square array $Y = \{y_{ij}\}$ of zeros and ones, where $y_{ij} = 1$ if competitor i won over j, and 0 if i lost (thus, $y_{ji} = 1 - y_{ij}$). The diagonal of the array is empty.

Tasks:

- Reduce the data to a lower-dimensional statistic in a way that reflects the idea that different competitors participate in the tournament or judgement with different individual winning abilities. Use Boltzmann's law to derive an exponential family distribution for the (primary) data Y.
- Find the order of the family (dimension of the parameter space).
- Interpret your model in terms of abilities and winning probabilities.

- Discuss the realism of the model for describing the probability structure behind each of the two types of data. Formulate some possible ways of regarding the given data as part of a much larger data set, and choose the one you would like to refer to.
- Think about extensions, for example a tournament in which each pair meets twice, and/or each competitor has home ground advantage once. An extension in the judgement type of data could be to imagine several judges independently evaluating all pairs of achievements.

Do *not* try to derive the structure function or the maximum likelihood estimates, they are not explicit. △

Discussion of Exercise 6.2 (Bradley–Terry type model)

Let us use the terminology of a tournament between players, each match ending in a win for one of the two players. Until later, assume each pair meets once, and that the concept of home ground is considered irrelevant. The outcome of the tournament is represented by the square array $Y = \{y_{ij}\}$. We should expect that the general winning abilities, and the variability between players in this respect, are reflected in the total number of wins achieved by the different players. These numbers are the row sums, $r_i = \sum_{j \neq i} y_{ij}$. Hence, the vector statistic $\boldsymbol{t}$ consisting of the row sums r_i of wins for each player i is a natural canonical statistic to try. The column sums are just complementary to the row sums, counting the number of losses, so they give no additional information.

Since the grand total number of wins, summed over all players, necessarily equals the total number of matches, $m(m-1)/2$, the number of linearly independent components of $\boldsymbol{t}$ is $m-1$.

Referring to Bolzmann's law, and in doing so, also (explicitly or implicitly) assuming that all data tables corresponding to a fixed vector $\boldsymbol{t}$ are equally probable, we can write down the probability for the data table Y as

$$\Pr(Y;\, \theta) \propto \exp\left\{\sum_i \theta_i\, r_i\right\} = \exp\left\{\sum_{ij,\, j\neq i} \theta_i\, y_{ij}\right\}$$

The proportionality factor is just the norming constant, a function of the parameters, that makes the probabilities sum to 1.

The exponent can be rewritten as (remember that $y_{ji} = 1 - y_{ij}$)

$$\exp\left\{\sum_{ij,\, j\neq i} \theta_i\, y_{ij}\right\} \propto \prod_{i<j} \exp\{(\theta_i - \theta_j)\, y_{ij}\}$$

This represents the data table as the outcome of $m(m-1)/2$ independent Bernoulli trials, each with its own odds, represented by the logit $\theta_i - \theta_j$. The probability $\pi_{ij} = \Pr(y_{ij} = 1)$ can be written

$$\pi_{ij} = \frac{e^{\theta_i-\theta_j}}{e^{\theta_i-\theta_j}+1} = \frac{e^{\theta_i}}{e^{\theta_i}+e^{\theta_j}}. \tag{6.15}$$

Note that the parameters θ_i are determined only up to a common additive constant. The model is called the Bradley–Terry model for pairwise comparisons. Models of this type can be fitted for example within **R**, using the package `BradleyTerry2` (Turner and Firth, 2012).

If tournament data should be regarded and interpreted as part of a much larger system, let us think in particular in terms of a large number of players who hypothetically could have participated in the tournament. If each player (still hypothetically) played against all these players, the average number of wins would tell the player's relative overall winning ability with much less uncertainty. It might appear as if the winning ability would depend on the average weakness (or strength) of the many added players, and therefore would be misleading as concerns the ability to win in the actual tournament. This is not so, however, because data only allow inference about differences of θ-values, and they will not be affected. We could always add a common constant to all θ-values without changing the logits.

As a model for random outcomes the preceding model appears more reasonable for matches played than for judgements of achievements. What is the randomness in the judgement outcome? In particular this is unclear if we consider replicates of the data table. If the same judge evaluates once more the same set of achievements the variation is likely to be much less than binomial, whereas if different judges evaluate them, the variation might be much higher than binomial. However, if a new set of achievements by the same competitors is judged by the same person, the model is perhaps applicable.

Some natural extensions in terms of a tournament is to allow players to meet twice, and to allow for a home ground advantage. Consider first the case of players meeting twice under identical conditions. The data array is now taken to be $\{y_{ijk}\}$, where $k = 1, 2$ is an index for the two matches. The canonical statistic will still be the total number of wins for each player, i.e. the numbers $r_i = \sum_{j \neq i}(y_{ij1} + y_{ij2})$. By the same rearrangement as before we find that the probability for the observed data array is proportional to

$$\prod_{i<j} \exp\{(\theta_i - \theta_j)(y_{ij1} + y_{ij2})\}$$

For each pair of players this represents two independent Bernoulli trials with the same probability of winning for the two matches.

For modelling a home ground advantage there are two main alternatives. Either we could have a general advantage, the same for all players, or we could have an advantage specific for each player. We just consider the former possibility here, but it is not more difficult to write down the model for the other situation. Suppose each pair of players meet twice, each having home ground once. If there is a general home ground advantage we should add to the canonical statistic a component being the total number of home wins. The number of away wins is just the complementary number. If we sort the data such that in $\{y_{ij1}\}$, player i has home ground if $i < j$, and vice versa in $\{y_{ij2}\}$, the total number of home wins is easily written down as $\sum_{i<j}(y_{ij1} + y_{ji2})$ (note indices!). Let the corresponding canonical parameter be λ. Then the effect on the probability model for the matches is that the logit for the match result y_{ijk} is changed from $\theta_i - \theta_j$ to $\theta_i - \theta_j + \lambda$ when i has home ground and $\theta_i - \theta_j - \lambda$ when j has home ground. Hence, the two matches between players i and j are still modelled as two Bernoulli trials, but now with different odds formulas.

The Bradley–Terry model is closely related to a model for rank data, called the Bradley–Terry/Mallows model. For a discussion of such data and the latter model in particular, see Section 12.7.

7

Curved Exponential Families

A full exponential family has a canonical statistic $\boldsymbol{t}$ and a parameter space $\boldsymbol{\Theta}$ of the same dimension, k say. It is also denoted a (k, k) exponential family, to tell that these two dimensions are both k, which is also the order of the family. If we impose a linear restriction on the canonical parameter vector, the minimal sufficient canonical statistic can be reduced in a corresponding way, so the resulting exponential family is still full, but of smaller order. A nonlinear restriction on the canonical parameter vector, on the other hand, will reduce the dimension of the parameter, but not the minimal sufficient statistic. We then talk of a *curved exponential family*, and denote it a (k, l) family, where $l < k$ is the dimension of the parameter vector. Thus, a curved family can be constructed by starting from a full family and imposing parameter restrictions. Alternatively, we might start with the restricted model and find that it may be regarded as a curved subfamily of one or several full families. Along this line, curved exponential families have become both standard practical choices and theoretical examples of models which are not full exponential families. Nor are they complete, for dimensional reasons, cf. Proposition 3.23. Here is first an examples section, also indicating some of the problems, followed by theory sections. More examples are found in later chapters, see the Index.

Classic references on curved families are Efron (1975, 1978). For a more recent discussion, see Barndorff-Nielsen and Cox (1994, Sec. 2.10).

7.1 Introductory Examples

Example 7.1 ***Normal sample with known coefficient of variation***
Suppose the coefficient of variation σ/μ in a sample from a $\mathrm{N}(\mu, \sigma^2)$ distribution is (regarded as) known, with the value $c > 0$, say. This submodel is parametrically one-dimensional. Its normal density can be written

$$f(\boldsymbol{y}; \mu) = (c\,\mu\,\sqrt{2\pi})^{-n}\, e^{-n/2}\, e^{\frac{1}{c^2\mu}\sum y_i + \frac{-1}{2c^2\mu^2}\sum y_i^2}.$$

The exponential factor, which depends on both parameter and data, cannot be reduced to depend on only a one-dimensional statistic, so the model is still of order 2. In other words, the two statistics $\sum y_i$ and $\sum y_i^2$ are jointly needed to form the minimal sufficient statistic, not only in the full exponential family but also in this restricted model. In the two-dimensional canonical parameter space $\boldsymbol{\Theta}$, the model with known coefficient of variation forms a parabola, that is, a one-dimensional curve in $\boldsymbol{\Theta}$: $\theta_2 = -(c^2/2)\,\theta_1^2$. The model is a (2, 1) curved exponential family. The discussion continues in Example 7.12. △

Example 7.2 ***Two normal samples with $\mu_1 = \mu_2$ ('Behrens–Fisher')***
Two independent samples from $N(\mu_1, \sigma_1^2)$ and $N(\mu_2, \sigma_2^2)$, respectively, with no parameters in common, are jointly represented by a full exponential family of order 4, with canonical statistic formed by all the sample sums and sums of squares, $(\sum_i y_{1i}, \sum_i y_{1i}^2, \sum_i y_{2i}, \sum_i y_{2i}^2)$, and with canonical parameters $\theta_1 = \mu_1/\sigma_1^2$, $\theta_2 = -1/2\sigma_1^2$, $\theta_3 = \mu_2/\sigma_2^2$, $\theta_4 = -1/2\sigma_2^2$. Suppose we specify the mean values to be the same, $\mu_1 = \mu_2$, denoted μ_0 say, for example as a hypothesis model to be tested. We then have an exponential family with a 3-dimensional parameter vector but still of order 4, that is, the minimal sufficient statistic remains the same. The restriction is indeed linear in the mean value parameter space $\boldsymbol{\mu}_t(\boldsymbol{\Theta})$, but not in the canonical parameter space $\boldsymbol{\Theta}$. The curved surface in $\boldsymbol{\Theta}$ is specified by the nonlinear relationship $\theta_1\,\theta_4 = \theta_2\,\theta_3$. The model is a (4, 3) exponential family. The problem to construct a test for the hypothesis $\mu_1 = \mu_2$ (or a confidence interval for $\mu_1 - \mu_2$) involves a lot of complications, see Section 7.2.2 and Example 7.10. It is known under the name of the *Behrens–Fisher problem*. △

Example 7.3 ***Probit regression***
In Example 2.2, logistic regression was introduced, with applications to dose–response and similar relationships for binary response data. Another relationship form that has been used for this type of data is the so called probit regression, when the Bernoulli 'success' probability $\pi_i = \pi(x_i)$ is linear in x via the so called *probit function*, nothing else than the *inverse normal probability function* Φ^{-1}, that is, $\Phi^{-1}(\pi(x_i)) = \alpha + \beta x_i$. As compared with one parameter $\pi(x)$ per dose value x, the probit regression forms a curved exponential family, whereas the logistic regression is linear in this canonical parameter space, and ony requires two parameters. This is the main reason logistic regression has largely taken over as a simple dose–response model. See further Example 9.3, but also Section 12.6 for another use of the probit. △

Example 7.4 ***SUR models in multivariate linear regression***
A *SUR* (*Seemingly Unrelated Regressions*) model is a special type of multivariate regression model, so let us first have a look at the general Gaussian multivariate linear regression model. This is given by

$$y = B\,x + \epsilon, \qquad \epsilon \sim \mathrm{N}(\mathbf{0}, \Sigma), \tag{7.1}$$

where B is a completely unspecified coefficients matrix, relating the response vector y to the vector x of explanatory variables. This is a regular exponential family with canonical parameter components being the elements of $B^T \Sigma^{-1}$ and Σ^{-1}, cf. the Gaussian linear model Example 2.9 and the multivariate normal distribution, Example 2.10. Interestingly, the ML estimates of B are identical with the set of univariate ML estimates, that is, with the ordinary least squares estimates for each response variable separately. This is true even though the noise terms are allowed to be correlated. By (3.14) the likelihood equations for B are in both cases $\sum_j y_j x_j^T = B \sum_j x_j x_j^T$, where j is the sampling unit index.

Turning now to the seemingly unrelated regressions (SUR) model, let each response variable (component of y) be regressed on its own set of explanatory variables, thought to be relevant for that particular response variable. If these regression formulas have no regressors in common, the regressions might appear to be mutually unrelated. However, if they have correlated noise terms, the regressions are not really unrelated, and even more, the ML estimates are not the simple least squares estimates. This model was introduced by Zellner (1962) and is quite popular in econometrics. In applications, the correlation is often due to time being the same. As an example, the response vector components could be the demands for some consumption goods, each demand explained specifically by variables related to that particular type of goods, and observed for a number of years or months. In a different study, responses could be consumptions registered for each of a number of consumer units, explained by variables measured separately for each of the units.

A SUR model can be regarded as a special form of a multivariate regression model (7.1). The SUR model says that the components of y depend on disjoint subsets of x-variables, and this can be expressed by a block-diagonal matrix B, obtained by inserting a zero for every combination of components y and x that is not in any regression equation. Unfortunately, this reduction of the parameter dimension is not accompanied by a reduction of the minimal sufficient statistic. The model imposes linear restrictions on B, but the canonical parameters of the full model cannot be expressed as linear functions of B only. For details in a simple case, let us consider a bivariate SUR model, with response variables y_1 and y_2:

$$\begin{cases} y_1 = \beta_1^T \boldsymbol{x}_1 + \epsilon_1 \,, \\ y_2 = \beta_2^T \boldsymbol{x}_2 + \epsilon_2 \,, \end{cases} \tag{7.2}$$

where $\boldsymbol{x}_1$ and $\boldsymbol{x}_2$ are disjoint vector variables and ϵ_1 and ϵ_2 are allowed to be correlated with an arbitrary covariance matrix $\boldsymbol{\Sigma}$ (2×2). In terms of (7.1), $\boldsymbol{x}^T = (\boldsymbol{x}_1^T \; \boldsymbol{x}_2^T)$ and

$$\boldsymbol{B} = \begin{pmatrix} \beta_1^T & 0 \\ 0 & \beta_2^T \end{pmatrix}. \tag{7.3}$$

For any reduction of the canonical statistic of (7.1) to take place, there must be a linear form in the canonical parameters that is identically zero in model (7.2). Let $\boldsymbol{C}$ be the concentration matrix $\boldsymbol{C} = \Sigma^{-1}$ corresponding to the unknown arbitrary covariance matrix Σ, and write (note $c_{21} = c_{12}$)

$$\boldsymbol{C} = \begin{pmatrix} c_{11} & c_{12} \\ c_{12} & c_{22} \end{pmatrix}.$$

Any linear form in the canonical parameters $\boldsymbol{BC}$ and $\boldsymbol{C}$ has the form

$$c_{11}(\kappa_{11} + \lambda_1^T \beta_1) + c_{12}(\kappa_{12} + \lambda_2^T \beta_1 + \lambda_3^T \beta_2) + c_{22}(\kappa_{22} + \lambda_4^T \beta_2)$$

for some constants κ and vectors λ. This can be identically zero, as function of the parameters β and c, only if κ and λ are zero. Hence, there is no nontrivial form that is zero, and this shows that no reduction is possible.

Drton and Richardson (2004) pointed out that econometric literature has often been vague about the character of the SUR model, and sometimes even erroneous by saying that the log-likelihood is globally concave. Concavity is a property for full exponential families in canonical parameters, and thus for the full multivariate regression model, but typically it does not hold for curved families. Drton and Richardson (2004) demonstrated that the log-likelihood for a SUR model can have several local maxima, so it is not true that the SUR log-likelihood is necessarily concave. This has consequences for the reliability of iterative ML search programs. Has the program found the global max or only a local extreme point? △

Example 7.5 *Covariance structures*

Think of a multivariate normal with an unspecified covariance matrix $\boldsymbol{\Sigma}$. We saw in Example 2.11 that after insertion of off-diagonal zeros in the concentration matrix $\boldsymbol{\Sigma}^{-1}$ we still have a full model, but of reduced dimension, characterized by conditional independences (see further Section 10.2). If we instead insert zeros in $\boldsymbol{\Sigma}$ itself, the resulting model is typically a curved family. Thus, this seemingly simpler procedure leads to a more complicated model. △

Example 7.6 ***Exponential life-times, with censored data***
When modelling life-time data (failure time data, or survival data) we typically must allow right-censoring. This means that we have only a lower limit to the life-time of some observational units. For example, the experiment or observation study might be stopped after some time, because we do not want to wait for failure or death of all items or individuals. Another complication is that individuals may disappear during the study, for example by moving from the area or by dying from a cause unrelated to the one under study.

Suppose a sample of size n is taken from the parametric life-time density $f(y; \lambda)$, with distribution function $F(y; \lambda)$. Two common principles for deliberate censoring are type I and type II censoring. In the former, all remaining items are censored at some fixed time, whereas in the latter case all items are followed until a fixed number of items remain, which are then all censored. It is important to register if and when an item has been censored, but we make the standard assumption that the censoring event contains no additional information about λ. This is obvious in type I censoring, and true but less obvious in type II.

An item not affected by censoring contributes to the likelihood as usual, by its density $f(y_i; \lambda)$, where y_i is the observed value for item i. If item i is censored at time y_i, however, we only know that its life-time exceeds y_i, so the item contributes by the tail probabiliity $1 - F(y_i; \lambda)$. The likelihood for the censored sample thus takes the form

$$L(\lambda; \text{data}) = \prod_{i=1}^{n} f(y_i; \lambda)^{\delta_i} \left(1 - F(y_i; \lambda)\right)^{1-\delta_i}, \tag{7.4}$$

where $\delta_i = 0$ if item i is censored, $\delta_i = 1$ if it is not.

After this general introduction, let us now consider an *exponential* life-time distribution, with

$$f(y; \lambda) = \lambda e^{-\lambda y},$$

and

$$1 - F(y; \lambda) = e^{-\lambda y}.$$

In life-time and failure-time contexts, λ is called the *hazard rate*, the conditional event rate given survival until y, being time-constant for the exponential. The likelihood (7.4) simplifies to

$$L(\lambda) = \lambda^d \prod_{i=1}^{n} e^{-\lambda y_i} = \lambda^d e^{-\lambda \sum_i y_i}, \tag{7.5}$$

where d is the number of uncensored life-times (failures or deaths). The statistic $\sum_i y_i = n\overline{y}$ is called the *total time at risk*. On log-scale we have

$$\log L(\lambda) = d \log \lambda - (n\overline{y})\,\lambda\,. \tag{7.6}$$

The form of the likelihood (7.5) or (7.6) shows that in general the model is a curved (2, 1) exponential family, with the two-dimensional canonical statistic $(d, \overline{y})$, but the single parameter λ. If d is fixed in advance, as in type II censoring, the model simplifies to a full exponential family. This also holds if the study is stopped at a predetermined total time at risk, but that situation is certainly less common in practice.

We will return to this set-up in Section 7.2, Example 7.9, but see also Cox and Oakes (1984, Ch. 3). Another perspective on censored data is as incomplete data from the underlying distribution, in this case the exponential distribution, see Example 8.1. △

Example 7.7 ***Binary fission, or Yule process***

Here is an example taken from Cox (2006, Ex. 2.4), somewhat related to the previous Example 7.6. Suppose that in a population of cells (or other elements), the cells divide in two, identical with their predecessors. For each cell (old and newborn) this occurs after an exponentially distributed life-time, Exp(ρ), so the probability for the life-time to exceed length $t > 0$ is $\exp(-\rho\, t)$. All cell life-times are assumed mutually independent. The exponential distribution lack of memory property implies that we may regard all cells existing at the start as being newborn. Similarly to the number of failures in the previous example, the total number of cells in a time period of fixed length is random and contributes information about ρ, in addition to the average life-time observed.

A linear pure birth process, called a *Yule process*, is essentially the same model. The only difference is that the cells now are assumed to give birth to additional cells instead of dividing in two, with birth intensity corresponding to division intensity. As an exponential family this curved model was first discussed by Keiding (1974) and Beyer et al. (1976). The full model canonical statistic consists of the total number of births and the total (or average) life-time in the fixed length time interval. △

Example 7.8 ***'Shaved dice' inference***

An instructive and entertaining, albeit constructed example of curved models was presented by Pavlides and Perlman (2010). First imagine a die has possibly been 'shaved' in one direction, so faces 1 and 6 have a higher probability than the other faces, α versus β, under the constraint $2\alpha + 4\beta = 1$. A sample of multinomial data from such a die is easily analyzed. By sufficency, the multinomial is reduced to a binomial, with 'success´ probability 2α. The estimated binomial probability is thus a linear function of α, and vice versa; no problems. However, in their scenario the authors now introduce the complication that data are from two such dice, used in a casino that has registered only the sum of the two dice. So if x_1 and x_2 are the two iid multinomial outcomes, the observed variate is $y = x_1 + x_2$. The variate y is of course also multinomially distributed, but with relatively complicated cell probabilities, $f_y(2) = \alpha^2$, $f_y(3) = 2\alpha(1 - 2\alpha)/4$, etc. Pavlides and Perlman show for example that if the shaved faces are 1 and 6, data follow an (6, 1) family, whose MLE solves a 6th degree polynomial equation. They also calculate formulas for the Fisher information, discuss ancillary information, etc, see their article. We will discuss the scenario again in Example 8.2, but then from an incomplete data point of view. △

Exercise 7.1 ***Normal correlation coefficient as single parameter***
Show that a mean- and variance-standardized bivariate normal distribution, i.e. a $N_2(0, 0, 1, 1, \rho)$, which has the correlation coefficient ρ as its only parameter, forms a (k, l) curved family, and specify the full family (smallest k). See further Exercise 7.5, Barndorff-Nielsen and Cox (1994, Ex. 2.38), and Small et al. (2000). △

Exercise 7.2 ***Poisson table under additivity***
Consider a 2×2 table of independent Poisson counts. Show that additivity of mean values, that is, $\lambda_{00} + \lambda_{11} = \lambda_{10} + \lambda_{01}$, is represented by a curved exponential family. △

7.2 Basic Theory for ML Estimation and Hypothesis Testing

Curved exponential families pose problems both in estimation and for hypothesis testing and confidence intervals. On the other hand such models have stimulated and motivated much development of statistical theory, both conceptually and technically, such as 'recovery of ancillary information', and conditional saddlepoint approximations, respectively (see Section 7.5). Quoting Efron (1978, p. 365), the curved exponential family is

> a paradigm for all smoothly defined statistical problems in which the

minimal sufficient statistic has higher dimension than the parameter space.

7.2.1 ML Estimation

Let the curved family be parameterized by ψ, to be estimated. This parameter is of dimension $\dim\psi = l < k = \dim\boldsymbol{\theta}$, where $\boldsymbol{\theta}$ is the full family canonical parameter vector. We assume that all components of $\boldsymbol{\theta} = \boldsymbol{\theta}(\psi)$ are smooth functions of ψ, with the $k \times l$ matrix of partial derivatives $\left(\frac{\partial\theta}{\partial\psi}\right)$. By the chain rule for differentiation, the score function U_ψ for ψ can be written

$$U_\psi(\psi) = \left(\frac{\partial\boldsymbol{\theta}}{\partial\psi}\right)^T U_\theta(\boldsymbol{\theta}(\psi)) = \left(\frac{\partial\boldsymbol{\theta}}{\partial\psi}\right)^T \{t - \mu_t(\boldsymbol{\theta}(\psi))\}, \tag{7.7}$$

see also the Reparameterization lemma, Proposition 3.14. Hence the scores for ψ are linear combinations of the scores for $\boldsymbol{\theta}$, with coefficients that depend on the parameters. The matrix $\left(\frac{\partial\theta}{\partial\psi}\right)$ gives the tangent directions along the curve (or surface) in $\boldsymbol{\Theta}$, and this allows the more precise interpretation that $U(\psi) = 0$ is achieved by finding a point $\boldsymbol{\theta}(\hat{\psi})$ on the curve in $\boldsymbol{\Theta}$ for which the vector $\boldsymbol{t} - \mu_t(\boldsymbol{\theta}(\hat{\psi}))$ in the observation space is orthogonal to the tangent vectors in the point $\boldsymbol{\theta}(\hat{\psi})$ in $\boldsymbol{\Theta}$. Note however that $\boldsymbol{t} - \mu_t(\boldsymbol{\theta}(\psi))$ and $\boldsymbol{\theta}(\psi)$ are not vectors in the same space, so the geometric interpretation is somewhat intricate, see Barndorff-Nielsen and Cox (1994, Fig. 2.1).

The observed information matrix is obtained by differentiating (7.7). This yields two types of terms, cf. the proof of Proposition 3.14:

$$\begin{aligned} J_\psi(\psi) &= -D_\psi^2 \log L(\psi) = -\left(\frac{\partial\boldsymbol{\theta}}{\partial\psi}\right)^T D_\theta^2 \log L(\boldsymbol{\theta}) \left(\frac{\partial\boldsymbol{\theta}}{\partial\psi}\right) - \left(\frac{\partial^2\boldsymbol{\theta}}{\partial\psi^2}\right)^T U_\theta(\boldsymbol{\theta}(\psi)) \\ &= \left(\frac{\partial\boldsymbol{\theta}}{\partial\psi}\right)^T I_\theta(\boldsymbol{\theta}(\psi)) \left(\frac{\partial\boldsymbol{\theta}}{\partial\psi}\right) - \left(\frac{\partial^2\boldsymbol{\theta}}{\partial\psi^2}\right)^T \{t - \mu_t(\boldsymbol{\theta}(\psi))\}. \end{aligned} \tag{7.8}$$

For $\dim\psi > 1$, the last term should be interpreted as the matrix obtained when a three-dimensional array is multiplied by a vector, cf. (3.19). This term has to do with the curvature of the surface $\boldsymbol{\theta} = \boldsymbol{\theta}(\psi)$ in $\boldsymbol{\Theta}$, and it will not be zero even in the MLE $\hat{\psi}$. However, its expected value is zero, since $E(\boldsymbol{t}) = \mu_t$, so the Fisher information matrix for ψ is given by

$$I_\psi(\psi) = \left(\frac{\partial\boldsymbol{\theta}}{\partial\psi}\right)^T I_\theta(\boldsymbol{\theta}(\psi)) \left(\frac{\partial\boldsymbol{\theta}}{\partial\psi}\right). \tag{7.9}$$

When $J_\psi(\psi)$ is not positive definite, this is because the last term of (7.8) is too large.

Example 7.9 ***Exponential life-times with censored data, continued***
We saw in Example 7.6 that a sample of exponential life-times, with censoring, has the log-likelihood

$$\log L(\lambda) = d \log \lambda - (n\,\overline{y})\,\lambda. \tag{7.10}$$

The score function is easily derived by direct differentiation, and (untypically) it admits an explicit ML estimator:

$$U(\lambda) = d/\lambda - n\,\overline{y} \quad \Rightarrow \quad \hat{\lambda} = d/(n\,\overline{y}). \tag{7.11}$$

How does this relate to the theory presented, and in particular to formula (7.7)? The role of ψ is taken by λ, and θ has components $\log\lambda$ and $-\lambda$, so

$$\left(\frac{\partial\theta}{\partial\lambda}\right) = \begin{pmatrix} 1/\lambda \\ -1 \end{pmatrix}.$$

The remaining calculation will be restricted to the case of Type I censoring at $y = y_0$. Then

$$E(d) = n\,(1 - e^{-\lambda y_0}),$$
$$E(\textstyle\sum y_i) = (n/\lambda)\,(1 - e^{-\lambda y_0}). \tag{7.12}$$

This makes the two mean value contributions cancel in the score expression (7.7), with (7.11) as the simple result.

Continued differentiation of the score function in (7.11) yields the observed information

$$J(\lambda) = d/\lambda^2, \qquad J(\hat{\lambda}) = (n\,\overline{y})^2/d \tag{7.13}$$

The expected information $I(\lambda)$ will differ from $J(\lambda)$ if d is random, as for example in type I censoring. When no censoring occurs, $d = n$, and the formulae simplify to those of an ordinary exponential sample. Note that this holds even if censoring might have occurred, i.e., when d is random. In particular, $I(\lambda) \neq J(\lambda)$ even when $d = n$. The observed information is a more relevant measure of information for the sample.

For a comparison of confidence interval constructions, some of them just corresponding to different information quantities, see Sundberg (2001). △

Asymptotically, when information increases and estimation becomes precise, we need only a local representation of the curved surface by its linear tangent plane, and both estimation and hypothesis testing problems approach those of linear submodels. In particular, ML estimation in the curved family is locally asymptotically equivalent to ML estimation in the full family of order l obtained when replacing the surface by its tangent plane through the true point $\theta(\psi)$ (provided of course that the true parameter point belongs to the curved model). The second term of J_ψ will be negligible, being only $\propto t - \mu_t = O(\sqrt{n})$ when the first term is $O(n)$. Hence, with probability tending to 1 with increasing information, there will be a root of the likelihood equations that will be a consistent estimator of ψ, normally distributed with variance-covariance matrix asymptotically given by I_ψ^{-1}. However, as soon as there are parts of the parameter space where $J_\psi(\psi)$ is not positive definite, we must be prepared for problems connected with multiple roots, see further Section 7.4.

For the numerical solution of the likelihood equations the choice between the Newton–Raphson and Fisher scoring methods (see Section 3.4) can be more crucial than in full families. There is a risk that, if the starting point is not close enough, the Newton–Raphson procedure will not converge to a maximum. On the other hand, since $I_\psi(\hat{\psi}) \neq J_\psi(\hat{\psi})$, the Newton–Raphson method will converge faster in a vicinity of $\hat{\psi}$. In critical cases one might consider trying the more robust scoring method for the first iterations, before changing method. See also Section 9.2.1 for further remarks.

Exercise 7.3 ***Mean values under type I censoring***
Verify formulas (7.12) for $E(d)$ and $E(\sum y_i)$. △

7.2.2 Hypothesis testing

Model reduction in connection with curved families depends on the character of the two models involved:

(i) From full to curved model
(ii) From curved to curved model
(iii) From curved to full model

We briefly comment on hypothesis testing in these situations.

(i) Consider first a dimension reduction from θ in a full model to ψ in a curved model, with the canonical statistic t in the full model being minimal sufficient also in the curved model. One example is the Behrens–Fisher problem, Example 7.2, when we want to test for equal mean-values in two samples without specifying equal variances. Another is testing the goodness of fit of the (2, 1) curved model of a censored exponential distribution, Example 7.6, against a model allowing a different censored proportion of data than motivated by the exponential distribution.

In Section 5.1 an exact test was constructed by a conditioning procedure for reductions from a full model to another full model. Such a conditioning procedure now fails, because there is no reduction of the minimal sufficient statistic. Thus we cannot obtain similarity (independence of nuisance parameters) in that way. Are there other similar tests? If the family under H_0 had been complete we could have concluded that every similar test must be a conditional test, conditioning on the minimal sufficient statistic under H_0 (Neyman structure; see Section 5.3). However, for dimensionality reasons this family cannot be complete, and the test cannot have Neyman structure. This makes it a generally open question whether there exists any nontrivial similar tests at all. A much investigated example is the classic Behrens–Fisher problem, where it is known since long that no such similar test exists that also satisfies any reasonable type of regularity condition.

(ii) Also if the model reduction considered is from one curved model to another, with no reduction in the minimal sufficient statistic, there is little chance that a nontrivial similar test can be constructed. Presumably we have to rely on large sample tests and large sample properties, that is, the deviance (LR), score, or Wald tests. No details will be given here, but the reader is referred to the general theory of Section 5.5.

(iii) Testing reduction from a curved model to a full model is different. Examples of this situation is testing whether the regression coefficient in probit regression is zero (H_0: no influence of dose), or testing within the curved Behrens–Fisher model if the common mean is specified (e.g. H_0: $\mu = 0$), or if the two variances are equal (H_0: $\sigma_1^2 = \sigma_2^2$). Whatever test statistic, calculation of p-values is made in the hypothesis model, which is full. The role of the curved model is to indicate the alternatives of interest.

A general procedure in such cases is to find the *best local test*, by locally linearizing the curved relationship, that is, to replace the curved model with a full model of parameter dimension dim $\boldsymbol{\psi}$ that best approximates the curved relationship. This would make possible a locally constructed exact test. An example is replacing the probit regression by logistic regression for testing H_0: no influence of dose. There may be alternative ways to avoid the curved model. To keep the hypothesis model but extend the alternative to a full model with minimal sufficient statistic $\boldsymbol{t}$ cannot often be recommended (but perhaps for testing H_0: $\mu = 0$ in the Behrens–Fisher example). Sometimes both the hypothesis model and the alternative model can be relaxed without changing the character of the model reduction. If asked to test H_0: $\sigma_1^2 = \sigma_2^2$ in the Behrens–Fisher model, a reasonable reaction would be to relax the mean value restriction and make the corresponding test in the usual two-sample model with no constraint on μ_1 and μ_2.

7.3 Statistical Curvature

It is intuitively reasonable that the more curved a curve or surface $\boldsymbol{\theta}(\boldsymbol{\psi})$ is in $\boldsymbol{\Theta}$, the further away we are from a nice full submodel, and the more problems or complications we may encounter. However, it is not immediately apparent that there exists a single characteristic making it possible to quantify this in a meaningful way. For simplicity we restrict the discussion to proper curves, that is, to families with $\dim \boldsymbol{\psi} = 1$ (not surfaces). For such models, the desirable characteristic was introduced by Efron (1975) under the name of *statistical curvature*. It is based on the concept of geometrical curvature of a curve, This concept is introduced in calculus courses as the rate of change of the (tangential) direction of a (smooth) curve in $\boldsymbol{\Theta}$ with respect to arc-length, as we move along the curve. Equivalently, the curvature is the inverse of the radius of curvature, that is, the radius of the circle arc locally best fitting the curve. The statistical concept is not quite so simple, however, because we must specify how distances and angles should be measured. For example, if a component of $\boldsymbol{\theta}$ is changed by a scale factor (together with the corresponding component of $\boldsymbol{t}$), the statistical situation is the same, so the statistical measure of curvature should be invariant to scale changes. Efron's statistical curvature satisfies this demand.

Definition 7.1 ***Statistical curvature***
The *statistical curvature* $\gamma_\psi \geq 0$ of a one-dimensional curved exponential family is the geometric curvature of $\boldsymbol{\theta}(\psi)$ in the $(\dim \boldsymbol{\theta})$-dimensional Euclidean space for which the Fisher information $I(\boldsymbol{\theta})$ defines the inner product.

The curvature squared, γ_ψ^2, is given by either of several equivalent formulas. Here is one of them:

$$\gamma_\psi^2 = \frac{\det \mathbf{M}(\psi)}{M_{20}(\psi)^3}, \tag{7.14}$$

with

$$\mathbf{M}(\psi) = \begin{pmatrix} M_{20}(\psi) & M_{11}(\psi) \\ M_{11}(\psi) & M_{02}(\psi) \end{pmatrix} = \begin{pmatrix} \left(\frac{\partial\theta}{\partial\psi}\right)^T I_\theta \left(\frac{\partial\theta}{\partial\psi}\right) & \left(\frac{\partial\theta}{\partial\psi}\right)^T I_\theta \left(\frac{\partial^2\theta}{\partial\psi^2}\right) \\ \left(\frac{\partial^2\theta}{\partial\psi^2}\right)^T I_\theta \left(\frac{\partial\theta}{\partial\psi}\right) & \left(\frac{\partial^2\theta}{\partial\psi^2}\right)^T I_\theta \left(\frac{\partial^2\theta}{\partial\psi^2}\right) \end{pmatrix},$$

where $\left(\frac{\partial\theta}{\partial\psi}\right)$ and $\left(\frac{\partial^2\theta}{\partial\psi^2}\right)$ are the vectors of first and second derivatives, respectively, of $\boldsymbol{\theta}$ by ψ along the curve. Note that $M_{20}(\psi) = i(\psi)$, see (7.9).

Here is an equivalent but statistically more informative expression, where we use lower case letters j, i and u with ψ to indicate that ψ is a scalar:

$$\gamma_\psi^2 = \frac{\operatorname{var}(j_\psi)\,(1-\rho^2(j_\psi,u_\psi))}{i_\psi^2} = \frac{\text{resid.var}(j_\psi|u_\psi)}{i_\psi^2}, \tag{7.15}$$

where ρ^2 is the squared correlation between the j_ψ and u_ψ, and resid.var is the residual variance in the theoretical regression of j_ψ on u_ψ. This characterization will be used in Section 7.4.

Proof of the equivalence of (7.14) and (7.15) Since $M_{20}(\psi) = i(\psi)$, two of the three factors of $M_{20}(\psi)^3$ in the denominator of (7.14) correspond to the denominator of (7.15). From (7.7), which says that u_ψ is a linear form in U_θ, and (7.8), which says that j_ψ is also a linear form in U_θ (except for a constant), it further follows that $\operatorname{var}(j_\psi) = M_{02}(\psi)$ and that $\operatorname{cov}(j_\psi, u_\psi) = M_{11}(\psi)$. Insertion yields the desired result. □

Here are some properties of the statistical curvature γ_ψ:

- The statistical curvature γ_ψ is invariant under reparameterization of the curved family. This can be seen from the geometrical interpretation, as argued by Efron (1975). A statistically oriented proof based on the representation (7.15) gives some additional insight, see Exercise 7.4. A straight calculus type of proof is also possible.
- When the curved model is not curved, but a straight line through $\boldsymbol{\Theta}$, then $\gamma_\psi = 0$. This is because in this case $\left(\frac{\partial^2\theta}{\partial\psi^2}\right) = \mathbf{0}$, so for example $\det \mathbf{M}(\psi) = 0$.
- If data form a sample of size n, the curvature $\gamma_\psi(n)$ will decrease with n according to the formula $\gamma_\psi(n) = \gamma_\psi(1)/\sqrt{n}$. Thus, the statistical curvature will tend to zero with n. This reflects the fact that large n yields high precision, so we can restrict the model to a local model in a neighbourhood of the true ψ-value, and locally the curved model looks like a full model of the lower dimension, $\dim\psi = 1$.
- Classic conjectures and provisional results by Fisher and Rao about second order efficiency can be expressed and better understood in terms of the curvature γ_ψ, as shown by Efron: Asymptotically as $n \to \infty$, $\gamma_\psi(1)^2\, i_\psi$ is the loss of information in the curved model MLE, relative to the upper bound $n\, i_\psi$ (information in sample of size n) in the denominator of the Cramér–Rao inequality (Proposition 3.25). This can be interpreted as a reduction in effective sample size from n to $n - \gamma_\psi(1)^2$. See also (7.21).

Exercise 7.4 ***Invariance of statistical curvature***
Verify, using representation (7.15), that the statistical curvature γ_ψ is invariant under reparameterizations. Let $\psi = \psi(\lambda)$, with $\dim\psi = \dim\lambda = 1$.

(a) First show that $U_\lambda = b\,U_\psi$ (cf. the Reparameterization lemma, Proposition 3.14) with $b = \psi'(\lambda)$, and next that

$$j_\lambda = b^2\, j_\psi - c\, u_\psi$$

with $c = \psi''(\lambda)$.
(b) The numerator of γ_ψ^2 in (7.15), and likewise of γ_λ^2, can be interpreted as the residual variance in a theoretical linear regression of j_ψ on u_ψ (regression coefficient $\mathrm{cov}(j,u)/\mathrm{var}(u)$). Use the formula in (a) to conclude that the numerator changes by the factor b^4.
(c) Use the Reparameterization lemma to conclude that the denominator of (7.15) is also changed by b^4, and hence that $\gamma_\lambda = \gamma_\psi$, as desired. △

7.4 More on Multiple Roots

Likelihood equations for curved families can have several roots. The bivariate normal with only correlation unknown (Exercise 7.1) is an artificial but relatively simple example, used by Barndorff-Nielsen and Cox (1994, Ex. 2.38) and Small et al. (2000). Outcomes with sample variances smaller than expected can yield three roots, see Exercise 7.5 for details. Examples 7.2 (Behrens–Fisher) and 7.4 (SUR) provide other illustrations, with models of more applied interest. The first of these is considered in some detail next. More examples are found in Efron (1978). The theory will be provided only for the case $\dim \psi = 1$, as in Efron (1978) and most of Sundberg (2010).

Example 7.10 ***Behrens–Fisher model, continued from Example 7.2***
In the two-sample situation with one and the same mean value parameter μ_0, the possibility of multiple roots to the likelihood equations was pointed out by Sugiura and Gupta (1987). When σ_1 and σ_2 have been solved for, as functions of μ_0, the resulting (i.e. profile) likelihood equation for μ_0 turns out to be a third degree polynomial equation. For data sets with sample means differing 'too much', this equation has three real roots, the middle of which represents a local minimum of the profile likelihood (saddlepoint of the original likelihood).

For simplicity, suppose the sample sizes are the same, $n_1 = n_2 = n$, and that the sample variances happen to be equal, $s_1^2 = s_2^2 = s^2$. Then, by symmetry, there is a (central) root at $\overline{y} = (\overline{y}_1 + \overline{y}_2)/2$. Typically this is the only root, but when the two mean values are considerably separated, $|\overline{y}_1 - \overline{y}_2| > 2\,s$, the likelihood will also have two maxima located towards $\overline{y}_1$ and $\overline{y}_2$, respectively, whereas the root in between represents a saddlepoint.

For very small n the probability for such outcomes is substantial. Already for moderately large n (≥ 10, say), the corresponding probability is negligible, as calculated under the Behrens–Fisher model (see Drton (2008) for an upper bound). If the model assumption of equal means is false, however, the probability can be large. Hence, multimodality of the likelihood is typically an indication and consequence of a bad fit between data and the model assumption of equal means.

In between the cases of a single root and three different roots we have the case of a triple root at $\overline{y}$, corresponding to a flat likelihood maximum. The profile log likelihood functions for μ_0 and for (μ_1, μ_2) in such a case, normalized to have maximum zero in the full model, are shown in Figure 7.1. Specifications are $n_1 = n_2 = 10$, $\overline{y}_1 = -1$, $\overline{y}_2 = 1$, $s_1 = s_2 = 1$, the symmetry implying a central root at $\mu_0 = 0$. The curved model in parameterization by (μ_1, μ_2) is the straight line $\mu_2 = \mu_1$. The flatness at the maximum is clear, and a normal approximation likelihood-based confidence interval for μ_0, albeit inadequate, would include both sample means, (-1 and $+1$).

This diagram differs from Figure 5 in Sundberg (2010) only in axis scaling. The latter has $\overline{y}_1 = -1/\sqrt{2}$, $\overline{y}_2 = 1/\sqrt{2}$, $s_1 = s_2 = 1/\sqrt{2}$. This is connected with a misprint in Sundberg (2010, Ex. 3.3), erroneously stating that multimodality occurs when $|\overline{y}_1 - \overline{y}_2| > \sqrt{2}\, s$, with $\sqrt{2}$ instead of 2. $\triangle$

Formula (7.7) can be used to elucidate the occurrence of flat likelihoods and multiple roots of the likelihood equations. As before, we restrict the presentation to the case $\dim \psi = 1$. For simplicity, we think of t as being of continuous type. Not surprisingly, the statistical curvature has a role to play. Suppose $\tilde{\psi}$ is a root for an observed $\tilde{t}$. Since the sample space for t is of higher dimension than the parameter space Ψ, a continuum of sample points t will have the same $\tilde{\psi}$ as a root. More precisely, let $\tilde{\mathcal{L}}$ be the line (when $\dim t = 2$) or hyperplane through $\tilde{t}$, characterized by having $\left(\frac{\partial \theta}{\partial \psi}\right)$ as normal vector, that is, formed by all points t satisfying

$$\tilde{\mathcal{L}}: \quad \left(\frac{\partial \boldsymbol{\theta}}{\partial \psi}\right)^T_{\tilde{\psi}} (t - \tilde{t}) = 0. \tag{7.16}$$

By the form of the score function (7.7), the same $\tilde{\psi}$ must be a root for any t in $\tilde{\mathcal{L}}$. By (7.8), if t is moved within $\tilde{\mathcal{L}}$ in a direction having positive scalar product with the vector $\left(\frac{\partial^2 \theta}{\partial \psi^2}\right)$, the observed information will decrease, linearly in t. If only the sample space for t extends far enough in this direction, we can achieve $j_\psi(\tilde{\psi}) < 0$, that is, for such t the root $\tilde{\psi}$ is a local minimum (and then certainly not the only root). The following result connects this feature with the statistical curvature $\gamma^2_{\tilde{\psi}}$ and with the

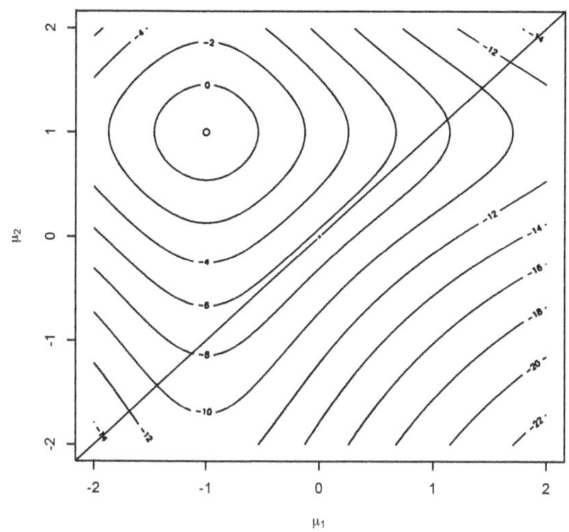

Figure 7.1 Behrens–Fisher model with flat profile log likelihood (triple root maximum), for μ_0 within the full model (μ_1, μ_2). The contours represent the two-dimensional profile log-likelihood for (μ_1, μ_2) when sample sizes are equal (=10), sample variances happen to be equal (=1), and sample means (±1) differ precisely much enough to yield a root of multiplicity 3. The circle point shows the sample means. The Behrens–Fisher model is the 45° line $\mu_2 = \mu_1 (= \mu_0)$, with maximum at $\mu_0 = 0$ (marked).

score test statistic $W_u(\tilde{\psi})$ for testing the adequacy of a reduction from the full model to the curved model. The result is taken from Sundberg (2010), but in essence the result goes back to Efron (1978, Th. 2), albeit in a less general setting there, with a different proof and not the same interpretation. The result is expressed in terms of the critical point where $J_\psi(\tilde{\psi}) = 0$, so $\tilde{\psi}$ is a multiple root (typically global or local maximum or saddle point).

Proposition 7.2 ***Multiple roots can indicate misfit of the curved model*** *If a root $\tilde{\psi} = \tilde{\psi}(t)$ of the one-dimensional likelihood equations (7.7) is a root of multiplicity ≥ 2 or a single root representing a local minimum, then the corresponding score test statistic $W_u(\tilde{\psi})$ for testing the curved model versus the full model will satisfy $W_u \geq 1/\gamma_{\tilde{\psi}}^2$ $(= O(n)$ for sample size $n)$.*

Letting $\boldsymbol{t}$ move in $\tilde{\mathcal{L}}$ from $\boldsymbol{\mu}_t(\boldsymbol{\theta}(\tilde{\psi}))$ along the projection of $\left(\frac{\partial^2\theta}{\partial\psi^2}\right)$, equality is attained at the turning point where the root is a multiple root (so the *lower bound for W_u is best possible), and further away from $\boldsymbol{\mu}_t(\boldsymbol{\theta}(\tilde{\psi}))$ in this direction, W_u is even larger and the root is no longer a maximum.*

Proof The following proof differs from the one in Efron (1978), and allows arbitrary dimension of $\boldsymbol{\theta}$. For ψ-values with nonpositive observed information, i.e. $j_\psi \leq 0$, it follows from (7.8) that

$$\left(\frac{\partial^2\boldsymbol{\theta}}{\partial\psi^2}\right)^T U_\theta(\boldsymbol{\theta}(\psi)) \geq \left(\frac{\partial\boldsymbol{\theta}}{\partial\psi}\right)^T I_\theta(\boldsymbol{\theta}(\boldsymbol{\psi}))\left(\frac{\partial\boldsymbol{\theta}}{\partial\psi}\right) = i_\psi \ (\geq 0). \tag{7.17}$$

We want to relate the left-hand side expression to the score test statistic W_u. It is a scalar product, on which we can use Cauchy–Schwarz inequality, and then to get W_u, we use the inequality in i_θ-norm version,

$$\begin{aligned}
&\left(\left(\frac{\partial^2\boldsymbol{\theta}}{\partial\psi^2}\right)^T U_\theta(\boldsymbol{\theta}(\psi))\right)^2 \\
&\leq \left(\left(\frac{\partial^2\boldsymbol{\theta}}{\partial\psi^2}\right)^T I_\theta(\boldsymbol{\theta}(\boldsymbol{\psi}))\left(\frac{\partial^2\boldsymbol{\theta}}{\partial\psi^2}\right)\right)\left(U_\theta(\boldsymbol{\theta}(\psi))^T I_\theta(\boldsymbol{\theta}(\boldsymbol{\psi}))^{-1} U_\theta(\boldsymbol{\theta}(\psi))\right) \\
&= \mathrm{var}(j_\psi)\, W_u(\psi),
\end{aligned} \tag{7.18}$$

where the last factor of (7.18) is the score test statistic W_u before inserting the root $\tilde{\psi}$ for ψ. When ψ is a root $\tilde{\psi}$, the inequality can also be sharpened. Since $\left(\frac{\partial\theta}{\partial\psi}\right)$ is orthogonal to U_θ in any root of the likelihood equation, the left-hand side of (7.17) in $\tilde{\psi}$ does not change if $\left(\frac{\partial^2\theta}{\partial\psi^2}\right)$ is replaced by $\left(\frac{\partial^2\theta}{\partial\psi^2}\right) - b\left(\frac{\partial\theta}{\partial\psi}\right)$, for any b. As b-value we choose the theoretical univariate regression coefficient of $\left(\frac{\partial^2\theta}{\partial\psi^2}\right) t$ on $\left(\frac{\partial\theta}{\partial\psi}\right) t$. and then we get the residual variance resid.var$(j_\psi \mid u_\psi)$ instead of var(j_ψ) in (7.18). Now combining the inequalities from (7.17) and (7.18), we obtain

$$W_u(\tilde{\psi}) \geq i_\psi(\tilde{\psi})^2/\mathrm{resid.var}(j_\psi(\tilde{\psi}) \mid u_\psi(\tilde{\psi}) = 1/\gamma_\psi(\tilde{\psi})^2,$$

the last expression, $1/\gamma_{\psi}^2$, following from (7.15).

Finally, to see that equality is possible, consider first the simplest case $\dim\boldsymbol{\theta} = 2$. Then there is only one direction orthogonal to $\left(\frac{\partial\theta}{\partial\psi}\right)$ so U_θ in $\tilde{\mathcal{L}}$ and the residual vector have the same direction, and we get equality in the Cauchy–Schwarz inequality. When $\tilde{\psi}$ is a root of multiplicity ≥ 2, there is also equality in (7.17). For $\dim\boldsymbol{\theta} > 2$, there is one direction in $\tilde{\mathcal{L}}$ that is the same as the direction of the residual vector, and in that particular direction, equality is attained. This finishes the proof. □

The components of the inequality $W_u \geq 1/\gamma_{\tilde{\psi}}^2$ may be interpreted in terms of distances. W_u is alternatively expressed as the Mahalanobis (squared) distance $D^2_M(t; \tilde{\psi})$ from t to its hypothesis model mean value,

$$D^2_M(t; \tilde{\psi}) = (t - \mu_t(\theta(\tilde{\psi})))^T \operatorname{var}(t; \theta(\tilde{\psi}))^{-1}(t - \mu_t(\theta(\tilde{\psi}))),$$

On the other side of the inequality, $1/\gamma^2_{\tilde{\psi}}$ has an interpretation as the squared radius of curvature in the point $\tilde{\psi}$, corresponding to the curvature $\gamma_{\tilde{\psi}}$. Note that for a sample of size n, $1/\gamma_{\psi} = O(\sqrt{n})$, so the lower bound for W_u increases with the amount of data proportionally to n.

Proposition 7.2 tells that if a root $\tilde{\psi}$ is a multiple root (implying a locally flat likelihood) or a local minimum, we should expect the curved model to fit badly, according to the large sample score test. Note however that this does not preclude the simultaneous presence of another root $\hat{\psi}$ that represents the global maximum and provides a good fit between model and data. In Proposition 7.2 the idea was to consider the same root $\tilde{\psi}$ for different sample points t, but a dual, generally more common point of view would be to regard t as fixed and study the log-likelihood surface for varying ψ.

The case $\dim \psi > 1$, certainly more frequent in practice, is more complicated to describe, see Sundberg (2010) for some discussion.

Exercise 7.5 ***Correlation parameter model, continued from Exercise 7.1***
In Exercise 7.1 the bivariate normal distribution $N_2(0, 0, 1, 1, \rho)$ model was considered, with only the correlation coefficient ρ unknown. Let $u = (\overline{y_1^2} + \overline{y_2^2})/2$ and $v = \overline{y_1 y_2}$, $t = (u, v)$.
(a) Derive the likelihood equation for ρ and show that
(i) when $u = E(u) = 1$, $\hat{\rho} = v$ is the one and only root,
(ii) when $v = 0$ and $u < 0.5$, $\hat{\rho} = 0$ is not the only root, but the likelihood equation also has the roots $\hat{\rho} = \pm\sqrt{1 - 2u}$ (triple root when $u = 0.5$). The middle root $\hat{\rho} = 0$ is then (necessarily) a local likelihood minimum.
(b) Derive the lines of type $\tilde{\mathcal{L}}$ (see (7.16)) through $\tilde{t} = (1, \tilde{v})$ and show that (provided $\tilde{v} \neq 0$) they cross the line $v = 0$ at $u = (1 - \tilde{v}^2)/2$, in agreement with the results in part (a). △

Exercise 7.6 ***Bivariate normal with mean on a parabola***
Let $\boldsymbol{y} = (y_1, y_2)$ be a single observation from $N_2(\boldsymbol{\mu}, I_{2\times 2})$, with independent components and $\boldsymbol{\mu}$ on the parabola $\mu_1 = \psi$, $\mu_2 = b\psi^2$, b known. This artificial model can be found among the examples in Efron (1975), in Barndorff-Nielsen and Cox (1994, Ex. 2.35), and in Sundberg (2010).
(a) Write down the log-likelihood and show that the likelihood equation is a third degree equation that can have three real roots. Consider in particular the case $y_1 = 0$ and various values of y_2.

(b) Calculate the statistical curvature γ_ψ, and in particular find its worst possible value.
(c) Extend these calculations to the case of a sample of size n. △

7.5 Conditional Inference in Curved Families

In a full exponential family, provided only that the MLE $\hat{\theta}(\boldsymbol{t})$ exists, we can let the MLE represent the minimal sufficient statistic, since it is a one-to-one function of the canonical $\boldsymbol{t}$, for $\boldsymbol{t} \in \mu_t(\boldsymbol{\Theta})$. For sufficiency in a curved family, $\hat{\boldsymbol{\psi}}(\boldsymbol{t})$ must be supplemented by a statistic of dimension $\dim \boldsymbol{t} - \dim \boldsymbol{\psi}$. Sometimes this supplementary statistic can be chosen such that the inference about $\boldsymbol{\psi}$ can be made conditional on it, resulting in increased relevance. The basis for this is a version of the conditionality principle different from the one in Section 3.5, but not without similarities and connections. The present principle is often phrased 'Condition on ancillary statistics', 'Recovery of ancillary information', or similarly. We start by specifying what we here mean by an ancillary statistic.

Definition 7.3 ***Ancillary statistic***
A statistic $\boldsymbol{a} = \boldsymbol{a}(\boldsymbol{t})$, a function of the minimal sufficient $\boldsymbol{t}$, is called *ancillary* if it is *distribution-constant*, that is, if its distribution is parameter-free.

When $\boldsymbol{a} = \boldsymbol{a}(\boldsymbol{t})$ is an ancillary statistic, the likelihood for $\hat{\boldsymbol{\psi}}$ and the density for $\boldsymbol{t}$ factorizes as follows, cf. (3.29) for S-ancillarity:

$$L(\boldsymbol{\psi};\, \boldsymbol{t}) = f(\boldsymbol{t};\, \boldsymbol{\psi}) = f(\boldsymbol{t}\,|\,\boldsymbol{a};\, \boldsymbol{\psi})\, f(\boldsymbol{a}), \tag{7.19}$$

so at least for ML estimation only the first factor is needed. This is not to say that $\boldsymbol{a}$ is irrelevant, because $\boldsymbol{a}$ appears also in the first factor. Of interest are the typical situations when the ancillary statistic $\boldsymbol{a}$ has the role of a *precision index* for $\hat{\boldsymbol{\psi}}$. A precision index yields information about the precision of $\hat{\boldsymbol{\psi}}$, but not about its location. For example, $\boldsymbol{a}$ is often a sort of random sample size, as in Example 7.11, and intuitively we should (of course) condition on sample size, to make the inference more relevant.

Proposition 7.4 ***Conditionality principle for curved families:***
Statistical inference (confidence statements, tests, etc) about the parameter $\boldsymbol{\psi}$ in a curved exponential family with ancillary statistic $\boldsymbol{a} = \boldsymbol{a}(\boldsymbol{t})$ should be made conditional on $\boldsymbol{a}$, that is, in the conditional model for $\boldsymbol{t}$ given $\boldsymbol{a}$.

Example 7.11 ***Two parametrically related Poisson variates***
We consider two variations of basically the same situation. Suppose first that y_1 and y_2 are independent observations from respectively $\text{Po}(c\,\psi)$ and

$\mathrm{Po}(c\,(1-\psi))$, where c is a known constant and ψ is a parameter, $0 < \psi < 1$. (Reader, check that this is a curved (2, 1) exponential family with (y_1, y_2) as canonical statistic in the full family!) The sum $a = y_1 + y_2$ is Poisson distributed with the known intensity c, so a is an ancillary statistic. Hence we should condition on a and make the inference in the binomial distribution $\mathrm{Bin}(a, \psi)$ for y_1 (or the complementary binomial for y_2). The statistic a plays the role of a random sample size for inference about ψ, whereas the value of c does not have a role in the inference about ψ.

In a less restricted variation of this model, suppose the two Poisson distributions are $\mathrm{Po}(\lambda\,\psi)$ and $\mathrm{Po}(\lambda\,(1-\psi))$, with ψ as before but λ as a free parameter on the real line. Suppose ψ is still the parameter of interest, so λ is a nuisance parameter, of secondary interest. In this case $a = y_1 + y_2$ has a distribution $\mathrm{Po}(\lambda)$ that only depends on the nuisance parameter λ. Note also that λ and ψ are variation independent. Hence, $a = y_1 + y_2$ is an S-ancillary statistic (see Section 3.5), and the factorization corresponding to (7.19) is a cut. By conditioning on a we eliminate the nuisance parameter. In this case, it is the conditionality principle of Section 3.5 that states we should do so. Note, however, that this conditional inference is formally the same as in the previous model, when $\lambda = c$ was assumed known in advance.

In the two-parameter full family, a mixed parameterization is instructive. Use y_1 and $a = y_1 + y_2$ as components to form the canonical statistic, and use the mean value for a, $E(a) = \lambda$, as one of the two parameters. The supplementary parameter is the canonical parameter for y_1, which is now $\log(\lambda\,\psi) - \log(\lambda\,(1-\psi)) = \log(\psi/(1-\psi)) = \mathrm{logit}(\psi)$, a one-to-one function of ψ. It is obvious that λ and ψ are variation independent, but this follows also from the general theory of the mixed parameterization, see Proposition 3.20. △

When and how could we find an ancillary statistic? Kumon (2009) has shown that only two types of curved exponential models can have exact ancillary statistics. One type is that of an *underlying cut* in an extended model, as illustrated by the preceding Example 7.11. In the extended two-parameter full family, $a = y_1 + y_2$ was an S-ancillary cut, that formed an ancillary statistic (in the strict sense) in the curved family corresponding to a known λ-value, $\lambda = c$. When it makes a difference, the known parameter should be the mean value of the cutting statistic (see Section 3.5).

The second case is when the curved family is a *transformation model* or a subfamily of such a model. The ancillary statistic is a maximal invariant under those transformations, see Kumon (2009) for details. Example 7.12, coming next, is a simple example of this type.

When no exact ancillary statistic can be found in any of these ways, there is still a possibility to form an approximate or 'locally' ancillary statistic, by large sample arguments, see further down.

Example 7.12 ***Normal sample with known coefficient of variation***
The (2, 1) curved exponential family $N(\mu, c^2\mu^2)$ of Example 7.1 is a transformation model under scale factor changes, that is, after a change of scale of y its distribution still belongs to the same family. A (maximal) invariant under such transformations is

$$a = \frac{(\sum y_i)^2}{\sum y_i^2}, \tag{7.20}$$

which is then also an ancillary statistic in this model. As an alternative, we can use the equivalent $a = s^2/\overline{y}^2$. That the distribution of such an a does not depend on μ is seen by normalizing both numerator and denominator by μ^2, since y/μ has a known distribution. See further Hinkley (1977). △

There is no guarantee that the ancillary statistic is unique. Here is an example from Basu (1964), that has been repeatedly discussed since then.

Example 7.13 ***Multinomial model with nonunique ancillary statistic***
Consider a fixed size sample from a multinomial with $2 \times 2 = 4$ cells and probabilities π_{ij} given by

p_{ij}	$j = 0$	$j = 1$	row sum
$i = 0$	$(1 + \psi)/6$	$(1 - \psi)/6$	1/3
$i = 1$	$(2 - \psi)/6$	$(2 + \psi)/6$	2/3
col. sum	1/2	1/2	1

This is a curved (3, 1) family (reader, check this!). The difference 2 in dimension from the full family opens up for two essentially different ancillary statistics. Any of the two row sums is one such ancillary statistic (binomial with known probability), and anyone of the two column sums is the second one. A natural reaction might be 'so what, why not condition on both of them'. However, they are not jointly ancillary, so such a conditioning is not justified, at least not without some loss of information.

In the light of Kumon's result, we should look for an extended model with a cut. This is obtained by allowing a row sum (or a column sum) to be binomially distributed, $\text{Bin}(n, \pi_0)$ say, for an unrestricted $0 < \pi_0 < 1$. This model extension makes the row (or column) sum S-ancillary. This does not work jointly for rows and columns, however. △

We now turn to approximate, local ancillary statistics, 'local' referring to the parameter space in a large sample setting, so the MLE is assumed to be relatively precise. When the curved family is locally approximated by its l-dimensional tangent space, it can locally be represented by a full l-dimensional family. The remaining $k - l$ components of $\boldsymbol{t}$ may be chosen to be locally ancillary, by choosing them information orthogonal to the l-dimensional canonical statistic (that is, the corresponding information matrix is blockdiagonal). A particular local ancillary statistic can be based on the difference $J(\hat{\boldsymbol{\psi}}) - I(\hat{\boldsymbol{\psi}})$ between observed and expected information in the MLE $\hat{\boldsymbol{\psi}}$, the *Efron–Hinkley ancillary* (Efron and Hinkley, 1978). More precisely, assuming $k = 2$, $l = 1$ for simplicity, they showed that asymptotically, for increasing sample size n,

$$a_{E-H} = \sqrt{n}\left\{J(\hat{\psi})/I(\hat{\psi}) - 1\right\}/\gamma_{\hat{\psi}} \sim \mathrm{N}(0, 1) \tag{7.21}$$

so a_{E-H} is approximately ancillary. Note the reappearance of the statistical curvature. Moreover, a_{E-H} recovers most of the ancillary information in $\boldsymbol{t}$. Even simpler, conditioning on a_{E-H} yields the approximate conditional variance $\mathrm{var}(\hat{\psi} \mid a_{E-H}) \approx J(\hat{\psi})^{-1}$, with multiplicative approximation error $\{1 + O(n^{-1})\}$, whereas the corresponding error factor for the expected information I is not better than $\{1 + O(n^{-1/2})\}$. That is, *the observed information is more relevant than the expected information* as a measure of the precision of the MLE. Alternative ancillary statistics that have been suggested are based on the (directed) likelihood ratio or the score statistic for testing the curved model within the full model. Note that we must trust the curved model. In practice, an extreme value of the ancillary statistic may rather indicate that something is wrong with the curved model, cf. Proposition 7.2. Therefore, a_{E-H} is of applied interest also in large sample situations, even though a_{E-H} is asymptotically uncorrelated with $\hat{\psi}$ itself. See Barndorff-Nielsen and Cox (1994, Ch. 7) and Ghosh et al. (2010, Sec. 4) for more details and more references.

The p^* formula (Barndorff-Nielsen's formula) provides a good approximation to the conditional distribution of the MLE $\hat{\boldsymbol{\psi}}$, also when given an exact or local ancillary statistic $\boldsymbol{a}$. A strict and general proof, that keeps full control of the magnitude of the approximation errors, is technically complicated, see Skovgaard (1990), Barndorff-Nielsen and Cox (1994, Ch. 7), or Jensen (1995, Ch. 4.2). We give here a somewhat heuristic proof in the spirit of Skovgaard (2001).

Proposition 7.5 ***Conditional saddlepoint approximation for the MLE***
The conditional saddlepoint approximation, given an ancillary statistic $\boldsymbol{a}$,

for the density $f(\hat{\psi}\,|\,\boldsymbol{a};\psi_0)$ of the MLE $\hat{\psi}$ in a curved exponential family is

$$f(\hat{\psi}\,|\,\boldsymbol{a};\psi_0) \approx \sqrt{\frac{\det J(\hat{\psi};\,\hat{\psi},\,\boldsymbol{a})}{(2\pi)^{\dim\psi}}}\,\frac{L(\psi_0)}{L(\hat{\psi})}. \tag{7.22}$$

Here $J(\hat{\psi};\,\hat{\psi},\,\boldsymbol{a})$ is the observed information with data $(\hat{\psi},\,\boldsymbol{a})$, calculated in the parameter point $\psi = \hat{\psi}$.

Proof In the proof, the ancillary statistic $\boldsymbol{a}$ will be assumed to supplement $\hat{\psi}$ such that $(\hat{\psi},\,\boldsymbol{a})$ represents the minimal sufficient statistic, that is a one-to-one function of $\boldsymbol{t}$. We first write $f(\hat{\psi}\,|\,\boldsymbol{a};\psi_0) = f(\hat{\psi},\boldsymbol{a};\psi_0)/f(\boldsymbol{a})$. Note that the density for $\boldsymbol{a}$ does not depend on the parameter ψ. This is utilized by considering the same relation for a different ψ-value and writing

$$f(\hat{\psi}\,|\,\boldsymbol{a};\psi_0) = \frac{f(\hat{\psi},\boldsymbol{a};\psi_0)}{f(\hat{\psi},\boldsymbol{a};\psi)}\,f(\hat{\psi}\,|\,\boldsymbol{a};\psi).$$

Note that the density ratio on the right is identical with the corresponding likelihood ratio $L(\psi_0)/L(\psi)$, since $(\hat{\psi},\,\boldsymbol{a})$ is minimal sufficient. As in previous derivations of the saddlepoint approximation we choose the arbitrary ψ to be $\hat{\psi}$. This implies that the last factor is the density for the argument $\hat{\psi}$ in the centre of its Gaussian approximation, where this approximation is good. Hence we can approximate the last factor by the usual product of a power of $1/\sqrt{2\pi}$ and the determinant of the inverse conditional variance-covariance matrix for $\hat{\psi}$. This inverse matrix can be replaced by the corresponding information matrix, as before. Note however, that we are in the conditional distribution given $\boldsymbol{a}$, so it is important that we do not use the expected information I, but a conditional information, that reflects the dependence on $\boldsymbol{a}$. The best and most convenient choice is the observed information J, calculated in the parameter point $\hat{\psi}$, but in principle an alternative could be the conditional expected information, given $\boldsymbol{a}$. □

Example 7.14 ***Two Poisson variates, continued***

As a simple illustration of Proposition 7.5, it is here applied to the curved, first version of the two Poisson variates, in Example 7.11. That is, y_1 and y_2 are mutually independent and $\text{Po}(c\,\psi)$ and $\text{Po}(c\,(1-\psi))$, respectively, with $a = y_1 + y_2$ as ancillary statistic to be conditioned on. It is easily checked that $\hat{\psi} = y_1/a$, or $y_1 = a\,\hat{\psi}$ and $y_2 = a\,(1-\hat{\psi})$, so the two Poisson densities yield the likelihood ratio

$$\frac{L(\psi_0)}{L(\hat{\psi})} = \left(\frac{\psi_0}{\hat{\psi}}\right)^{a\hat{\psi}} \left(\frac{1-\psi_0}{1-\hat{\psi}}\right)^{a(1-\hat{\psi})}$$

The observed information is

$$J(\psi) = y_1/\psi^2 + y_2/(1-\psi)^2$$

where y_1 and y_2 may be expressed in terms of $\hat{\psi}$ and a. In the parameter point $\psi = \hat{\psi}$ this expression simplifies to

$$J(\hat{\psi};\, \hat{\psi}, a)) = \frac{a}{\hat{\psi}\,(1-\hat{\psi})}.$$

This expression can also be obtained as the conditional mean value $E(J\,|\,a)$ in the point $\hat{\psi}$. Combination of the likelihood ratio and the information yields the approximation formula

$$f(\hat{\psi}\,|\,a;\,\psi_0) \approx \frac{1}{\sqrt{2\pi}\,\sqrt{a\,\hat{\psi}\,(1-\hat{\psi}}} \left(\frac{\psi_0}{\hat{\psi}}\right)^{a\hat{\psi}} \left(\frac{1-\psi_0}{1-\hat{\psi}}\right)^{a(1-\hat{\psi})}. \tag{7.23}$$

How should this formula be understood? In this case the exact conditional distribution of $y_1 = a\hat{\psi}$, given a, is exactly known, being Bin(a, ψ). Formula (7.23) is simply the result of a Stirling's formula approximation (Section B.2.2) of the binomial coefficient in the exact density. △

Remark 7.6 ***Locally ancillary and S-ancillary statistics.*** Proposition 7.5 holds approximately also if $\boldsymbol{a}$ is an (in a suitable sense) approximately ancillary statistic, of course. Asymptotically, this is a locally ancillary statistic. The proposition holds also in a full exponential family, with $\boldsymbol{t} = (\boldsymbol{u}, \boldsymbol{v})$, if $\boldsymbol{a}$ is replaced by the component $\boldsymbol{u}$ for conditional inference about $\boldsymbol{\theta}_v$. The proof works in precisely the same way. The statistic $\boldsymbol{u}$ is S-ancillary with a distribution depending only on the parameter $\boldsymbol{\mu}_u$ (note: $\boldsymbol{\theta}_v$ and $\boldsymbol{\mu}_u$ are variation independent), and in the proof it makes no difference if the value of $\boldsymbol{\mu}_u$ is fixed and known or fixed and unknown. △

Exercise 7.7 ***Bivariate normal with mean on a circle***
The following curved model is discussed by Fisher (1956, Sec. 5.8) and by Efron (1978). Let $\boldsymbol{y} = (y_1, y_2)$ be a single observation from $\mathrm{N}_2(\boldsymbol{\mu}, \mathrm{I}_{2\times 2})$, with independent components and $\boldsymbol{\mu}$ on the circle $|\boldsymbol{\mu}| = \rho$, with ρ known, Choose polar coordinates ρ and ψ for $\boldsymbol{\mu}$, so the likelihood is

$$L(\psi; \boldsymbol{y}) = (2\pi)^{-1} e^{-|\boldsymbol{y}-\boldsymbol{\mu}(\psi)|^2/2}$$

(a) Derive the likelihood equation and the observed information for ψ. Show that to each observed $\boldsymbol{\mu} \neq \mathbf{0}$ the likelihood equation has two roots, one maximum and one minimum.
(b) Note that by symmetry arguments the distribution of $r = |\boldsymbol{y}|$ must be

free from the parameter ψ. In other words, r is ancillary for ψ and the inference should be made conditional on the value of r. Therefore it is enough to know the observed information in (a), and its expected value is not needed.
(c) Derive the conditional distribution of z given r. Make use of the von Mises distribution, Example 2.14.
(d) Derive the statistical curvature of the family and show that it is everywhere $\gamma = 1/\rho$, so the circle radius ρ is also the radius of curvature.
(e) Extend the results in (a)–(d) from a single bivariate observation to a sample of n such observations. △

Exercise 7.8 ***Why is Basu's theorem not applicable?***
In Example 7.12, what is wrong in the following argument: 'Since the (ancillary) statistic (7.20) has a distribution that is free of the parameters, it is independent of the minimal sufficient statistic $(\overline{y}, s^2)$'? △

Exercise 7.9 ***Conditional versus unconditional inference***
In the curved (2, 1) exponential family in Example 7.11, compare conditional and unconditional inference (ML estimation and test of the hypothesis $\text{logit}(\psi) = 0$, for example). Also, make a diagram with the ψ-curve in $\boldsymbol{\Theta}$ and the corresponding mean value curve in the observation space, and check geometrically for some imagined observation the orthogonality property mentioned in Section 7.2. △

8

Extension to Incomplete Data

Sometimes it is not possible to observe a variate $\boldsymbol{x}$ that should have a distribution of exponential type, but only some function of $\boldsymbol{x}$, $\boldsymbol{y} = \boldsymbol{y}(\boldsymbol{x})$, is available for observation. The distribution for $\boldsymbol{y}$ is then only rarely an exponential family, but the relationship with $\boldsymbol{x}$ can be theoretically and computationally exploited. We say that $\boldsymbol{y}$ are *incomplete data* from the exponential family for $\boldsymbol{x}$. Here are first some examples, mostly taken from Sundberg (1974a).

8.1 Examples

Example 8.1 ***Grouped or censored continuous type data***
If the actual outcome of a continuous type variate x is not registered when x falls in a certain interval $I \subset \mathbb{R}$, but only its presence in I (a categorical indicator variate), the observed data may be characterized as incomplete data from the distribution for x. Examples range from fully grouped data, when the whole real line $\mathbb{R}$ is divided in intervals, via partially grouped data to censored (type I) data, when outcomes exceeding a specified value c are only registered as exceeding c, so $I = \{c, \infty\}$.

Censored data of type I typically appear in technical applications, when a set of new items are exposed until they fail, and their life-times are registered, provided they fail before a predetermined time c. At time c, the whole experiment is stopped, even if some items have not yet failed.

An alternative view of censored data, as following a curved family within a family of higher dimension, was introduced in Example 7.6, cf. also Example 8.2.

Truncation is different, and does not bring us outside the class of exponential families. If data are truncated for $x > c$, say, such outcomes leave no trace of having existed. This corresponds to setting $h(x) = 0$ for $x > c$, with a corresponding adjustment of $C(\theta)$. However, the parameter space may be affected, see for example Section 14.3. △

Example 8.2 ***Aggregated multinomial data***
Discrete data may also be grouped, exemplified by aggregated multinomials. The multinomial distribution with all probabilities unknown is a simple exponential family, and so are also many more specified multinomial models, with fewer parameters. One illustration is Example 7.8, where the underlying situation is that dice have been shaved such that two of their six faces (might) have a different probability than the other four faces. Genotype distributions in genetics provide many examples, and in Section 8.6 we will discuss in detail the structure of blood group genotype data.

Now suppose only aggregated such data are observable, obtained by merging cells of the multinomial in some way. Phenotype data instead of genotype data, when all different genotypes cannot be distinguished, is such an example. The shaved dice is another, in which the underlying multinomial for two dice, with $6 \times 6 = 36$ possible outcomes, is aggregated by observing only the sum of the two dice, with 11 possible outcomes. For aggregated data the simplicity is often lost, and we get a more or less complicated curved exponential family for this multinomial, to be seen in Example 7.8. But an alternative to be considered now is to regard such aggregated data as incomplete data from the original, simple multinomial. This is usually preferable to the curved model. In the shaved dice example, Sundberg (2018) shows that an analysis as incomplete data is not only simpler, in some respects, but also provides more insight, in particular elucidating the information content in ancillary statistics to be conditioned on, and opens up for use of the EM algorithm (Section 8.3). △

Example 8.3 ***The folded normal distribution***
Suppose that instead of having a sample of x-values from $\mathrm{N}(\mu, \sigma^2)$, we observe only the modulus $y = |x|$. The resulting distribution for y is obtained by folding the distribution for x around $x = 0$, hence the name 'folded normal distribution'. The distribution has been used in applications. Leone et al. (1961) give several examples from industrial practice where the model could be reasonable, for example for measurements of flatness of lenslike objects placed convex side down on a flat surface. Another application is when we have a sample of paired data, but have lost the identity of the items within pairs (Hinkley, 1973). Nevertheless we might want to make inference about the size of the systematic difference from the pairwise differences, in particular to test the hypothesis $\mu = 0$.

A notable feature of the model is that the sign of μ is unidentifiable from the incomplete y-data. It is natural to imagine the parameter space for (μ, σ^2) folded along the line $\mu = 0$ and regard $|\mu|$ as parameter instead of

μ. Otherwise expressed, we replace the original parameters by the equivalence classes of nondistinguishable parameters. A peculiar effect of this folding is that parameter points with $\mu = 0$, often being of central applied interest, are no longer in the centre of the parameter space, but instead are found on the boundary of the folded parameter space. For inferential consequences, see Example 8.9. △

Example 8.4 ***Mixture of two normal distributions***
Many anatomic characteristics in animal poulations may be taken to be normally or log-normally distributed, with parameters being different for males and females, or depending on species or age. For example, in some bird species, male and female adults appear indistinguishable (to humans), but simple measurements of bill length and/or wing length may indicate the sex. If males and females each have a normal distribution, our data $\boldsymbol{y}$ will be from a mixture of these normal distributions, with some unknown mixing proportion. A univariate model for such a mixed sample will have four or five parameters: A binomial mixing parameter π_0, population mean values μ_1 and μ_2, and one common or two separate population variances, σ^2, or σ_1^2 and σ_2^2. If the measured variate clearly depends on sex, its mixed distribution will have two modes, whereas if the variate is an inefficient separator, the distribution will have only one mode, flattened in comparison with the normal. Suppose we also had observed the sex. Then we would simply divide up the mixture sample into two Gaussian samples, a simple exponential family model. Hence we could regard the actually observed data as incomplete data from an exponential family, $x = \{y, z\}$, where z is an unobserved sex indicator variable (0–1 type), and y is the measured characteristic.

Mixture models is one of the standard tools when dealing with classification problems. When the model parameters have been estimated, an individual can be classified, with more or less precision, by calculating and comparing the conditional expected class probabilities given the observed values for that individual.

There is, however, also an identifiability problem, analogous to that for the folded normal distribution (Example 8.3). From mixture data we might be able to distinguish the two components of a mixture, but without further (external) information we cannot say which label (e.g. male/female) should be attached to one or the other of the components. This type of information is lost in the mixture. △

Example 8.5 ***Random data with measurement errors***
Suppose exponentially distributed life-times x are registered with additive

$N(\mu, \sigma^2)$ measurement errors z, independent of the x-values, so the observed data are represented by $y = x + z$. Life-times with nonnegligible measurement errors, could be an adequate description of data from physical instruments registering very short life-times. Since x and z separately follow distributions of exponential type, the joint distribution of (x, z) is of exponential type. However, the convolution of the two distributions is not of exponential type, so the actually observed variate y, that is a function of (x, z), is not modelled by an exponential family. △

Example 8.6 ***Missing data in multivariate analysis***
When components of multivariate data vectors are missing, it can sometimes be realistic to regard the actually observed parts of data as incomplete data corresponding to the observation of a function of the complete data. However, the question why data are missing is important, and the missingness may contain information about the missing value itself, or more generally affect the interpretation of some model parameters. If we trust that data are *missing completely at random* (MCAR), we can safely regard missing data from a multivariate normal distribution as incomplete data in the sense we use the term here. For different types of missingness, and the consequences for inference, see Little and Rubin (2002). △

Example 8.7 ***The negative binomial distribution***
The negative binomial distribution with two free parameters is sometimes used as a distribution to model count data showing Poisson overdispersion, that is, excess of variance as compared with a Poisson distribution. An example of such data are accident counts for a heterogeneous group of car drivers. The negative binomial probability for observing a count $y = 0, 1, 2, \ldots$ is

$$f(y;\, \psi, \kappa) = \frac{\Gamma(\kappa + y)}{y!\,\Gamma(\kappa)}\, \psi^y\, (1 - \psi)^\kappa\,.$$

For fixed κ this was discussed in Exercise 3.20 as a one-parameter exponential family, only with slightly different notations ($\psi = 1 - \pi_0$). With both ψ and κ as parameters, however, it is easy to see that this is not a two-parameter exponential family distribution. However, it may be regarded as the distribution for incomplete data from an exponential model in several different ways.

One way is to assume that y_i is conditionally Poisson distributed, $\mathrm{Po}(\lambda_i)$, given a set of individual latent *frailty* or *risk proneness* type parameters λ_i, which are assumed to be iid gamma distributed. The complete data would be the y_i-values together with their λ_i-values. The observed data y are from a gamma mixture of Poisson distributions.

A different interpretation of the negative binomial is obtained if z is taken as Poisson and y is taken as the sample sum of z logarithmically distributed iid variates (see Exercise 2.3). In this setting, the complete data are $x = (y, z)$, and they form an exponential family since the logarithmic distribution is a linear family, that is, with $t(y) = y$.

The negative binomial can be taken as the basis for a generalized linear model with dispersion parameter, see Example 9.6. △

Example 8.8 ***The Cauchy distribution family***
The Cauchy distribution family, with density

$$f(y;\, \lambda, \mu) = \frac{1}{\pi} \frac{\lambda}{\lambda^2 + (y - \mu)^2},$$

can be characterized as incomplete data from an exponential family in at least two ways. One is as the distribution of the ratio of two variates having a two-parameter bivariate normal distribution, with means zero and equal variances. Another is as a mixture of normal $\mathrm{N}(\mu, \sigma^2)$ distributions, with a fixed μ but random σ^2, generated by sampling from a $\lambda^2 \chi^2(1)$ for $1/\sigma^2$. △

Exercise 8.1 ***Folded binomial distribution***
The following model, proposed by Urbakh (1967), might be called a folded binomial. Pairs of different but morphologically indistinguishable types of chromosomes were studied. Within each pair, m labels were randomly allocated as if by Bernoulli trials, with the unknown probability depending on chromosome type. For each pair, the number of labels were counted in the two indistinguishable chromosomes, say y and $m-y$, $y \leq m/2$. Characterize this as a situation with incomplete data from an exponential family. △

Exercise 8.2 ***Wrapped normal distribution***
The wrapped normal is a distribution on the circle (unit circumference, say), obtained by winding a normal density around the circle in consecutive rounds. The model has some applications in statistics, see e.g. Mardia and Jupp (2000). Characterize this as a model with incomplete data from an exponential family. △

8.2 Basic Properties

We first derive a useful expression for the probability density for $\boldsymbol{y} = \boldsymbol{y}(\boldsymbol{x})$, that reflects the exponential type origin of the complete data $\boldsymbol{x}$.

The marginal probability or density for a certain outcome $\boldsymbol{y}$ is obtained by summing or integrating the distribution for the complete data $\boldsymbol{x}$ over the region $\boldsymbol{y}(\boldsymbol{x}) = \boldsymbol{y}$. Let us, without restriction, assume that $\boldsymbol{\Theta}$ includes $\boldsymbol{\theta} = \boldsymbol{0}$, so $h(\boldsymbol{x}) = h_x(\boldsymbol{x})$ represents a probability distribution. We factorize $h_x(\boldsymbol{x}) = h(\boldsymbol{x} \mid \boldsymbol{y})\, h_y(\boldsymbol{y})$. Then we can express the marginal density for $\boldsymbol{y}$ as

$$f(y;\,\theta) = \int_{y(x)=y} \frac{1}{C(\boldsymbol{\theta})}\, e^{\theta^T t(x)}\, h_x(\boldsymbol{x})\, \mathrm{d}\boldsymbol{x}$$

$$= \frac{1}{C(\boldsymbol{\theta})} \left(\int_{y(x)=y} e^{\theta^T t(x)}\, h(\boldsymbol{x} \mid \boldsymbol{y})\, \mathrm{d}\boldsymbol{x} \right) h_y(\boldsymbol{y}) = \frac{C_y(\boldsymbol{\theta})}{C(\boldsymbol{\theta})}\, h_y(\boldsymbol{y}),$$

where

$$C_y(\boldsymbol{\theta}) = \int_{y(x)=y} e^{\theta^T t(x)}\, h(\boldsymbol{x} \mid \boldsymbol{y})\, \mathrm{d}\boldsymbol{x}.$$

Like $C(\boldsymbol{\theta})$, $C_y(\boldsymbol{\theta})$ can be regarded as a norming constant in an exponential family, but now for the conditional distribution of $\boldsymbol{x}$, given $\boldsymbol{y}$, which is the motivation for the notation C_y. Only the basic probability measure differs from that of the original family, being $h(\boldsymbol{x} \mid \boldsymbol{y})\, \mathrm{d}\boldsymbol{x}$ for a given $\boldsymbol{y}$ instead of $h(\boldsymbol{x})\, \mathrm{d}\boldsymbol{x}$. Since the canonical statistic is the same, we can immediately conclude from the general exponential family theory that the following basic properties hold:

$$D \log C_y(\boldsymbol{\theta}) = E_\theta(\boldsymbol{t} \mid \boldsymbol{y}) = \mu_{t|y}(\boldsymbol{\theta}), \tag{8.1}$$

$$D^2 \log C_y(\boldsymbol{\theta}) = \mathrm{var}_\theta(\boldsymbol{t} \mid \boldsymbol{y}) = V_{t|y}(\boldsymbol{\theta}). \tag{8.2}$$

The corresponding factorization of the complete data likelihood via conditioning on $\boldsymbol{y}$ yields

$$\log L(\boldsymbol{\theta}\,;\, \boldsymbol{y}) = \log L(\boldsymbol{\theta}\,;\, \boldsymbol{x}) - \log L(\boldsymbol{\theta}\,;\, \boldsymbol{x} \mid \boldsymbol{y}). \tag{8.3}$$

As we have just seen, both terms on the right correspond to exponential families with the same canonical statistic $\boldsymbol{t}$, which cancels, and (8.3) simplifies to

$$\log L(\boldsymbol{\theta}\,;\, \boldsymbol{y}) = \log C_y(\boldsymbol{\theta}) - \log C(\boldsymbol{\theta}).$$

From this expression we first get the Fisher score function

$$U(\boldsymbol{\theta}\,;\, \boldsymbol{y}) = D \log L(\boldsymbol{\theta}\,;\, \boldsymbol{y}) = E_\theta(\boldsymbol{t} \mid \boldsymbol{y}) - E_\theta(\boldsymbol{t}). \tag{8.4}$$

Compare with the complete data score function! The two functions differ only in that (8.4) has the conditional expected value of the canonical statistic $\boldsymbol{t}$, given the observed data $\boldsymbol{y}$, instead of the unobservable $\boldsymbol{t} = \boldsymbol{t}(\boldsymbol{x})$ itself. Unfortunately, this implies a number of complications. First, some work is required to calculate $E_\theta(\boldsymbol{t}\,|\,\boldsymbol{y})$, and to solve the likelihood equations $E_\theta(\boldsymbol{t}\,|\,\boldsymbol{y}) - E_\theta(\boldsymbol{t}) = \mathbf{0}$. Next, the likelihood equations will often have multiple roots, and there need not even be a finite global maximum but only local maxima.

From the second derivatives of $\log C$ and $\log C_y$ we obtain the observed information for $\boldsymbol{\theta}$ as

$$J(\boldsymbol{\theta}\,;\,\boldsymbol{y}) = -D^2 \log L(\boldsymbol{\theta}\,;\,\boldsymbol{y}) = \mathrm{var}_\theta(\boldsymbol{t}) - \mathrm{var}_\theta(\boldsymbol{t}\,|\,\boldsymbol{y}).$$

The expected Fisher information $I(\boldsymbol{\theta})$ is obtained as its expected value.

In the following proposition we summarize what we have found:

Proposition 8.1 ***Likelihood equations and information matrices***
When $\boldsymbol{y}$ *are incomplete data from an exponential family with canonical statistic* $\boldsymbol{t}$*, the likelihood equations can be written*

$$E_\theta(\boldsymbol{t}\,|\,\boldsymbol{y}) - E_\theta(\boldsymbol{t}) = \mathbf{0}, \tag{8.5}$$

and the observed and expected information matrices for the corresponding canonical parameter $\boldsymbol{\theta}$ *are given by*

$$J(\boldsymbol{\theta}\,;\,\boldsymbol{y}) = \mathrm{var}_\theta(\boldsymbol{t}) - \mathrm{var}_\theta(\boldsymbol{t}\,|\,\boldsymbol{y}), \tag{8.6}$$

$$I(\boldsymbol{\theta}) = \mathrm{var}_\theta(\boldsymbol{t}) - E_\theta(\mathrm{var}_\theta(\boldsymbol{t}\,|\,\boldsymbol{y})) = \mathrm{var}_\theta(E_\theta(\boldsymbol{t}\,|\,\boldsymbol{y})). \tag{8.7}$$

Consequently, to have some information about all components of the parameter vector, we need a positive definite variance matrix for the conditional expected value of $\boldsymbol{t}$, given our data $\boldsymbol{y}$. This is in accordance with intuition: If the expected value of some component of $\boldsymbol{t}$, given data $\boldsymbol{y}$, does not depend on $\boldsymbol{y}$, we will not be able to estimate the whole parameter vector $\boldsymbol{\theta}$, due to the one-to-one relationship between the latter and $\boldsymbol{\mu}_t(\boldsymbol{\theta})$.

By utilizing the underlying exponential family properties, the following large-sample property can be shown to hold, without any other further regularity condition imposed on the model, than that the Fisher information should be positive definite (denoted $I > 0$); a $\sqrt{n}$-consistency condition:

Proposition 8.2 ***Large sample properties of the MLE***
Suppose we have a sample of size n *from a distribution representing incomplete data from an exponential family, and that the* $\sqrt{n}$*-consistency condition* $I_1(\boldsymbol{\theta}_0) > 0$ *is satisfied in the true point* $\boldsymbol{\theta}_0$*, where* I_1 *is the information (8.7) for* $n = 1$*. Then, with probability increasing to 1 as* $n \to \infty$*, the likelihood equations will have a root* $\hat{\boldsymbol{\theta}}$ *in any (fixed and open) neighbourhood of* $\boldsymbol{\theta}_0$*. In a small enough neighbourhood of* $\boldsymbol{\theta}_0$ *this root* $\hat{\boldsymbol{\theta}}$ *is unique with probability tending to 1. The estimator* $\hat{\boldsymbol{\theta}}$ *is asymptotically efficient and asymptotically normally distributed,* $\mathrm{N}(\boldsymbol{\theta}_0,\, I(\boldsymbol{\theta}_0)^{-1}) = \mathrm{N}(\boldsymbol{\theta}_0,\, I_1(\boldsymbol{\theta}_0)^{-1}/n)$.

Proof See Sundberg (1974a). □

Remark 8.3 ***The $\sqrt{n}$-consistency condition*** $I(\theta_0) > 0$. This condition implies $\sqrt{n}$-consistency, but is not necessary for consistency. Situations can occur in practice characterized by a slower convergence than $1/\sqrt{n}$, due to an extreme loss of information, typically in connection with unidentifiability phenomena. A relatively simple example is provided by the folded normal distribution (Example 8.3), when $\mu_0 = 0$, see Example 8.9. △

Example 8.9 ***The folded normal distribution, continued***
Here we observe, as in Example 8.3, a sample of values of $y = |x|$, where x is distributed as $N(\mu, \sigma^2)$. The sign of μ is unidentifiable, so we assume $\mu \geq 0$. In this case it is very difficult to distinguish different positive μ-values close to zero, in the sense that very large samples will be required. As soon as the true value μ is nonzero, the Fisher information is positive and the MLE of $|\mu|$ is unique and $\sqrt{n}$-consistent. However, if the true value is $\mu = 0$, the Fisher information fails to be positive definite. Nevertheless, the MLE of $|\mu|$ is still consistent, even though the $\sqrt{n}$-consistency criterion cannot be satisfied. More precisely, if σ is known, the MLE goes towards $\mu = 0$ at the slow rate $n^{-1/4}$ instead of $1/\sqrt{n}$, and for estimated σ the rate is even slower, $n^{-1/8}$ (Sundberg, 1974b). The reason for the problem is connected with the fact that the allowed value $\mu = 0$ forms the boundary of the folded parameter space corresponding to the folded data. △

In a normal mixture, Example 8.4, an analogous situation appears when the two normal components are identical or close to identical. The corresponding lack of identifiability was discussed by the end of that example. This has consequences not only for parameter estimation, but also for hypothesis testing, see Section 8.4.

8.3 The EM Algorithm

Only exceptionally the likelihood equations for an incomplete data model will have an explicit solution. The EM algorithm is an iterative solution algorithm that uses the underlying model for the complete data, and which is therefore natural to discuss here. It has a particularly simple form when the complete data model is of exponential type, as assumed here.

The likelihood equations $\boldsymbol{\mu}_t(\boldsymbol{\theta}) = \boldsymbol{t}$ for complete data $\boldsymbol{x}$ has the unique solution $\hat{\boldsymbol{\theta}}(\boldsymbol{t}) = \boldsymbol{\mu}_t^{-1}(\boldsymbol{t})$. Frequently, $\boldsymbol{\mu}_t^{-1}$ is explicit. With incomplete data $\boldsymbol{y} = \boldsymbol{y}(\boldsymbol{x})$, $\boldsymbol{t}$ is replaced in the score function by its expected value $E_\theta(\boldsymbol{t}|\boldsymbol{y})$, see (8.4), and the MLE must satisfy the equation system

$$\hat{\boldsymbol{\theta}} = \boldsymbol{\mu}_t^{-1}(E_{\hat{\theta}}(\boldsymbol{t} \mid \boldsymbol{y})). \tag{8.8}$$

This suggests the following alternating iterative algorithm, after the selection of a starting value $\boldsymbol{\theta}^{(0)}$:

(1) Calculate $E_{\boldsymbol{\theta}^{(0)}}(\boldsymbol{t} \mid \boldsymbol{y})$.
(2) Insert $E_{\boldsymbol{\theta}^{(0)}}(\boldsymbol{t} \mid \boldsymbol{y})$ on the right-hand side of (8.8), as a proxy for $\boldsymbol{t} = \boldsymbol{t}(\boldsymbol{x})$, thus obtaining a new $\boldsymbol{\theta} = \boldsymbol{\theta}^{(1)}$ by application of $\boldsymbol{\mu}_t^{-1}(.)$.

Repeat the two steps until convergence. In the terminology of the EM-algorithm the first step is the E-step (*E*xpectation step), and the second is called the M-step (*M*aximization step), representing the complete data likelihood maximization by $\boldsymbol{\mu}_t^{-1}$.

Note that the two steps jointly can be described by $\boldsymbol{\theta}^{(k+1)} = \boldsymbol{g}(\boldsymbol{\theta}^{(k)})$ for a specific vector-valued function $\boldsymbol{g}$, given as the right-hand side of (8.8). A root $\hat{\boldsymbol{\theta}} = \hat{\boldsymbol{\theta}}(\boldsymbol{y})$ is characterized by $\boldsymbol{g}(\hat{\boldsymbol{\theta}}) = \hat{\boldsymbol{\theta}}$. Differently expressed, $\hat{\boldsymbol{\theta}}$ is a *fixed point* of the function $\boldsymbol{g}$.

To understand heuristically the convergence properties of the algorithm, we consider the one-parameter case. Whether the algorithm converges or not, with a starting point close to $\hat{\theta}$, depends on the slope g' of g in $\hat{\theta}$. For $\theta^{(0)}$ close to $\hat{\theta}$, the deviation $\theta^{(0)} - \hat{\theta}$ will change by the factor $g'(\hat{\theta})$ (Draw a diagram of some $g(\theta)$ versus θ, crossing the 45° line $g(\theta) = \theta$ at some $\theta = \hat{\theta}$, and try the algorithm!) If $|g'(\hat{\theta})| < 1$, the algorithm will converge, and the smaller it is, the faster is the convergence. On the other hand, if $|g'(\hat{\theta})| > 1$, it will diverge. The convergence is called geometric, because after the initial phase the remaining error tends to decrease geometrically at each step, by the factor $g'(\hat{\theta})$. If $|g'(\hat{\theta})| = 1$, very slow convergence is possible but depends on higher-order derivatives.

What is then g' in this case, with its particular composite function g? The inner function has derivative $\mathrm{var}_\theta(t \mid y)$ and the outer function μ_t^{-1} has derivative $1/\mathrm{var}_\theta(t)$. Hence, the composite function has derivative

$$\frac{\mathrm{var}_\theta(t \mid y)}{\mathrm{var}_\theta(t)} = 1 - \frac{\mathrm{var}_\theta(t) - \mathrm{var}_\theta(t \mid y)}{\mathrm{var}_\theta(t)} = 1 - \frac{J(\theta; y)}{J(\theta; t)}, \tag{8.9}$$

where $J(\theta; y)$ and $J(\theta; t) = I(\theta; t)$ are the observed information quantities with incomplete and complete data, respectively, see (8.6). Clearly the expression is nonnegative, being a ratio of two variances. Also, in the MLE point $\hat{\theta}$, with $J(\hat{\theta}; y) \geq 0$, (8.9) is ≤ 1, and typically < 1. Hence, the algorithm can be generally expected to converge, and the convergence rate depends on the relative amount of information about t retained in y. Note also that the left-hand side of (8.9) does not depend on the chosen parameterization, so the convergence rate is invariant under reparameterization.

For planning purposes we may of course replace the numerator of (8.9) by its expected value, to obtain $1 - I_y(\theta)/I_x(\theta)$, where the Fisher informations are for incomplete and complete data, respectively, and in which we insert a provisional value for θ.

The multiparameter version of the previous argument can be expressed in terms of eigenvalues, as follows (Sundberg, 1972, 1976):

Proposition 8.4 ***Local convergence of the EM algorithm***
EM convergence is determined by the eigenvalues of the matrix

$$\mathrm{var}_{\hat{\theta}}(\boldsymbol{t})^{-1}\,\mathrm{var}_{\hat{\theta}}(\boldsymbol{t}|\boldsymbol{y}) = I - J(\boldsymbol{\theta};\, \boldsymbol{t})^{-1} J(\boldsymbol{\theta};\, \boldsymbol{y}). \tag{8.10}$$

If $\hat{\boldsymbol{\theta}}$ is a local maximum of the likelihood function, then all eigenvalues of (8.10) are nonnegative and ≤ 1, and they represent the estimated relative loss of information about $\boldsymbol{t}$ in different directions. The geometric rate of convergence of the algorithm, after some initial steps, is given by the largest of these eigenvalues. The eigenvalues are invariant under reparameterization of the model, hence, so is also the rate of convergence.

Suppose the maximal eigenvalue of (8.10) in $\hat{\boldsymbol{\theta}}$ is < 1, so there is not complete loss of information about any parameter component. Proposition 8.4 then tells that the algorithm is bound to converge geometrically to $\hat{\boldsymbol{\theta}}$, provided that it starts close enough. Local maxima $\hat{\boldsymbol{\theta}}$ are *points of attraction* of the EM algorithm. Local minima and saddlepoints of $\log L$ also yield roots to the likelihood equations, but they are not points of attraction, because such roots would have a maximal eigenvalue > 1. EM will not converge to such points.

The theory presented was expressed in terms of finding roots of the likelihood equations, but alternatively it can be formulated in terms of likelihood maximization. The complete data log-likelihood is $\log L(\boldsymbol{\theta}; \boldsymbol{t}) = \boldsymbol{\theta}^T \boldsymbol{t} - \log C(\boldsymbol{\theta})$. Its conditional mean value for $\boldsymbol{\theta} = \boldsymbol{\theta}^{(0)}$ is

$$E_{\theta^{(0)}}\{\log L(\boldsymbol{\theta};\, \boldsymbol{t}) \,|\, \boldsymbol{y}\} = \boldsymbol{\theta}^T E_{\theta^{(0)}}(\boldsymbol{t} \,|\, \boldsymbol{y}) - \log C(\boldsymbol{\theta}). \tag{8.11}$$

Maximization of this expression with respect to $\boldsymbol{\theta}$ for fixed $\boldsymbol{\theta}^{(0)}$ is equivalent to the M-step as given, and the calculation of the conditional mean value is equivalent to the previous E-step. Reader, check yourself, by differentiation, that when $\boldsymbol{\theta}^{(0)} = \hat{\boldsymbol{\theta}}$, $\boldsymbol{\theta} = \hat{\boldsymbol{\theta}}$ will maximize (8.11)!

The fact that locally the EM algorithm will converge to a (nearby) log-likelihood maximum, and only to a maximum, is strongly connected with a global property of the EM algorithm, that the likelihood is bound to increase in each step. This result is formulated and proved here:

Proposition 8.5 ***Globally, each EM step will increase the likelihood***
Whatever starting value $\boldsymbol{\theta}^{(0)}$*, the next value* $\boldsymbol{\theta}^{(1)}$ *after an EM step satisfies* $\log L(\boldsymbol{\theta}^{(1)}; \boldsymbol{y}) \geq \log L(\boldsymbol{\theta}^{(0)}; \boldsymbol{y})$.

Proof By (8.3) we can write

$$\log L(\boldsymbol{\theta}; \boldsymbol{y}) = \log L(\boldsymbol{\theta}; \boldsymbol{t}) - \log L(\boldsymbol{\theta}; \boldsymbol{t}|\boldsymbol{y})$$

Here $\boldsymbol{t}$ must cancel on the right-hand side, so we can choose any $\boldsymbol{t}$ there. Corresponding to the E-step we use the $\boldsymbol{t}$-value $\boldsymbol{t} = E_{\theta^{(0)}}(\boldsymbol{t}|\boldsymbol{y})$. Then the first term $\log L(\boldsymbol{\theta}; \boldsymbol{t})$ has the form (8.11) and will be maximized by $\boldsymbol{\theta}^{(1)}$, according to the M-step. On the other hand, the second term in the difference will be maximized by $\boldsymbol{\theta}^{(0)}$, since we are in the conditional family, given $\boldsymbol{y}$, with the observed $\tilde{\boldsymbol{t}} = E_{\theta^{(0)}}(\boldsymbol{t}|\boldsymbol{y})$ and the likelihood equation $\tilde{\boldsymbol{t}} = E_{\theta}(\boldsymbol{t}|\boldsymbol{y})$ for $\boldsymbol{\theta}$. Thus, comparing $\log L(\boldsymbol{\theta}^{(1)}; \boldsymbol{y})$ with $\log L(\boldsymbol{\theta}^{(0)}; \boldsymbol{y})$, the first term is larger and the second term, to be subtracted, is smaller. Thus, the inequality holds. □

The term EM was coined by Dempster et al. (1977), when they introduced the algorithm in a general framework of incomplete and complete data from models not restricted to be exponential families. The E-step is then the calculation of the conditional mean value of $\log L(\boldsymbol{\theta})$ under $\boldsymbol{\theta}^{(0)}$, as in (8.11), and the M-step is the maximization of the resulting conditional mean of $\log L$ with respect to $\boldsymbol{\theta}$. The convergence rate can be expressed in terms of the complete and incomplete data information matrices in the same way as in Proposition 8.4, and they also showed that the likelihood increases, Proposition 8.5. However, the general results are less explicit than the exponential family results.

Before the studies for exponential families by Sundberg (1972, 1976) and in EM generality by Dempster et al. (1977), the algorithm had been around in special cases for a long time, see Meng and van Dyk (1997) for history. In genetics it was used already in the 1950'es. For example, Ceppellini et al. (1955) used the idea in situations when different genotypes x correspond to the same observed phenotype y, such as in Section 8.6. Baum et al. (1970) developed the algorithm with applications to hidden Markov models in mind (Section 8.7).

There are alternatives to the EM algorithm. Possible alternative algorithms include the Newton–Raphson, the quasi-Newton, and the Fisher scoring algorithms (Section 3.4). When they work they are likely to be faster, often much faster than EM, but at least the Newton–Raphson and quasi-Newton methods will be more sensitive to the choice of starting value. In case of large parameter dimension, the matrix inversion required in the three related methods might be a problem, but on the other hand the dimension of the equation system can often be reduced by elimination of some equations and parameters before these methods are applied.

There are several other modifications of the EM algorithm. The M-step might require an iterative method by itself, but such iterations need not be full sequences, as exemplified by the ECM type of EM modifications introduced by Meng and Rubin (1993). Sometimes the EM not only needs but allows speeding up, see for example Meng and van Dyk (1997). One of many variants is the PX–EM method, see Section 8.8.

For all methods the difficult situations are when the likelihood surface looks like a ridge that is very flat along the ridge, and such situations tend to appear frequently when data are interpreted as incomplete data. Note however that the direction of slowest EM convergence is not along the ridge of $L(\boldsymbol{\theta}; \boldsymbol{y})$, but in the only indirectly related direction of most information loss relative to the complete data (compare Figure 8.1 with Figure 8.3).

Remark 8.6 ***Choice of complete data model.*** The complete data model for $\boldsymbol{x}$ can be a purely hypothetical construct, and there might be several complete data models possible, with different forms of incompleteness but resulting in the same model for $\boldsymbol{y}$. Examples are when negative binomial data (Example 8.7) or Cauchy data (Example 8.8) are regarded as incomplete data. The EM algorithm may converge faster for one of the representations, but other aspects are the difficulty in deriving a formula for $E(\boldsymbol{t} \mid \boldsymbol{y})$ in the E-step and in solving the maximization problem in the M-step.

When the complete data are multinomially distributed, the Poisson trick from Section 5.6 can be utilized. This means the multinomial is replaced by a set of mutually independent Poisson variates, by regarding the total as a Poisson outcome. This can simplify the theoretical calculations of $\mathrm{var}(\boldsymbol{t})$ and $\mathrm{var}(\boldsymbol{t} \mid \boldsymbol{y})$, and thereby the EM convergence rate to be expected. A possible drawback is that the dimension is increased, by 1. However, this will not affect the convergence itself, because the added component will correspond to a zero eigenvalue, and the Poisson parameter for the sample total will converge to the actual sample size already in the first iteration. △

Exercise 8.3 ***Rate of convergence and sample size***

Suppose the complete data form a sample $\{\boldsymbol{x}_i\}$ of size n, and the incomplete data are $\boldsymbol{y}_i$, related by $\boldsymbol{y}_i = \boldsymbol{y}(\boldsymbol{x}_i)$, as e.g. for mixture data as in Example 8.4.

Show that the expected rate of convergence does not depend on the sample size (only on the degree of information loss in the function $y(\boldsymbol{x})$). △

Exercise 8.4 ***Observed components of t***
Suppose some components of $\boldsymbol{t}$ are observed, thus forming part of the observed data $\boldsymbol{y}$. Show that (8.10) then has the corresponding eigenvalues zero, and draw conclusions about the behaviour of the EM algorithm. △

8.4 Large-Sample Tests

As in Chapter 5, consider a model-reducing hypothesis. In Section 5.4 it was shown that the so-called exact test has a number of locally asymptotically ($n \to \infty$) equivalent large sample tests, including the likelihood ratio test, the score test, and Wald's test. For the exponential families of Chapter 5 it was sufficient that the family was regular (or at least that the parameter $\boldsymbol{\theta}_0$ under H_0 was not on the boundary of $\boldsymbol{\Theta}$). The mutual local asymptotic equivalence between these large sample tests, and their approximate χ^2 distribution, hold much more generally than for exponential families, as can be shown under a set of regularity conditions, see Section 5.5. With incomplete data from an exponential family we are in an intermediate situation. All regularity conditions needed for the general result, except one, are automatically satisfied through the underlying exponential family. The necessary condition remaining is that the information matrix be positive definite in θ_0. Equivalently expressed, this is the $\sqrt{n}$-consistency condition for the MLE introduced in connection with Proposition 8.2. For precise formulations and proofs of test properties, see Sundberg (1972, 1974a).

Remember that $I(\boldsymbol{\theta}) = \mathrm{var}_\theta(E_\theta(\boldsymbol{t} \,|\, \boldsymbol{y}))$, so the condition is a demand on data $\boldsymbol{y}$ to yield substantial information on all components of $\boldsymbol{t}$. A simple nonstandard example when the condition breaks down is testing H_0: $\mu = 0$ in the folded normal distribution, see Example 8.9, and Sundberg (1974b) for more details. Another is the identifiability problem for a mixture of normals, see Example 8.4, which makes it nonstandard and complicated to test if the number of mixture components can be reduced.

8.5 Incomplete Data from Curved Families

It is not difficult to extend the previous results on the likelihood equations and the EM algorithm to curved exponential families. For example, this will be needed in Section 8.8. Suppose $\boldsymbol{\theta} = \boldsymbol{\theta}(\boldsymbol{\psi})$ for a lower-dimensional

ψ. The likelihood for ψ is of course only the restriction of the likelihood for θ, but expressed in terms of ψ. that is $L(\theta(\psi))$. Thus, the score function can be written

$$U_\psi(\psi) = \left(\frac{\partial\theta}{\partial\psi}\right)^T U_\theta(\theta(\psi)) = \left(\frac{\partial\theta}{\partial\psi}\right)^T \left\{E_{\theta(\psi)}(t \mid y) - \mu_t(\theta(\psi))\right\}, \qquad (8.12)$$

where $\left(\frac{\partial\theta}{\partial\psi}\right)$ is the Jacobian matrix of the function $\theta(\psi)$), compare with (7.7) for complete data. Its variance is the Fisher information,

$$I_\psi(\psi) = \left(\frac{\partial\theta}{\partial\psi}\right)^T \operatorname{var}_{\theta(\psi)}\left(E_{\theta(\psi)}(t \mid y)\right) \left(\frac{\partial\theta}{\partial\psi}\right).$$

Since also for curved families, the only difference between the score functions for incomplete data and for complete data is that t has been replaced by $E(t \mid y)$, the EM algorithm works in the same way as for full families:

E-step Calculate $E_{\theta(\psi)}(t \mid y)$ for the current ψ-value.
M-step Solve the complete data likelihood equations after replacing t by $E_{\theta(\psi)}(t \mid y)$ from the E-step, This yields a new ψ for the new E-step.

For a preliminary calculation of what EM rate of convergence to expect, we are free to use the local character of this property and approximate the Jacobian in the anticipated point $\theta(\psi)$. This corresponds to approximating the curved model by a (local) linear reduction of the given full model to a model with a smaller-dimensional canonical statistic.

8.6 Blood Groups under Hardy–Weinberg Equilibrium

Human blood groups can be related to a gene existing in three different forms (alleles), denoted *A*, *B* and 0. Since each individual has two genes, (one per chromosome), there are 6 different combinations: *AA*, *A*0, *AB*, 00, *B*0 and *BB*. A complication is that the alleles *A* and *B* are dominant (codominant) over the recessive allele 0, which implies that the gene combinations *AA* and *A*0 are indistinguishable. Likewise, *BB* and *B*0 are indistinguishable. In genetic terminology the two *genotypes* *AA* and *A*0 are represented by the same observed *phenotype*, and likewise for *BB* and *B*0. If genotypes are the complete data, phenotypes are incomplete data.

We assume a random sample of size n has been taken from some large population of individuals (so no need to distinguish sampling with or without replacement). The population is assumed to be in Hardy–Weinberg equilibrium, see further Section 14.1. This implies that the sampling of individuals is probabilistically equivalent to independent sampling of the two genes per individual from the total population of genes (with the alleles in proportions p, q and r, respectively, $p+q+r = 1$). Thus, complete genotype data for the 6 blood group genotypes under Hardy–Weinberg equilibrium follow the special multinomial distribution shown in Table 8.1a. This reduces to a simple trinomial family (order 2) for the total number of genes of type A, B and 0 (sum $= 2n$), with respective probabilities p, q and r (sum $p + q + r = 1$), see Exercise 8.5. We are free, however, to extend the trinomial family to have order 3 by using the Poisson trick, assuming n to be a Poisson outcome, see Remark 8.6 in Section 8.3.

As mentioned, the multinomial probabilities for the 6 blood group genotypes are given in Table 8.1a, with one cell for each of the 6 genotypes. The neighbouring Table 8.1b shows how phenotype counts y_i, ($\sum y_i = n$) will be denoted. Phenotype data will of course also be multinomially distributed, but no longer following a full exponential family, and we cannot expect an explicit MLE. *Note:* An alternative interpretation of phenotype data is as a curved family; to see this is left as part of Exercise 8.5.

As an illustration some fictive data with $n = 100$ will be used, shown in Table 8.1c, and perfectly corresponding to $p = 0.5$, $q = 0.3$, $r = 0.2$. However, remember that the sample size has no effect on the EM algorithm per se, so the fictive data may as well be regarded as fictive percentages.

	A	0	B
A	p^2	$2pr$	$2pq$
0		r^2	$2qr$
B			q^2

Table 8.1a.
Genotype probabilities

	A	0	B
A	y_1		y_3
0		y_4	
B			y_2

Table 8.1b.
Phenotype counts

	A	0	B
A	45		30
0		4	
B			21

Table 8.1c.
Fictive data

Figures 8.1 and 8.2 show the behaviour of the EM algorithm for these data, with three different starting points, and in two different magnifications. It is clearly seen how all three sequences rapidly approach a certain almost linear curve through the MLE. Along this curve (line) the deviation from the MLE appears to be reduced by a factor 1/2 for each iteration. The eigenvalues of the matrix $\text{var}(\boldsymbol{t})^{-1}\text{var}(\boldsymbol{t}|\boldsymbol{y})$ are in fact 0.09 and 0.51, which perfectly explains the behaviour seen (Proposition 8.4). Figure 8.3 shows the log-likelihood surface, for comparisons. Note that the line of relatively slow convergence appears unrelated to the local form of the likelihood surface.

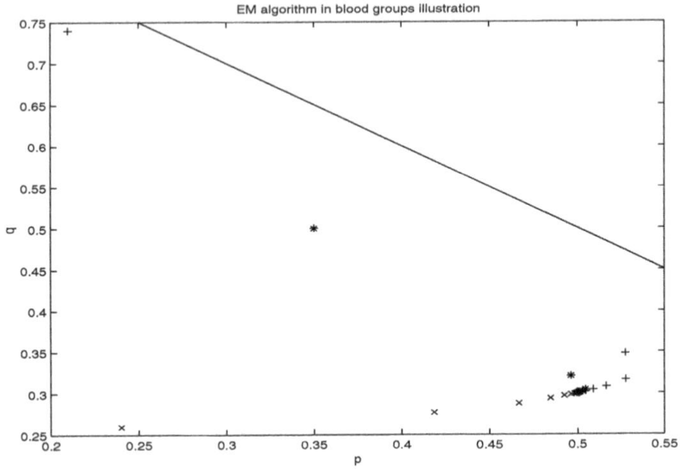

Figure 8.1 EM algorithm illustrated for three different starting points, indicated by $\times$, $*$, $+$. The solid line is part of the parameter space boundary. The MLE is $(\hat{p}, \hat{q}) = (0.5, 0.3)$.

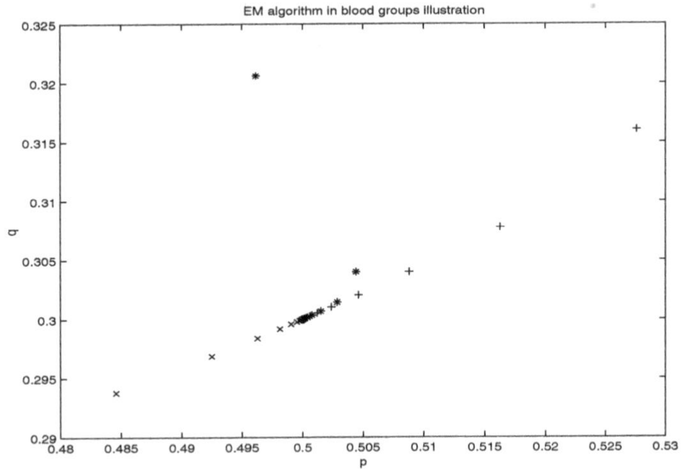

Figure 8.2 EM algorithm illustrated for three different starting points. Shown in a closer vicinity of the MLE than in Figure 8.1.

Exercise 8.5 ***Geno- and phenotype data under H–W equilibrium***

(a) Show that the genotype data multinomial model of Table 8.1a is a full exponential family of order 2, and characterize its canonical statistic.

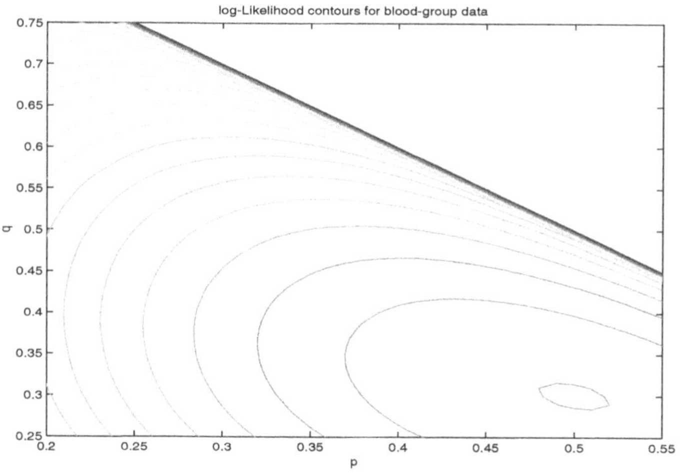

Figure 8.3 Log-likelihood surface contours for the blood groups example, corresponding to Figures 8.1 and 8.2. The MLE is $(\hat{p}, \hat{q}) = (0.5, 0.3)$.

(b) Show that the phenotype data multinomial model may alternatively be seen as a $(3, 2)$ curved exponential model. △

8.7 Hidden Markov Models

In Example 8.4, the concept of mixtures was discussed as an important example of incomplete data. In Hidden Markov Models (HMMs), observations are from a mixture of distributions when, additionally, the components of the mixture are represented by the unobserved states of an underlying random process. HMMs are popular in a wide range of applications, including speech recognition, digital communication, ion channel modelling, DNA sequence alignment, and other computational biology. See Zucchini et al. (2016) for an introduction with additional examples and extensions, and efficient use of **R** packages, and see Cappé et al. (2005) for a rigid treatment of the statistical theory of HMMs. HMMs also lead on to other dynamic models, with underlying, latent random processes, see Section 13.2.

The basic HMM consists of a latent part and an observed part. The latent (hidden) part is a finite state homogeneous Markov chain $\{x_t, t = 1, 2, \dots, T\}$ (t for time), fully specified by its transition matrix, cf. Example 2.13. The observed part is a matching sequence $\{y_t\}$, where the distribution of y_t is state dependent and, given $x_t = x$, belongs to a state dependent parametric family $f(y; \theta_x)$.

The number K of hidden states will be assumed known, but in reality K must also be assessed, of course. The architecture of the HMM specifies K and the transitions possible between the K states. To each possible transition, from state i to state j, corresponds a transition probability, a parameter α_{ij}. Hence, the full model parameter vector consists of all unknown transition probabilities and the set of all K characteristics θ_x. There can also be an unknown initial distribution for the Markov chain, in particular if the chain is recurrent, so every state can be reached from every other state. In other HMMs the chain is transient, moving in only one direction.

Note that x_t and the bivariate (x_t, y_t) form Markov chains, whereas the observable y_t typically does not. A schematic picture of the structure is shown in Figure 8.4.

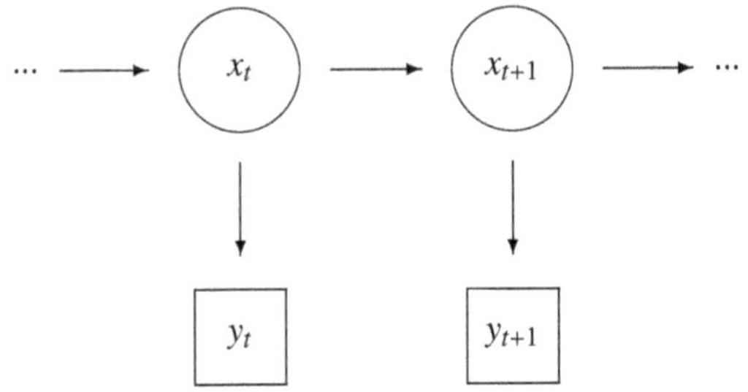

Figure 8.4 Schematic picture of HMMs, showing an excerpt of the latent Markov chain $\{x_t\}$ and the corresponding observed $\{y_t\}$, depending only on x_t for the same t.

Here is a relatively simple example. Consider a recurrent chain that alternates between the two states $x = 0$ and $x = 1$, and where the observed output y is normally distributed, $N(\mu_0, \sigma_0^2)$ or $N(\mu_1, \sigma_1^2)$ depending on x. This could be a very simplistic model for an ion channel that is open or closed, with Y_t representing a measured electrical current. The observed data model is a mixture of two Gaussian distributions, but not the standard one discussed in Example 8.4, but with membership dependent between successive observations. The parameter vector now also includes two transition probabilities α_{01} and α_{10} for the hidden Markov chain. A two-state example with discrete output distribution is obtained if y is instead Poisson distributed. Zucchini et al. (2016, Ch. 15) illustrate its use for modelling long individual series of daily seizure counts for epileptic patients.

Statistical inference for HMMs can be divided in different task types. One is estimation etc. of the model parameters, another is the reconstruction of the hidden chain $\{x_t\}$. Other aspects are goodness of fit and model selection. We comment here on estimation of the model parameters.

ML estimation is not a simple task, because the likelihood is a complicated function. Frequently it will have multiple local maxima (and then of course also other extremal points). For iterative procedures, the choice of starting values can be both difficult and crucial. The EM algorithm has a long history in HMMs, even before the name EM was coined, and it is still much used for HMM likelihood maximization (Baum–Welch procedure), but there are competitors more specific to HMM models (Zucchini et al., 2016, Ch. 3–4). The choice of K-value from data is generally difficult – a crucial question being if the latent structure can be motivated and interpreted, or if it just provides a flexible descriptive fit to data.

8.8 Gaussian Factor Analysis Models

A factor analysis (FA) model is a multivariate model in which the observed vector $\boldsymbol{y}$ is regarded as more or less determined by a smaller number of *latent* (unobservable) random variables or *factors*, $\boldsymbol{z}$ say. For example, individual results on a set of mental tests may be interpreted in terms of one or several individual intelligence factors (logic, creativity, ...). In the most classic FA model, the p-dimensional Gaussian vector $\boldsymbol{y}$ is modelled as

$$\boldsymbol{y} = \boldsymbol{\mu} + \Lambda\, \boldsymbol{z} + \boldsymbol{\epsilon}, \qquad (8.13)$$

where $\boldsymbol{\mu}$ is the mean value vector, $\boldsymbol{z}$ is a q-dimensional latent Gaussian vector of mean zero, $q < p$ (typically $q \ll p$), Λ is a $p{\times}q$ coefficient matrix (*loadings matrix* – to be determined), and $\boldsymbol{\epsilon}$ is a Gaussian pure noise vector, $\mathrm{N}_p(\mathbf{0}, \Psi)$, where Ψ is an unknown diagonal matrix. That Ψ is diagonal means that the latent $\boldsymbol{z}$ is intended to explain all correlation among the components of the response vector $\boldsymbol{y}$. We will soon below regard $\boldsymbol{z}$ as an unobserved part of complete data $\boldsymbol{x} = (\boldsymbol{y}, \boldsymbol{z})$.

With a sample of observations $\boldsymbol{y}_i$, each will have its own values $\boldsymbol{z}_i$ and $\boldsymbol{\epsilon}_i$. Without restriction, $\boldsymbol{z}_i$ may be assumed distributed as $\mathrm{N}_q(\mathbf{0}, \mathrm{I}_{q\times q})$, which implies that the observed data is a sample from a $\mathrm{N}_p(\boldsymbol{\mu}, \Sigma)$, with

$$\Sigma = \Lambda\Lambda^T + \Psi.$$

The sample mean $\bar{\boldsymbol{y}}$ and the sample covariance matrix S of $\boldsymbol{y}$ are jointly minimal sufficient statistics for the model parameters $(\boldsymbol{\mu}, \Lambda, \Psi)$, and the usual method to estimate them nowadays is by maximum likelihood. This is not quite trivial, because the parameter dimension is often high, and the model is a curved family (provided $\dim(z) = q$ is small enough, so the model is identifiable). When $\dim(z) > 1$ there is an additional lack of uniqueness due to the possibility of rotating z. In ML estimation this is most conveniently handled by prescribing that $\Lambda^T\Psi^{-1}\Lambda$ should be diagonal.

There are several computer packages for factor analysis. ML estimation is typically carried out by an algorithm iteratively alternating between ML estimation of Λ for given Ψ and of Ψ for given Λ. The standard procedure, deriving from Jöreskog in the 1960s, is to solve an eigenvalue–eigenvector equation to obtain Λ in the former step, and to maximize the log-likelihood numerically with respect to Ψ in the latter step. That the latter step is preferable in this form is not obvious, because the final solution should satisfy the simple relation $\widehat{\Psi} = \mathrm{diag}\{S - \widehat{\Lambda\Lambda^T}\}$, where diag represents diagonalization, replacing all elements outside the diagonal by zeros. However, when this formula for $\widehat{\Psi}$ was used during the iteration process, it was found unstable, sometimes yielding negative variances, and it was abandoned. For more details, see for example Bartholomew et al. (2011, Ch. 3).

An alternative idea was proposed by Rubin and Thayer (1982), namely to compute the MLE by the EM algorithm, regarding $\boldsymbol{x} = (\boldsymbol{y}, z)$ as complete data and the observed $\boldsymbol{y}$ as incomplete. The distribution of the complete $(\boldsymbol{y}, z)$ data can be seen as a standard normal for z (i.e. fully specified), combined with a conditional multivariate linear regression of $\boldsymbol{y}$ on z. The latter has regression coefficients and a residual variance matrix as parameters. The restriction $\mathrm{var}(z) = \mathrm{I}_{q\times q}$ on the distribution of the regressor z is not a linear hypothesis in the canonical parameters, however. Thus the complete data follow a curved family, but this is not a matter of principle for the EM algorithm, see Section 8.5.

We first consider the M-step, given complete $(\boldsymbol{y}, z)$ data. We can utilize that the distribution of z is free of parameters, being a standard normal, that could and should be conditioned on. The resulting conditional factor is the regression of $\boldsymbol{y}$ on z. This is an ordinary multivariate Gaussian regression, see first part of Example 7.4, with parameters $(\boldsymbol{\mu}, \Lambda, \Psi)$, and ML estimation in this model is an almost trivial extension of univariate regression (component-wise least squares and Ψ estimated from residual sums of squares and products). Thus the M-step is simple and explicit.

The E-step requires formulas for $E(z_i \,|\, \boldsymbol{y}_i)$ and $E(z_i z_i^T \,|\, \boldsymbol{y}_i)$. They follow from the joint normality of $(\boldsymbol{y}, z)$ by theoretical regression of z on $\boldsymbol{y}$. For simplicity of writing we assume $\boldsymbol{y}$-data here already centred (eliminating $\hat{\mu} = \overline{\boldsymbol{y}}$):

$$E(z_i \mid \boldsymbol{y}_i) = \Lambda^T \Sigma^{-1} \boldsymbol{y}_i \,, \tag{8.14}$$

$$\begin{aligned} E(z_i z_i^T \mid \boldsymbol{y}_i) &= \mathrm{var}(z_i \mid \boldsymbol{y}_i) + E(z_i \mid \boldsymbol{y}_i) E(z_i \mid \boldsymbol{y}_i)^T \\ &= \mathrm{I}_{q\times q} - \Lambda^T \Sigma^{-1} \Lambda + \Lambda^T \Sigma^{-1} (\boldsymbol{y}_i \boldsymbol{y}_i^T) \Sigma^{-1} \Lambda \,. \end{aligned} \tag{8.15}$$

Remembering from Example 2.10 that the conditional variance is obtained by inverting the joint variance matrix, selecting the part in question and inverting it back again, the variance term of (8.15) is obtained by using (B.3) of Proposition B.1. This term can alternatively and simpler be expressed in terms of the diagonal matrix $\Lambda^T \Psi^{-1} \Lambda$, as follows (Exercise 8.6):

$$\mathrm{I}_{q\times q} - \Lambda^T \Sigma^{-1} \Lambda = \left(\mathrm{I}_{q\times q} + \Lambda^T \Psi^{-1} \Lambda \right)^{-1} , \tag{8.16}$$

In the FA model (8.13) for $\boldsymbol{y}$ we assumed $\mathrm{var}(z) = \mathrm{I}_{q\times q}$, for parameter identifiability. Note that we can easily transform from a z with $\mathrm{var}(z) = A$, say, to another z with $\mathrm{var}(z) = \mathrm{I}_{q\times q}$, by scaling z by $A^{-1/2}$ and Λ by $A^{1/2}$. Now imagining that we had complete data, we could actually estimate both the regression coefficients matrix Λ and $A = \mathrm{var}(z)$, the latter from the sample covariance matrix of z. If the complete data model is extended from curved to full model by expanding the parameter set, allowing arbitrary $\mathrm{var}(z)$, the typically quite slow EM algorithm is actually speeded up considerably. This idea was introduced under the name PX-EM by Liu et al. (1998). The FA model is one of their examples of the usefulness of the idea.

The computations needed in the EM algorithm yield as a by-product one of the variants proposed for prediction of the factor scores z_i, namely the so-called *regression scores* (another variant, the Bartlett scores, is obtained by generalized least squares). The regression scores predictor is simply $E(z_i \mid \boldsymbol{y}_i)$ as given by formula (8.14), but with the ML estimates inserted for Λ and $\Sigma = \Lambda\Lambda^T + \Psi$. A somewhat less efficient but more robust version uses the sample variance-covariance matrix S instead of the MLE of Σ.

In this section we assumed data were Gaussian. In Section 12.6 we look at latent variable models and methods when data are binary.

Exercise 8.6 ***Matrix identity in factor analysis***
Prove the matrix identity (8.16) by verifying directly that if $\Sigma = \Lambda\Lambda^T + \Psi$, then $\mathrm{I}_{q\times q} + \Lambda^T \Psi^{-1} \Lambda$ is the inverse of $\mathrm{I}_{q\times q} - \Lambda^T \Sigma^{-1} \Lambda$ and vice versa. *Note:* The identity is a special case of the so-called *binomial inverse theorem*. △

9

Generalized Linear Models

The theoretical framework called *Generalized linear models* extracts the common features of *Linear models* and *Log-linear models*, but includes a wide variety of other models as well. The common theme is a linear exponential family, a linear structure ('linear predictor') for some function of the mean value, and the so-called link function that connects the linear structure with the exponential family via the mean value. The most important extension from the basic generalized linear model goes via *exponential dispersion models*, with a more liberal variance structure.

The generalized linear model framework and terminology was designed in the early 1970s by Nelder and Wedderburn. A classic text is by McCullagh and Nelder (1989, 2nd Ed.), but there are numerous other books and book chapters. Statistical packages typically have a rich toolbox of such methods. Fahrmeir and Tutz (2001) deal with multivariate generalized linear models, and will be referred to in Section 13.2.1.

9.1 Basic Examples and Basic Definition

We here briefly remind the reader about the presumably well-known structures of linear models, log-linear models for count data, and logistic regression, in order to point out the structure elements that will be brought into the unified theory of generalized linear models. In parallel with logistic regression (logit-based) we will look at probit-based regression models, in order to go outside the main subclasses.

Example 9.1 ***Gaussian linear models, continued from Example 2.9***
Consider again the standard linear model with normally distributed error term, saying that our response values y_i are $\mathrm{N}(\mu_i, \sigma^2)$ and mutually independent, $i = 1, \ldots, n$. The mean values μ_i are expressed linearly by a design matrix $A = \{a_{ij}\}$ or a set of regressors $X = \{x_{ij}\}$, and a smaller number of parameters, $\mu_i = \sum_j a_{ij}\beta_j$ or $= \sum_j x_{ij}\beta_j$. In vector notations: $\boldsymbol{\mu} = X\boldsymbol{\beta}$.

The model has a specified distribution type for the randomness in data (the normal distribution), a mean value structure, and a variance–covariance structure. The mean value structure is parametrically linear; the model expresses the mean values linearly in the smaller number of β-parameters. This linear function will be called the *linear predictor*, a term that is particularly natural in a regression context. The simple variance structure states that the observations have a common variance σ^2, and if it is unknown, it is in the present context called the *dispersion parameter*.

Here the mean value vector itself was modelled as a linear function of $\boldsymbol{\beta}$. Among possible alternatives we could imagine some other function of the mean value, $g(\mu)$, expressed linearly. If for example the *link function* g is the log function, the mean value structure is multiplicative and not additive, but the distribution type still normal, and in particular not log-normal.

A more common modification of the standard linear model is a change of the variance structure. For example, we might want to have a variance that is proportional to μ_i^2, $\mathrm{var}(y_i) = \sigma^2 v_y(\mu_i) = \sigma^2\mu_i^2$. The function $v_y(\mu)$ will be called the *variance function*. △

Example 9.2 ***Multiplicative Poisson and log-linear models for contingency tables, continued from Example 2.5***

In this example, observed numbers y_i are mutually independent Poisson distributed, $y_i \sim \mathrm{Po}(\mu_i)$, where μ_i is the mean value. The distribution type is now Poisson. Index i may represent the cells of a cross-tabulation.

A log-linear model, for example a multiplicative Poisson model, is characterized by some linear structure (linear predictor) for the log of the mean value, $\log \mu_i = \sum_j a_{ij}\beta_j$ or $= \sum_j x_{ij}\beta_j$, When $\log \mu$ is additive, μ itself has a multiplicative structure, and vice versa. We say that the model has *log-link*, when as here the linear predictor refers to the log of the mean value.

The variance is not constant, as in the standard linear model, but it is a known function of the Poisson mean value, $\mathrm{var}(y_i) = \mu_i$, so it does not require a dispersion parameter. However, a possibly desirable extension of the model could be to a model that allows under- or (more likely) over-dispersion relative to the Poisson distribution.

Log-linear models for multinomial data, for example in the form of contingency tables, can be treated within the same Poisson framework by using the Poisson trick of Section 5.6, that is, assuming the total to be Poisson distributed and the multinomial to be conditional on this total. In this way, the Poisson-based models can be used to cover the corresponding multinomial models. Extensions of this idea are found in survival analysis (Section 9.7.1), and in analysis of spatial point processes (Section 13.1.1). In the special case of the binomial, however, it is not needed, see Example 9.3.

Suppose we have data $\{y_i\}$ modelled as $\text{Po}(N_i \mu_i)$, with known weights (e.g. population sizes) N_i and a log-linear model for μ_i. This includes for example the weighted multiplicative Poisson model of Exercises 2.1, 3.21 and 5.14. Such models are fitted by software for generalized linear models by treating $\log N_i$ as a regressor with regression coefficient β_0 fixed at $\beta_0 = 1$. An alternative is to use $z_i = y_i/N_i$ as data (which is no longer Poisson) in combination with a weighted variance for z_i, see Proposition 9.6. Another example is the density estimate refinement by tilting, Example 2.12 and Exercise 3.26, with the frequencies of a discretized version of the original density estimate f_0 corresponding to N_i (Efron and Tibshirani, 1996). △

Example 9.3 ***Probit and logit regression models for binary data***
In a dose–response experiment a sample of individuals is exposed to some substance in different doses. We take the response variable y here to be binary, expressing whether or not an individual reacts in some specific way to the dose given. As a basic model we assume that the individuals respond mutually independently, and that individual i, who has received dose x_i, reacts with a probability that depends on the dose, $y_i \sim \text{Bin}(1, \pi_i)$, where $\pi_i = \pi(\boldsymbol{x}_i) = E(y_i)$. For the function $\pi(x)$, different forms are possible. In a logistic regression model the logit of the mean value, $\text{logit}(\pi_i)$, is assumed linear in dose x_i. The probit regression model differs only in that a different function than the logit is assumed linear in x, namely the *probit function*, which is defined as the inverse of the standard normal distribution function, $\text{probit}(\pi_i) = \Phi^{-1}(\pi_i)$.

The probit has an older history than the logit, and has been much used in dose–response situations back in time, see also Example 7.3. Notice that both functions map the unit interval on the real line. For another use of the probit in modelling, see Section 12.6.

Like in the Poisson case, the variance for the binomial is a fully known function of the mean, so no dispersion parameter is needed. △

Altogether this means that in the last example we have a new distribution type, the binomial or Bernoulli, as compared with the previous examples. We also have a choice between two different link functions, the logit and the probit, linking the linear regression with the mean value, and we have a completely known variance function, given as soon as we specify the distribution type. But common to all three examples is a specification of a distribution type, a linear stucture/predictor, and a link function that connects the mean value with the linear predictor.

The distribution types of the three examples have one feature in common, that is not quite apparent, but is required for them to be examples of generalized linear models. In the Gaussian model, let the dispersion parameter have a known value, for simplicity. Then all three of them form linear one-parameter exponential families, the term *linear* meaning that y itself is the canonical statistic, $t(y) = y$. The generalized linear model is fully specified by additionally selecting a set of regressors (or a design matrix) that will form the linear predictor of some function of the mean value parameter, and by selecting this particular link function of the mean value. Hence, there are three steps in the modelling by generalized linear models, and the natural order is often the following:

- **Linear predictor:** What are the regressors or class variables, collected in a vector $\boldsymbol{x}$, which will be tried in order to explain (linearly) the systematic part of the variability in data? Form a function in these regressors, $\eta = \boldsymbol{x}^T\boldsymbol{\beta}$ (the linear predictor), $\dim \boldsymbol{\beta} = k < n$.
- **Distribution type:** For a fixed mean value μ of y, what is the distribution type that describes the randomness in y? Select the distribution type among linear one-parameter exponential families, for example the Gaussian (with fixed variance = dispersion parameter), Poisson, binomial, exponential, gamma (with fixed dispersion parameter), inverse Gaussian (with fixed dispersion parameter).
- **Link function:** What is the link function g that will tie the mean value μ to the linear predictor η, $\eta = g(\mu)$? One such function is a particularly natural choice, the canonical link, see next paragraph.

As a result of the last two steps, we have to each observation y_i, $i = 1, \dots, n$, a triple of parameters to consider: θ_i, μ_i, and η_i. All these parameters are equivalent representations of the distribution for y_i. That there is a one-to-one relationship between the canonical parameter θ_i and the mean value parameter μ_i was established in Proposition 3.11. The relationship between μ_i, and η_i is given by the link function $\eta_i = g(\mu_i)$, and we simply demand that the link function g should be a one-to-one function. By combination we can express the relationship between θ_i and η_i through $\eta = g(\mu) = g(\mu(\theta))$. The model becomes particularly simple if this function of θ is the identity function, $\eta = \theta$, because then the canonical $\theta_1, \dots, \theta_n$ are linear functions of the smaller-dimensional $\boldsymbol{\beta}$. This link function is called the *canonical link.*

In order to see the implications for the model, consider the saturated exponential family for the whole $\boldsymbol{y}$-vector, with a corresponding $\boldsymbol{\theta}$-vector having one component per observation. In vector notations the exponent of this model is $\boldsymbol{\theta}^T\boldsymbol{y} = \sum \theta_i y_i$. When $\boldsymbol{\theta} = \boldsymbol{\eta} = X\boldsymbol{\beta}$, for a design matrix X

of regressor values, the exponent can be rewritten as $\boldsymbol{\beta}^T(X^T\boldsymbol{y})$. This means that the resulting model is a $\dim\boldsymbol{\beta}$-dimensional exponential family, with $\boldsymbol{\beta}$ as a reduced canonical parameter vector and the corresponding sums of products vector $X^T\boldsymbol{y}$ as canonical statistic. This is the same canonical statistic that we have in an ordinary linear model multiple regression of $\boldsymbol{y}$ on the columns of X. For the Poisson model distribution, the canonical link is the log-link. For the Bernoulli distribution, θ is the logit of π, so the logit function is the canonical link in this case. This is the main reason that logistic regression almost totally replaced probit regression during the second half of the 20th century. If we do not use the canonical link, the model is a curved exponential family, with a minimal sufficient statistic of higher dimension than $\boldsymbol{\beta}$, see Chapter 7.

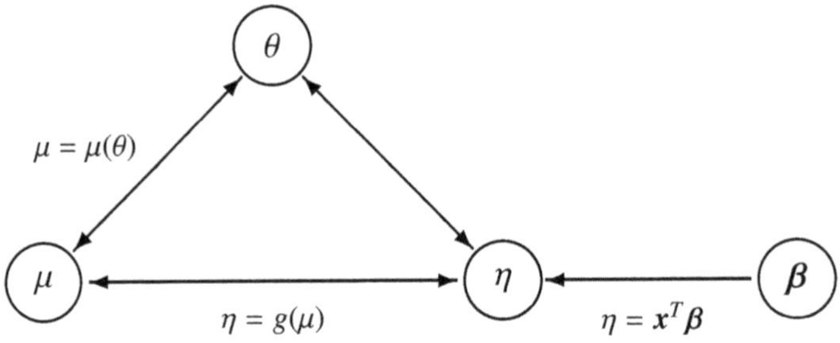

Figure 9.1 'Triangular diagram' for parameter relationships in generalized linear models. When link is canonical, $\eta \equiv \theta$

The definition of generalized linear models is far from complete without the incorporation of the dispersion parameter. In some models, for example the Gaussian linear model, it is natural also to allow a scale parameter in the model, and the model remains an exponential family, just with one additional parameter. Generally, however, the introduction of the dispersion parameter can make the model quite complicated, so we will first investigate more fully models without dispersion parameter. As we have just seen, this incorporates the important Poisson and Bernoulli (binomial) distribution families.

We assume there is a linear predictor in the models under consideration. If the relationship were intrinsically nonlinear, such that no link function to a linear predictor could be found, the theory would get more complicated and less explicit, of course. Such models are necessarily curved families. Wei (1998) provides some theory.

Table 9.1 *Common distributions used in generalized linear models.*

Distribution type	Canonical link function	Variance function	Dispersion parameter
Normal $N(\mu, \sigma^2)$	μ	constant, = 1	$\phi = \sigma^2$
Bernoulli $Bin(1, \pi_0)$	$\mathrm{logit}(\pi_0)$	$\pi_0\,(1 - \pi_0)$	constant, = 1
Poisson $Po(\mu)$	$\log(\mu)$	μ	constant, = 1
Exponential	$1/\mu$	μ^2	constant, = 1
Gamma (Example 9.5)	$1/\mu$	μ^2	$\phi = 1/(1+\beta)$

Exercise 9.1 ***Log-link for binary data***
For Bernoulli type data, the log-link is rarely used. What could be the reason? Consider in particular a case when $\boldsymbol{x}^T\boldsymbol{\beta} > 0$. △

Exercise 9.2 ***Generalized linear models for the exponential distribution***
Verify the exponential distribution line of Table 9.1. △

9.2 Likelihood Theory for Generalized Linear Models without Dispersion Parameter

Suppose we have independent scalar observations $y_1, \ldots, y_n$ from a linear exponential type distribution family, with density

$$f(y_i;\, \theta_i) = \frac{h(y_i)}{C(\theta_i)}\, e^{\theta_i\, y_i} \tag{9.1}$$

Equivalently, this family can be parameterized by its mean value μ_i for y_i. Because we have regression and similar situations in mind, the n observations have individual mean values, each being a specified function of a smaller-dimensional parameter vector $\boldsymbol{\beta}$, $\mu_i = \mu_i(\boldsymbol{\beta})$, and likewise $\theta_i = \theta_i(\boldsymbol{\beta})$. With a canonical link, θ_i is a linear function of $\boldsymbol{\beta}$, and we have a full exponential family for $\{y_i\}$ of dimension $\dim\boldsymbol{\beta}$. With a noncanonical link, θ_i is a nonlinear function of $\boldsymbol{\beta}$, and we have a curved exponential family, see Section 7.2 for the general form of such likelihood equations, etc.

The log-likelihood for $\boldsymbol{\beta}$ is obtained via the log-likelihood for $\boldsymbol{\theta}$ as

$$\log L(\boldsymbol{\beta};\, y_1, \ldots, y_n) = \sum_{i=1}^{n} \{\theta_i(\boldsymbol{\beta})\, y_i - \log C(\theta_i(\boldsymbol{\beta}))\} + \text{constant}, \tag{9.2}$$

where the function $\theta_i(\boldsymbol{\beta})$ is a composite function, via the successive relations of $\boldsymbol{\theta}$ with $\boldsymbol{\mu}$, $\boldsymbol{\mu}$ with $\boldsymbol{\eta}$, and $\boldsymbol{\eta}$ with $\boldsymbol{\beta}$. The score function is obtained by the chain rule, a special case of (3.15), cf. also (7.7):

$$U_j(\boldsymbol{\beta}) = \frac{\partial \log L}{\partial \beta_j} = \sum_{i=1}^{n} \{y_i - \mu_i(\boldsymbol{\beta})\} \left(\frac{\partial \theta_i}{\partial \beta_j} \right). \tag{9.3}$$

For simplicity, we first treat the case of canonical link, i.e. $\boldsymbol{\theta} \equiv \boldsymbol{\eta}$, and $\partial \theta_i / \partial \beta_j = x_{ij}$. Then the first part of the log-likelihood simplifies to

$$\sum_i \theta_i(\boldsymbol{\beta})\, y_i = \sum_i \eta_i(\boldsymbol{\beta})\, y_i = \sum_i \left(\sum_j x_{ij} \beta_j \right) y_i = \sum_j \beta_j \left(\sum_i x_{ij}\, y_i \right) = \sum_j \beta_j t_j\,, \tag{9.4}$$

with the β_j as canonical parameter components and $t_j = \sum_i x_{ij} y_i$ as canonical statistics, $j = 1, \ldots, k$ ($= \dim \boldsymbol{\beta}$). Hence the MLE, if it exists, is the necessarily unique root of the likelihood equation system $\boldsymbol{t} - E_\beta(\boldsymbol{t}) = \boldsymbol{0}$. Here $E_\beta(\boldsymbol{t})$ has the components $\sum_i x_{ij} \mu_i(\boldsymbol{\beta})$, so the equations can be written

$$\sum_i x_{ij} \{y_i - \mu_i(\boldsymbol{\beta})\}, \qquad j = 1, \ldots, k. \tag{9.5}$$

Typically this is a nonlinear equation system in $\boldsymbol{\beta}$, that must be solved by some iteration method. The exception that $\mu_t(\boldsymbol{\beta})$ is linear, implying a linear equation system, occurs when the canonical link function is the *identity link*, $\boldsymbol{\mu} \equiv \boldsymbol{\eta}$, as in the Gaussian linear model.

With a noncanonical link function, we have to use the chain rule repeatedly, to derive the score function for $\boldsymbol{\beta}$. The derivative $\partial \theta_i / \partial \beta_j$, that was simply x_{ij} under the canonical link, will now depend on $\boldsymbol{\beta}$, and this dependence must be expressed via $\boldsymbol{\mu}$ in the triangular diagram Figure 9.1:

$$\frac{\partial \theta_i}{\partial \beta_j} = \frac{\partial \theta_i}{\partial \mu_i} \frac{\partial \mu_i}{\partial \eta_i} \frac{\partial \eta_i}{\partial \beta_j}. \tag{9.6}$$

Here

$$\frac{\partial \theta_i}{\partial \mu_i} = 1 \Big/ \left(\frac{\partial \mu_i}{\partial \theta_i} \right) = \frac{1}{\mathrm{var}(y_i;\, \mu_i)}, \tag{9.7}$$

$$\frac{\partial \mu_i}{\partial \eta_i} = 1 \Big/ \left(\frac{\partial \eta_i}{\partial \mu_i} \right) = \frac{1}{g'(\mu_i)}, \tag{9.8}$$

and finally

$$\frac{\partial \eta_i}{\partial \beta_j} = x_{ij}\,. \tag{9.9}$$

The factors $\mathrm{var}(y_i)^{-1}$ and $g'(\mu_i)^{-1}$ typically both depend on the parameter

$\boldsymbol{\beta}$, via μ_i. Note that, except for the link function and the form of the predictor, we need only know how the variance is related to the mean value in the (one-parameter) distribution family, to be able to set up the likelihood equations. This has implications for the quasi-likelihood method (Section 9.5).

Summarizing in matrix terms, we have now derived the following result.

Proposition 9.1 ***Likelihood equations***
The likelihood equation system $U(\boldsymbol{\beta}) = \mathbf{0}$ *for a generalized linear model with canonical link function (full exponential family),* $\boldsymbol{\theta} \equiv \boldsymbol{\eta} = X\boldsymbol{\beta}$*, is*

$$X^T \{\mathbf{y} - \boldsymbol{\mu}(\boldsymbol{\beta})\} = \mathbf{0}. \tag{9.10}$$

For a model with noncanonical link (curved family), the corresponding equation system has the form

$$X^T G'(\boldsymbol{\mu}(\boldsymbol{\beta}))^{-1} V_y(\boldsymbol{\mu}(\boldsymbol{\beta}))^{-1} \{\mathbf{y} - \boldsymbol{\mu}(\boldsymbol{\beta})\} = \mathbf{0}, \tag{9.11}$$

where $G'(\boldsymbol{\mu})$ *and* $V_y(\boldsymbol{\mu})$ *are diagonal matrices with diagonal elements* $g'(\mu_i)$ *and* $v_y(\mu_i) = \mathrm{var}(y_i;\, \mu_i)$*, respectively.*

9.2.1 Numerical Methods for MLE Computation

The usual method for solving the likelihood equations for $\boldsymbol{\beta}$ in a generalized linear model is Fisher scoring method (Section 3.4). It has here the interpretation as a weighted least squares method, where the weights depend on the parameter values, and therefore must be updated iteratively. An alternative name in this context is therefore the Iterative Weighted (or reweighted) Least Squares method, IWLS.

For models with canonical link, the scoring method is identical with the Newton–Raphson method. This is obvious, since canonical link implies that the parameter vector $\boldsymbol{\beta}$ is canonical, and the expected and observed information matrices are identically the same for exponential families in canonical parameterization. The equivalence of Fisher scoring and Newton–Raphson implies both a lack of sensitivity to the choice of starting point and a quite fast (quadratic) convergence towards the MLE.

With a noncanonical link, Newton–Raphson and Fisher scoring (IWLS) are different methods. Most statistical packages allow a choice between them. However, note that even if the chosen method converges, this is not a guarantee that we have found the global maximum, since likelihood equations for curved families can have multiple roots.

9.2.2 Fisher Information and the MLE Covariance Matrix

To obtain the observed and expected information matrices for a general link function is essentially only an application of the curved family formulas (7.8) and (7.9). We will be somewhat more explicit here. The left-hand sides of (9.10) and (9.11) are the score vectors under canonical and noncanonical link, respectively. Differentiating by $\boldsymbol{\beta}$ once more, and if necessary taking the expected value, yields the following expressions for the observed and expected information matrices, see the proof that follows.

Proposition 9.2 ***Observed and expected information***
The observed and expected information matrices for a generalized linear model with canonical link function are identical and given by the formula

$$J(\boldsymbol{\beta}) = I(\boldsymbol{\beta}) = X^T V_y(\boldsymbol{\mu}(\boldsymbol{\beta}))\, X, \tag{9.12}$$

a weighted sum of squares and products matrix of the regressor $\boldsymbol{x}$.
With noncanonical link, the Fisher information matrix is given by

$$I(\boldsymbol{\beta}) = \left(\frac{\partial\theta}{\partial\beta}\right)^T V_y(\mu(\beta)) \left(\frac{\partial\theta}{\partial\beta}\right) = X^T G'(\boldsymbol{\mu}(\boldsymbol{\beta}))^{-1} V_y(\boldsymbol{\mu}(\boldsymbol{\beta}))^{-1} G'(\boldsymbol{\mu}(\boldsymbol{\beta}))^{-1} X \tag{9.13}$$

and, with some abuse of notation when $\dim \boldsymbol{\beta} > 1$*, see comment to formula (7.8), the observed information can be written*

$$J(\boldsymbol{\beta}) = I(\boldsymbol{\beta}) - \left(\frac{\partial^2 \boldsymbol{\theta}}{\partial \boldsymbol{\beta}^2}\right)^T \{\boldsymbol{y} - \boldsymbol{\mu}(\boldsymbol{\beta})\}. \tag{9.14}$$

Here G' *and* V_y *are the same diagonal matrices as in Proposition 9.1.*

Proof With canonical link, thus with $\boldsymbol{\beta}$ as canonical parameter, we have $J = I$, and the expression (9.12) is most easily and immediately obtained as var$(U(\boldsymbol{\beta})$ from the left-hand side of (9.10). With noncanonical link we instead get I in (9.13) as var$(U(\boldsymbol{\beta})$ from the left-hand side of (9.11). Note that one of the two factors V_y^{-1} from (9.11) cancels with the V_y from var$(\boldsymbol{y})$. The observed information J is obtained by differentiation of the score function, which now yields the additional, data-dependent term in (9.14) from differentiation of the factor $\left(\frac{\partial\theta}{\partial\beta}\right)$. Note that it has expected value zero, so the first term is the expected information I. The whole procedure is a special case of the derivation of (7.8). □

9.2.3 Deviance and Saturated Models

The trivial model obtained when letting each observation have its own mean value (i.e. $\hat{\mu}_i = y_i$) is called a *saturated model*. This model is essentially only of interest as a reference model. If a specific generalized linear model is fitted to data we can compare its likelihood maximum with that of the saturated model by their log likelihood ratio, or equivalently the *deviance*, relative to the saturated model. This deviance is also called the *residual deviance*, and in a Gaussian linear model (with $\sigma^2 = 1$), it is the residual sum of squares. A special case is the *null deviance*, when the null model (with a constant predictor) is compared with the saturated model. Here is the standard definition of deviance:

Definition 9.3 ***Deviance***
The *deviance* (or residual deviance) D for a generalized linear model is

$$D = D(y, \boldsymbol{\mu}(\hat{\boldsymbol{\beta}})) = 2\left\{\log L(y;\, \boldsymbol{y}) - \log L(\boldsymbol{\mu}(\hat{\boldsymbol{\beta}});\, \boldsymbol{y})\right\}, \tag{9.15}$$

where both log-likelihoods are written in terms of the mean value parameter vector $\boldsymbol{\mu}$, for the saturated model and the model in question, respectively.

As an example, for the binomial with data $\{n_i, y_i\}$, the deviance is

$$D = 2\sum_i \left\{ y_i \log\left(\frac{y_i}{n_i \pi_i(\hat{\beta})}\right) + (n_i - y_i)\log\left(\frac{n_i - y_i}{n_i - n_i \pi_i(\hat{\beta})}\right)\right\} \tag{9.16}$$

Note that each observation yields its contribution to the deviance, in the form of one term of the sum, $D = \sum D_i$.

The factor 2 in the deviance comes from the large sample result stating that for a model-reducing hypothesis, twice the log-ratio is approximately χ^2 distributed (Section 5.5). Often, however, when the larger of the two models is the saturated model, this distributional property should not be referred to, because the large sample approximation need not be valid for the saturated model involved. On the other hand, it can be used for the difference of two such deviances, because the contribution from the saturated model cancels. A deviance for testing a hypothesis model nested within another, wider model can always be expressed as the difference between the corresponding residual deviances. Even if none of two models compared is nested within the other (for example a logit versus a probit model), their residual deviances may be used to compare these models, but then without explicit reference to the χ^2 distribution.

Exercise 9.3 ***Poisson deviance***
Find an expression for the deviance of a generalized linear model with the Poisson distribution family, analogous to the binomial. △

9.2.4 Residuals

In ordinary Gaussian linear models, the vector of residuals, that may be written $\boldsymbol{y} - \hat{\boldsymbol{\mu}} = \boldsymbol{y} - \boldsymbol{\mu}(\hat{\boldsymbol{\beta}})$, is not only used for assessing and estimating the size of σ^2, but also for checking that the model is reasonable in its mean value structure and that the variance is constant, as it is assumed to be. For a generalized linear model based on the binomial or Poisson distribution, the corresponding residuals are not natural. The first complication to deal with is that var(y_i) depends on μ_i, as expressed by the variance function $v_y(\mu_i)$ (element of the diagonal matrix $V_y(\boldsymbol{\mu})$). Therefore the Pearson residuals for a generalized linear model are by definition normalized by $\sqrt{v_y(\mu_i)}$, that is, by $\sqrt{\pi_i(1-\pi_i)}$ for Bernoulli trials and by $\sqrt{\mu_i}$ for Poisson models.

An even better alternative might be to use the analogy with the Gaussian model deviance, to which the squared residuals all contribute additively. The analogue for the binomial or Poisson would then be the so-called *deviance residuals*, which quantifies how the residual deviance (9.15) is the sum of terms D_i, contributed from each observation. As an example, consider the binomial deviance (9.16), which is the sum (over i) of all individual squared deviance residuals. As the residual itself we take $\sqrt{D_i}$, with the same sign as $y_i - \hat{\mu}_i$ in the Pearson residual. When the ratio $y_i/(n_i\,\pi_i(\hat{\boldsymbol{\beta}}))$ is close enough to 1, so that a quadratic approximation of the log function is adequate, it can be shown (see Exercise 9.4) that such an approximation yields

$$D_i \approx \frac{(y_i - n_i\,\pi_i(\hat{\boldsymbol{\beta}}))^2}{n_i\,\pi_i(\hat{\boldsymbol{\beta}})\,(1-\pi_i(\hat{\boldsymbol{\beta}}))},$$

which is exactly the squared Pearson residual. The approximate equality of these deviance residuals has of course its analogue in the asymptotic equivalence of the deviance test and the score test (Section 5.5).

Exercise 9.4 ***Binomial residuals***
In the binomial case, check by expansion of the log function that the (squared) deviance residual is approximated by the (squared) Pearson residual. △

Exercise 9.5 ***Poisson residuals***
Derive the deviance residuals for the Poisson family. △

9.3 Generalized Linear Models with Dispersion Parameter

So far, the underlying distribution family was assumed to be a one-parameter linear exponential family. The normal distribution is an example showing the need to extend the theory to allow a scale parameter or a variance (=dispersion parameter) in the model. For later use we will also extend the results by allowing known 'weights', corresponding to an extension to the variance σ^2/w_i for the normal distribution, where the w_is are known numbers. We start by revisiting the normal distribution.

Example 9.4 ***Normal distribution with dispersion parameter***
Before we introduce a dispersion parameter, consider the density of the $\mathrm{N}(\mu, 1)$ family, which can be written

$$f(y;\, \mu, 1) = e^{\mu\, y\, -\, \mu^2/2}\, h(y), \qquad h(y) \propto e^{-y^2/2}. \tag{9.17}$$

Here $\mu = \theta$ is the canonical parameter, and $\mu^2/2$ is the $\log C(\theta)$ term, whose derivative with respect to $\theta = \mu$ is the expected value of y (Check!). When a variance parameter σ^2 is introduced, the density changes to

$$f(y;\, \mu, \sigma^2) = e^{\frac{\mu y - \mu^2/2}{\sigma^2}}\, h(y;\, \sigma^2), \qquad h(y;\, \sigma^2) \propto e^{-y^2/(2\sigma^2)}. \tag{9.18}$$

As we know, this is a two-parameter exponential family, but this property will not be utilized here. Instead we will be more general in that respect, for later reference. Note first that

(1) $E(y;\, \mu,\, \sigma^2) = \mu$ holds also in the model with a variance parameter σ^2;
(2) $\mathrm{var}(y;\, \mu,\, \sigma^2) = \sigma^2 = \sigma^2\, \mathrm{var}(y;\, \mu, 1)$

These relationships will be seen to hold in a wider generality after this example.

Next, for any specified value of $\sigma^2 > 0$, note that the score function and the Fisher information for μ are immediately obtained by differentiation of the exponent of (9.18):

(1) The score function for μ is $U(\mu) = \dfrac{y-\mu}{\sigma^2}$;
(2) $I(\mu) = 1/\sigma^2$.

Note also that as an equivalent alternative to differentiation of U, we may obtain $I(\mu)$ as $\mathrm{var}(U(\mu);\, \mu, \sigma^2) = \mathrm{var}(y;\, \mu, \sigma^2)/(\sigma^2)^2 = 1/\sigma^2$.

Alternatively we may use the fact that for any fixed value of $\sigma^2 > 0$, (9.18) represents an exponential family for which we can take $t(y) = y/\sigma^2$

as canonical statistic in combination with the (unchanged) canonical parameter $\theta = \mu$. The general theory for exponential families yields

$$U(\theta) = U(\mu) = t - E(t) = \frac{y-\mu}{\sigma^2},$$

and

$$I(\theta) = I(\mu) = \text{var}(t) = \text{var}(y/\sigma^2) = 1/\sigma^2,$$

that is, once more the same results.

Thus, the effects of introducing σ^2 in the denominator of the exponent as shown in (9.18) are quite simple: The likelihood equations for $\theta = \mu$ are not affected at all, whereas the information matrix is divided by σ^2.

Analogously, if we have a sample of size n from a normal distribution, the same form (9.18) holds with y replaced by $\bar{y}$ and σ^2 by σ^2/n in the exponent of (9.18). △

More generally than in the Gaussian example, we next assume we are given a two-parameter distribution family for y_i of the form

$$f(y_i;\,\theta_i,\phi) = e^{\frac{\theta_i y_i - \log C(\theta_i)}{\phi}}\, h(y_i;\,\phi), \qquad (9.19)$$

where $C(\theta)$ is the norming constant in the special linear exponential family when $\phi = 1$. In the Gaussian Example 9.4, we wrote μ for θ (since they were the same) and σ^2 for ϕ (since ϕ was the variance for y). The two-parameter family may be an exponential family, as in the Gaussian example and in the gamma Example 9.5 following next, but that is not generally required. In any case, for given ϕ, if we have a distribution family (9.19), it is a linear exponential family for y_i/ϕ with θ_i canonical and log-norming constant $(1/\phi)\log C(\theta_i)$ (for simplicity, think of the support set for y as being $\mathbb{R}$, $\mathbb{R}_+$ or $\mathbb{R}_-$, so it is the same for y/ϕ for all ϕ). Thus the score function is

$$U_\phi(\theta_i) = \frac{y_i - \frac{d\log C(\theta_i)}{d\theta_i}}{\phi}, \qquad (9.20)$$

and since the score has expected value zero, we conclude that

$$E(y_i) = \frac{d\log C(\theta_i)}{d\theta_i} = \mu(\theta_i)$$

holds also in the extended family, and not only when $\phi = 1$.

Differentiating $(1/\phi)\log C(\theta_i)$ once more with respect to the canonical θ_i we will get the variance for the canonical statistic y_i/ϕ. Thus $\text{var}(y_i/\phi;\,\theta_i,\,\phi) = \text{var}(y_i;\,\theta_i,\,\phi = 1)/\phi$, or in a simpler form $\text{var}(y_i;\,\theta_i,\,\phi) = \phi\, v_y(\mu_i)$, where

$v_y(\mu_i) = \mathrm{var}(y_i;\ \theta_i,\ \phi = 1)$ is the *variance function*. The results so far are collected in the following proposition.

Proposition 9.4 ***Moments for models with dispersion parameter***
In a distribution family with density

$$f(y;\ \theta, \phi) = e^{\frac{\theta y - \log C(\theta)}{\phi}}\ h(y;\ \phi),$$

the expected value and variance for y are given by

$$\begin{aligned} E(y;\ \theta,\ \phi) &= E(y;\ \theta,\ \phi = 1) = \mu(\theta),\ \ \textit{free from dependence on}\ \phi; \\ \mathrm{var}(y;\ \theta,\ \phi) &= \phi\, v_y(\mu),\ \ \textit{where}\ v_y(\mu) = \mathrm{var}(y;\ \theta,\ \phi = 1). \end{aligned} \tag{9.21}$$

Thus, both mean value and variance are easily expressed in terms of the exponential family for $\phi = 1$, for which they can be obtained by differentiating $\log C(\theta)$.

For the normal distribution these relationships are simply expressed as $E(y) = \mu$ being free of σ^2, $v_y(\mu) \equiv 1$, and $\mathrm{var}(y) = \sigma^2 = \phi$. We next take a look at the gamma family to see a somewhat more complicated example, but also its analogies with the Gaussian example.

Example 9.5 ***Exponential and gamma (with dispersion parameter)***
If we start from the exponential distribution,

$$f(y;\ \theta) = e^{\theta y + \log(-\theta)}, \qquad y > 0,\ \theta < 0,$$

instead of $\mathrm{N}(\mu,\ 1)$, and try to extend it by a free variance parameter we will see that we are lead to a two-parameter gamma family, having the exponential as a special case. The gamma should not be surprising, remembering that a sample sum or mean from the of exponential is gamma distributed.

The gamma density was introduced in Chapter 2, Example 2.7, as

$$f(y; \alpha, \beta) = \frac{\alpha^{\beta+1}}{\Gamma(\beta + 1)}\, y^{\beta}\, e^{-\alpha y}, \quad y > 0, \tag{9.22}$$

but now we want a different parameterization, in terms of a canonical or mean parameter for y together with a dispersion parameter or variance. From the form of (9.22) and the mean and variance calculations already made in Example 3.9, we have the following results:

$$\begin{aligned} \mu = E(y) &= (\beta + 1)/\alpha \quad && \text{from Example 3.9} \\ var(y) &= \mu^2/(\beta + 1) \quad && \text{from Example 3.9} \\ \alpha &= -\theta/\phi \quad && \text{from comparison of (9.22) with (9.19)} \end{aligned}$$

The exponential distribution, for which $\phi = 1$, has $\mu = -1/\theta$. Thus, $\phi = 1/(\beta+1)$, and we get the following alternative representation of the gamma distribution as a generalized linear model, with μ and ϕ as new parameters:

$$f(y;\, \mu, \phi) = e^{\frac{1}{\phi}\{(-1/\mu)\, y - \log \mu\}}\, h(y;\, \phi)\,.$$

Additionally, note that the variance function is $v_y(\mu) = \text{var}(y;\, \mu,\, \phi = 1) = \mu^2$. This is the well-known variance–mean relationship for the exponential distribution. It was here extended to the gamma distribution, reading then $\text{var}(y;\, \mu,\, \phi) = \phi\, \mu^2$.

We may also parenthetically observe that the function $h(y;\, \phi)$ depends on y through $y^{\beta} = y^{-1+1/\phi}$, thus verifying that we have in fact a two-parameter exponential family. △

From Example 9.5 we conclude that the gamma family is a two-parameter family that corresponds to the normal family, with the following analogies and differences: In both cases the score function for the original θ is $(y-\mu(\theta))/\phi$. This implies that the likelihood equations for β in the generalized linear model are free from the parameter ϕ. The Fisher information for θ is $I(\theta) = \text{var}(y;\, \theta, \phi)/\phi^2 = v_y(\mu)/\phi$, in both cases proportional to $1/\phi$, but with different variance functions of μ. Likewise, we simply divide $J(\beta)$ and $I(\beta)$ in Proposition 9.2 by ϕ to get the corresponding formulae with dispersion parameter. We formulate these general results for inference about $\boldsymbol{\beta}$ as a proposition:

Proposition 9.5 ***Estimation of $\boldsymbol{\beta}$ in models with dispersion parameter***
In a generalized linear model with dispersion parameter ϕ,

- *the likelihood equations for $\boldsymbol{\beta}$ are given by Proposition 9.1, whatever be $\phi > 0$;*
- *the information matrices $I(\boldsymbol{\beta})$ and $J(\boldsymbol{\beta})$ are given by Proposition 9.2 only when $\phi = 1$; for general ϕ, divide these expressions by ϕ.*

Occasionally we have reason to also introduce an individual weights factor in the model, such that $\text{var}(y_i;\, \mu_i, \sigma^2) = (\sigma^2/w_i)\, v_y(\mu_i)$, where w_i is a known 'weight' (that need not sum to 1), For example, if the y_i in a logistic regression situation is a frequency representing n_i Bernoulli trials with the same x-value x_i, we must reduce the Bernoulli variance function $\pi_i(1-\pi_i)$ to $\pi_i(1-\pi_i)/n_i$. For more examples, see Section 9.7.2. It is left as Exercise 9.10 to demonstrate the following extension of Proposition 9.5:

Proposition 9.6 ***ML estimation of β in models with weights***
In a generalized linear model with dispersion parameter ϕ/w_i, the modification required in likelihood equations and Fisher information for β is that the diagonal elements $1/v_y(\mu_i(\beta))$ of the variance factor diagonal matrix V_y^{-1} are replaced by $w_i/v_y(\mu_i(\beta))$ in Proposition 9.1 and by $(w_i/\phi)/v_y(\mu_i(\beta))$ in Proposition 9.2. In matrix form this is expressed by insertion on each side of V_y^{-1} of a diagonal matrix W with w_i on the diagonal.

Estimation of the Dispersion Parameter ϕ

For estimation of ϕ it is natural to use the residual sum of squares, divided by the appropriate degrees of freedom (residual mean square). As residuals we may take the Pearson residuals, and use

$$\hat{\phi} = \frac{1}{n-k} \sum_i \frac{(y_i - \mu_i(\hat{\beta}))^2}{v_y(\mu_i(\hat{\beta}))} \,. \tag{9.23}$$

An alternative is to use the deviance residuals, that is, to use

$$\hat{\phi} = D(y, \mu(\hat{\beta});\ \phi = 1)\ /\ (n-k),$$

D being the residual deviance with the saturated model as reference. Note that here D must be the so-called *unscaled deviance*, obtained for $\phi = 1$. Another alternative is the ML method for ϕ, preferably conditional on $\hat{\beta}$. In the Gaussian linear model, the alternatives makes no difference, but more generally, most experience seems to speak for (9.23). Perhaps the main argument is that (9.23) is not sensitive to errors in the distribution type for y, if only the predictor and the variance structure is correctly modelled.

Note that with dispersion parameter we lose the possibility to check the model by just looking at the size of the residual mean square. Instead, the residual mean square indicates whether data support the use of a model without dispersion parameter, thus answering questions about possible under- or over-dispersion in for example binomial and Poisson models. Finally, note that the deviance for testing hypotheses about β will be changed by a factor $1/\phi$ when a dispersion parameter is introduced into a model. For estimation of ϕ by deviance we must use the unscaled deviance, whereas for hypothesis testing we use the *scaled deviance*, that is, the unscaled deviance normalized by the true ϕ or an estimated $\hat{\phi}$. Thus, distinguish scaled and unscaled deviances in statistical program output!

Example 9.6 ***Negative binomial to allow Poisson overdispersion***
Poisson regression has simply variance equal to the mean μ. Often there is overdispersion, however, when the (residual) variance is seen to be higher

than the mean (e.g. due to heterogeneity). As an alternative distribution to the Poisson, we will here consider the negative binomial, that allows Poisson overdispersion (but not underdispersion). Another alternative is to avoid specifying a distribution and work with a quasi-likelihood representing the desired link and variance functions, see Section 9.5.

The two-parameter negative binomial probability function for counts, $y = 0, 1, 2, \ldots$, can be written (see also Example 8.7)

$$f(y; \psi, \kappa) = \frac{\Gamma(\kappa + y)}{y!\,\Gamma(\kappa)} \psi^y (1 - \psi)^\kappa .$$

Its mean value is $\mu = \kappa\psi/(1-\psi)$, and the variance is $\mathrm{var}(y) = \kappa\psi/(1-\psi)^2 = \mu + \mu^2/\kappa$. In the limit as $\kappa \to \infty$ and $\psi \to 0$, such that the mean value μ stays constant, we obtain the Poisson distribution. The standard interpretation is that the Poisson parameter is given heterogeneity via a gamma distribution, see Example 8.7, and the Poisson itself then corresponds to a degenerate gamma, of zero variance. Thus, as an extension of the Poisson for generalized linear models, the negative binomial is somewhat complicated, having the Poisson as a limit and not as a proper member.

A different approach to fitting the negative binomial, advocated in particular by Hilbe (2011), see also McCullagh and Nelder (1989, Ch. 11), is to use the fact that it is a linear exponential family for each fixed value of κ. Select a start value for κ, with later updating, and use the corresponding negative binomial in a generalized linear model. However, if it should be compared with a Poisson, note that Poisson and the negative binomial do not have the same canonical link. If we used the log-link with the Poisson-based model, we should use the same link when modelling with the negative binomial. However, the canonical link is $\log \psi = \log(1 + \kappa/\mu)$. The updating of κ could be to increase the log-likelihood maximum or to reduce the deviance (9.15) (sum of deviance residuals).

Note the implicit assumption that when μ varies, because η varies, κ should remain the same. If, in fact, κ varies with μ, the variance function formula, given a fixed κ, $v_y(\mu) = \mu + \mu^2/\kappa$, would be misleading. △

Exercise 9.6 ***Binomial distribution***

In Example 9.3 about logit and probit regression we assumed that each unit was an individual Bernoulli trial. Sometimes the observation unit is instead the frequency in a number of Bernoulli trials with the same value of the regressor x, a number that is $\mathrm{Bin}(n, \pi(x))$. Show, by starting from $\mathrm{Bin}(1, \pi_0)$, that if a generalized linear model is based on the distribution of the relative frequency of successes in n Bernoulli trials with the same π_0, this distribution can be characterized as having the same variance function as for $n = 1$, but dispersion parameter $\phi = 1/n$. △

Exercise 9.7 ***Inverse Gaussian distribution, continued from Exercise 4.6***
The Inverse Gaussian density (Exercise 2.6 and Exercise 3.25) is

$$f(y;\, \mu, \alpha) = \sqrt{\frac{\alpha}{2\pi y^3}}\, e^{-\frac{\alpha(y-\mu)^2}{2\mu^2 y}} \qquad y, \mu, \alpha > 0.$$

In Exercise 3.25 it was told that $E(y) = \mu$ and $\mathrm{var}(y) = \mu^3/\alpha$. Show that this is a linear exponential family for each fixed value of α, and that in a generalized linear model it has canonical link $1/\mu^2$, variance function μ^3, and dispersion parameter $1/\alpha$. Finally, show that the family is closed with respect to rescaling, that is, for any $c > 0$, $z = cy$ has a distribution in the same family, cf. Section 9.4. △

Exercise 9.8 ***Information orthogonality***
Show that in any generalized linear model of type (9.19), and with canonical link, say, the parameters β and ϕ are information orthogonal, that is, they have a diagonal Fisher information matrix. △

Exercise 9.9 ***Proposition 9.4***
As an alternative proof of the variance proportionality to ϕ in Proposition 9.4, without reference to exponential family properties, start from (9.19) and use instead the fact that the variance of the score function equals minus the expected second derivative of the loglikelihood. △

Exercise 9.10 ***Proposition 9.6***
Prove the modification stated in Proposition 9.6. △

9.4 Exponential Dispersion Models

In the previous section we considered models with a dispersion parameter from the point of view of generalized linear models. A more general treatment of the underlying two-parameter distribution type has been given in particular by Bent Jørgensen, who introduced the term *exponential dispersion models*, but the ideas go much further back in time. An exhaustive account is found in his book (Jørgensen, 1997), to which we refer for details and further references. Here is a brief look at some of the interesting properties of these families.

An exponential dispersion (ED) model (Jørgensen: reproductive ED), has density

$$f(y;\,\theta,\phi) = e^{\frac{\theta y - \kappa(\theta)}{\phi}}\, h(y;\,\phi). \tag{9.24}$$

We denote this family $\mathrm{ED}(\mu,\phi)$ (where $\mu = \mu(\theta)$). The parameter ϕ is called the dispersion parameter. For each given ϕ, an $\mathrm{ED}(\mu,\phi)$ represents a linear exponential family, with canonical statistic y/ϕ and $\log C(\theta) = \kappa(\theta)/\phi$. Only three ED models are at the same time two-parameter exponential families, namely the normal, gamma and inverse Gaussian families. If the support of a linear exponential family is on a lattice (i.e. the integers Z, or a set of type $a + b\,Z$), we must let the ED model have a ϕ-dependent support, an artificial construction. Thus, the ED family generated by the Poisson family is a 'scaled Poisson' family, that might be denoted $\phi\,\mathrm{Po}(\mu/\phi)$.

When forming a generalized linear model, we use data y_i that are modelled by $\theta_i = \theta_i(\beta)$, and either a common ϕ, or more generally a factor of type $\phi_i = \phi/w_i$, where the w_is are known weights, see Proposition 9.6. In this generality the ED models have a remarkable convolution property. If $y_i \sim \mathrm{ED}(\mu,\,\phi/w_i)$ and they are mutually independent, then

$$\frac{\sum_i w_i y_i}{\sum_i w_i} \sim \mathrm{ED}(\mu,\,\phi/\textstyle\sum_i w_i). \tag{9.25}$$

This is a well-known property for the normal distribution, even when the mean value varies with i, but it also holds for the gamma and the inverse Gaussian distributions, as other examples.

As we have seen, the variance of an $\mathrm{ED}(\mu,\phi)$, can be written $\mathrm{var}(y) = \phi\, v_y(\mu)$, where the last factor is the variance function. Another remarkable ED property is that different EDMs cannot have the same variance function. (This would not be true if the variance function were expressed as function of θ!) For linear exponential families (i.e. for $\phi = 1$), see Exercise 3.9. For EDMs it follows as a consequence, since $\mathrm{var}(y) = \phi\,\mathrm{var}(y;\,\phi = 1)$.

ED models with $v_y(\mu) = \mu^p$, are of special interest for use in generalized linear models. These models are called *Tweedie models*, after M.C.K. Tweedie. We have already met the normal ($p = 0$), Poisson ($p = 1$), gamma ($p = 2$), and inverse Gaussian ($p = 3$; Exercise 9.7). Tweedie models with $1 < p < 2$ are *compound Poisson distributions*. These are distributions for sums of random variables with a Poisson number of terms. They are continuous distributions on the positive half-axis $\mathbb{R}_+$, except for an atom at $y = 0$. Tweedie models with $p \le 0$ or $p > 2$ are so-called stable distributions, on $\mathbb{R}$ and $\mathbb{R}_+$, respectively, which means they can occur as limit distributions for suitably normalized sums of iid variables. Tweedie models with $0 < p < 1$ do not exist, see Exercise 9.12.

It is well-known that the normal and gamma families are closed with respect to scale transformation, that is, a rescaling of y by a scale factor does not bring us outside the family. The same property for the inverse Gaussian family was part of Exercise 9.7. In fact, the property precisely characterizes the Tweedie models among all the exponential dispersion models.

Exercise 9.11 ***Tweedie models***
Verify by differentiation (or solving differential equations) that all Tweedie models, except the Poisson ($p = 1$) and the gamma ($p = 2$), have $\kappa(\theta) \propto \theta^b$ for some b. Show the relation $p = (b-2)/(b-1)$, or $b = (p-2)/(p-1)$. △

Exercise 9.12 ***Nonexistence of Tweedie models***
The result of Exercise 9.11 explains the nonexistence of Tweedie models for $0 < p < 1$. Show first that $0 < p < 1$ corresponds to $b > 2$. Show next that $\kappa(\theta) \propto \theta^b$ with $b > 2$ implies not only that $\theta = 0$ belongs to Θ, but also that $\text{var}(y;\ \theta = 0) = 0$. Why is the latter property impossible? △

9.5 Quasi-Likelihoods

Statistical packages typically have some built-in standard distribution families and their canonical links, and some other frequently chosen link functions. Taking the function `glm` in **R** as example, it allows only the canonical links for the Gaussian and the inverse-Gaussian distributions, whereas it allows also a couple of noncanonical links for the binomial, Poisson and gamma families. For example, the gamma family may have identity, inverse or log link. However, there is a road to more freedom, allowing any of these link functions for any distribution. This is quasi-likelihood, recalled in `glm` in **R** by specifying `quasi` instead of a standard distribution.

Note that for a generalized linear model, only the link function, the variance function and the linear predictor were needed to write down the likelihood equations (9.11). In the Fisher information for $\boldsymbol{\beta}$, the dispersion parameter was also needed (Proposition 9.4). For quasi-likelihood estimation the same characteristics are specified, but without prescribing that they were derived from some particular likelihood.

More precisely, when modelling by quasi-likelihood, a particular distribution for the response is *not specified*, but only the form of the link function and how the variance function depends on the mean. A response distribution fulfilling these model assumptions need not even exist. For example, if desired, we could fit an ‘impossible Tweedie model’ with $v_y(\mu) = \mu^p$ for some $0 < p < 1$, even though there is no linear exponential family that can have such a variance function. Note that for any linear exponential family, the variance of the response will depend on the mean in a characteristic way, see Exercise 3.9, and it will have the dispersion parameter as a multiplier. For example, $v_y(\mu) = \mu$ characterizes the Poisson distribution. However, when the distribution is unspecified, we are free to change the variance function.

As another example, quasi-likelihood can be used to allow overdispersion (or underdispersion) by inserting a dispersion parameter in the variance of the Poisson or binomial. For example, an 'overdispersed Poisson' quasi-likelihood would correspond to the variance $\phi\, v_y(\mu) = \phi\, \mu$. This function differs from the variance function of the negative binomial, $v_y(\mu) = \mu + \mu^2/\kappa$ (given κ), a real distribution that is also used to model overdispersion, see Example 9.6.

Quasi-likelihood estimation uses formally identical techniques to those for the Gaussian distribution, so it provides a way of fitting Gaussian models with nonstandard link functions or variance functions to continuous data. As an example, consider fitting the nonlinear regression

$$y = \frac{\theta_1 z_1}{z_2 + \theta_2} + \epsilon, \tag{9.26}$$

with $\epsilon \sim \mathrm{N}(0, \sigma^2)$. The function could for example represent a Michaelis–Menten relationship with additive error (for $z_1 = z_2 =$ substrate concentration). Alternatively, (9.26) is expressed

$$y = \frac{1}{\beta_1 x_1 + \beta_2 x_2} + \epsilon$$

where $x_1 = z_2/z_1$, $x_2 = 1/z_1$, $\beta_1 = 1/\theta_1$, and $\beta_2 = \theta_2/\theta_1$. This is linear for the inverse of the mean value, but the Gaussian does not allow the inverse link function (at least not in **R**: `glm`). Quasi-likelihood, that allows the inverse link, solves this problem.

9.6 GLMs versus Box–Cox Methodology

When a Gaussian linear model has been fitted to a set of data but appears not to have the right mean value structure, one can try a Gaussian model with another link function than the identity link. If y-data are positive, we could for example try a log-link or a square-root link function, that is, assume that y_i is $\mathrm{N}(\mu_i, \sigma^2)$ with a linear structure for $\eta_i = g(\mu_i)$, for a suitable nonlinear link function g. Note that such a change of link changes only the mean value, keeping the distribution type and variance structure unaffected.

An alternative to the change of link function is to transform the response variable. Assuming that $\log y$ or $\sqrt{y}$ follows the Gaussian linear model, instead of the original y, will imply approximately the same change of the mean value μ of y, but it will also change the distribution type and variance structure for y, perhaps to the better. For example, if $\log y_i$ is assumed normally distributed with constant variance, the original y_i must have been log-normally distributed with approximately constant coefficient of variation, and with a multiplicative mean value structure for y_i corresponding to an additive mean value for $\log y_i$. This often makes an improvement, from the modelling point of view, But how do we know what transformation to apply to y? The *Box–Cox methodology* (Box and Cox, 1964) provides an answer, for situations when y takes only positive values.

The Box–Cox technique aims at finding the best power-law transformation (the log-transform included). Thus, the transformations under consideration are essentially from y to y^λ, with λ as a kind of parameter, and we will go for the MLE of λ. Equivalently to y^λ we may use

$$y_\lambda = \frac{y^\lambda - 1}{\lambda}.$$

Note that we have a Gaussian linear model for y_λ precisely when we have such a model for y^λ ($\lambda \neq 0$), since they are related by a linear function. The advantage of y_λ is that it allows the log-transform to be included, being the limit of y_λ when $\lambda \to 0$ (because $y^\lambda = e^{\lambda \log y} \approx 1 + \lambda \log y$ for small λ).

The desired model for the transformed variable y_λ is the Gaussian linear model. Since y_λ depends on the parameter λ, we cannot regard y_λ as data, so we must maximize the likelihood for the original data $\{y_i\}$. By the chain rule, the densities for y_λ and y are related by

$$f_Y(y) = f_{Y_\lambda}(y_\lambda)\,\frac{dy_\lambda}{dy} = f_{Y_\lambda}(y_\lambda)\, y^{\lambda-1}$$

This yields the log-likelihood (additive constant deleted)

$$\log L(\boldsymbol{\beta}, \sigma^2, \lambda; \boldsymbol{y}) \tag{9.27}$$
$$= -\frac{1}{2\sigma^2}\sum_i (y_\lambda - \mu_i(\boldsymbol{\beta}))^2 - \frac{n}{2}\log\sigma^2 + (\lambda - 1)\sum \log y_i\,,$$

where $\mu_i(\boldsymbol{\beta})$ is the mean value in the linear model for y_λ, with parameter $\boldsymbol{\beta}$.

Right now we are primarily interested in the choice of λ, so a natural next step is to consider the profile likelihood for λ, by maximizing (9.27) with respect to $\boldsymbol{\beta}$ for fixed λ. This can be done explicitly, because for fixed λ we have a linear model in the data $\boldsymbol{y}_\lambda$. The profile likelihood is quite simple, even though this is not needed for the method to work. Note that $\sum_i (y_\lambda - \mu_i(\hat{\boldsymbol\beta}))^2$ is the residual sum of squares, alternatively written as $(n - \dim \boldsymbol\beta)\,\hat\sigma^2(\boldsymbol{y}_\lambda)$. It follows that the first term of (9.27) becomes a constant, so the profile log-likelihood L_p can be simplified to

$$\log L_p(\lambda) = -\frac{n}{2}\log \hat\sigma^2(y_\lambda) + (\lambda - 1)\sum \log y_i. \tag{9.28}$$

The value of λ maximizing (9.28) is found by plotting this function or by some other method. If the maximizing value is close to an interpretable value, for example $\lambda = 0$ or $\lambda = 0.5$, this value can be a more natural choice for the transformation.

In the early 1980s there was some debate in the literature whether or not the uncertainty in λ should be taken account of in inference for $\boldsymbol\beta$. The strongest arguments that it should not influence the inference for $\boldsymbol\beta$ are two:

- λ is not a parameter in the ordinary sense,
- $\boldsymbol\beta$ has no meaning and cannot be interpreted without reference to a fixed (selected) λ-value.

9.7 More Application Areas

9.7.1 Cox Regression Partial Likelihoods in Survival Analysis

Cox proportional hazards models are used to model survival data under log-linear influence of explanatory covariates, allowing right-censoring and an unspecified underlying hazard function. The data used in the *Cox partial likelihood* is at each time of 'death' a specification of which units are currently at risk (the *risk set*) and which of these units that 'dies'. For simplicity we assume all events are disjoint in time (i.e. no ties). The partial likelihood is the product over all such events. At a prescribed event with k units in the risk set, the corresponding factor of the partial likelihood is of type

$$L_i(\boldsymbol\beta) = \frac{e^{\boldsymbol\beta^T \boldsymbol{x}_i}}{\sum_{j=1}^k e^{\boldsymbol\beta^T \boldsymbol{x}_j}}, \tag{9.29}$$

where the $\boldsymbol{x}_j$s are the covariate vectors for the current risk set, and index i refers to the item that dies. This is a multinomial likelihood for the outcome of a single trial, and as usual we will introduce an auxiliary Poisson model to represent an equivalent likelihood. The development that follows comes from Whitehead (1980), see also McCullagh and Nelder (1989, Sec. 13.5).

Assume y_j is a Poisson number of deaths of item j, $j = 1, \ldots, k$, mutually independent with mean values $\mu_j = e^{\alpha + \boldsymbol{\beta}^T \boldsymbol{x}_j}$. The corresponding Poisson type likelihood (except for a constant factor) is

$$L_{\text{Po}}(\alpha, \boldsymbol{\beta}) = e^{\sum\{\log(\mu_j) y_j - \mu_j\}} = e^{\alpha \sum y_j + \boldsymbol{\beta}^T \sum \boldsymbol{x}_j y_j - \sum \mu_j}. \quad (9.30)$$

We will use it with all but one y_j being zero, and $\sum y_j = 1$, as observed. The parameter α will take care of the denominator of (9.29). More precisely, since the exponential family likelihood equations are $\boldsymbol{t} = E(\boldsymbol{t})$, we must in particular have $\sum y_j = \sum \mu_j$, which for given $\boldsymbol{\beta}$ yields $e^{\hat{\alpha}} = e^{\hat{\alpha}(\boldsymbol{\beta})} = 1/\sum e^{\boldsymbol{\beta}^T \boldsymbol{x}_j}$. When this is inserted in the Poisson likelihood (9.30), in combination with $y_i = 1$ and $y_j = 0$ for $j \neq i$, we get

$$L_{\text{Po}}(\hat{\alpha}(\boldsymbol{\beta}), \boldsymbol{\beta}) = e^{\hat{\alpha} + \boldsymbol{\beta}^T \boldsymbol{x}_i - 1} \propto \frac{e^{\boldsymbol{\beta}^T \boldsymbol{x}_i}}{\sum e^{\boldsymbol{\beta}^T \boldsymbol{x}_j}},$$

where the right-hand side is the likelihood factor (9.29). Thus, after insertion of the MLE of α for given $\boldsymbol{\beta}$, the two likelihoods are equivalent.

It only remains to combine the information from all the partial likelihood factors, each having its own risk set and its own death event, in a product as if they corresponded to independent experiments. The resulting partial likelihood is then equivalent to the profile likelihood for $\boldsymbol{\beta}$ in a combined Poisson model with a separate α parameter associated with each event. The equivalence holds both for ML estimation of $\boldsymbol{\beta}$ and for likelihood ratio considerations. The Poisson likelihood is easily handled as a Poisson generalized linear model.

9.7.2 Generalized Linear Models in Nonlife Insurance Pricing

Generalized linear models have become increasingly popular in actuarial practice in nonlife insurance companies since the 1990s, but already the first edition from 1983 of McCullagh and Nelder (1989) has examples of claim number analysis and accident severity analysis. Actuaries use the models for setting prices of insurance policies, based both on external data and on their own historical data of policies and paid claims. We will here take a brief look at the basic problems and model types in insurance pricing (premium rating), in a presentation based on the first half of the book by Ohlsson and Johansson (2010).

The premium (per policy year) to be paid by policy-holders should be in correspondence with the expected loss for the insurance company, so the expected loss must be estimated. The so-called *pure premium* is the product of the *claim frequency* (accident rate) and the corresponding *claim severity* (average damage cost per claim) for the policies under consideration. These characteristics are estimated for a group of similar policy-holders, a so-called *tariff cell*. This means that the pure premium will be specified as dependent on a number of selected predictor variables, called rating factors, which define the tariff cells. These could for example be age and gender when the policy-holder is a private person, or type, age and model of a car to be insured, or population density and other characteristics of the geographic area of the policy-holder. Typically all rating factors are made categorical, by dividing values into classes. Thus a tariff cell is a category combination in a cross-tabulation. The generalized linear model helps to estimate the characteristics of a tariff cell by using data from all cells to form and estimate a suitable linear predictor. This is particularly helpful when some tariff cells have very little data.

Consider a particular tariff cell. The claim frequency is, more precisely, the number of claims normalized by the duration, which is the total time exposed to risk (sum of policy years) within the tariff cell. The claim severity is an average cost per claim formed as the total cost normalized by the number of claims. The *pure premium* is their product,

$$\text{pure premium} = \text{claim frequency} \times \text{claim severity}.$$

The pure premium is the quantity of ultimate interest, but it is standard to analyse frequency and severity separately, before the results are combined. The analysis of claim severity is then made conditional on the observed number of claims, in accordance with the inferential principle of Section 3.5.

Models considered here will be based on three basic assumptions both on the claim frequency and on the claim severity:

- independence between individual policies (so catastrophes are excluded)
- independence between disjoint time intervals
- homogeneity within tariff cell (so policies within a cell are exchangeable)

Pure premium, claim frequency and average claim severity are 'key ratios', sums normalized by an exposure weight, either the duration (the two first) or the number of claims (severity). A consequence of the basic assumptions is that such key ratios y satisfy

$$E(y) = \mu, \qquad \text{var}(y) = \sigma^2/w,$$

where μ, σ^2 and w are characteristics of the tariff cell, and w is the total duration (total number of policy years), or the number of claims, respectively.

(Note that this fits with the convolution property of exponential dispersion models, (9.25), if tariff cells with the same μ and σ^2 are merged.)

For modelling claim frequency the Poisson is the natural first try as a distribution in a generalized linear model. For claim severity the gamma distribution has the corresponding role, because severity has a skew distribution on $\mathbb{R}_+$, with a variance in some way increasing with the mean. Before we discuss alternatives to these two distributions, we briefly consider the link function. Since half a century or so, the standard model is multiplicative, and several fitting methods were developed before the generalized linear models entered the scene. In the terminology of the latter, multiplicativity corresponds to use of the log-link. This is the canonical link for Poisson, but not for gamma.

Considering claim frequency data, the basic assumptions listed indicate that we might expect the claim numbers to be Poisson distributed, and the model for them is identical with the weighted multiplicative Poisson models of Exercises 2.1, 3.21 and 5.14. The duration factor in the Poisson parameter is then perhaps best treated as a known intercept on the log-linear scale. However, it is often found that the residual deviance is too large to fit the Poisson. The reasons for this *Poisson overdispersion* could be lacking homogeneity of means within tariff cells, or clustering effects, or that the multiplicative model for the means does not fit well enough. Traditionally, overdispersion has been accounted for by use of the negative binomial, see Example 9.6, but a flexible alternative is to use quasi-likelihood methods for a suitable variance function. This is often chosen as representing a Tweedie model, but in a quasi-likelihood there is no real need to bother about the existence of a distributional correspondence.

Turning to claim severity data, the gamma model has variance function $v_y(\mu_i) = \mu_i^2$ (Example 9.5). After scaling with a given number w_i of claims, we get $\mathrm{var}(y_i) = \phi\,\mu_i^2/w_i$. For two cells with equal claim numbers this implies they should have equal coefficients of variance $\sqrt{\mathrm{var}(y_i)}/\mu_i$, reflecting equal coefficient of variation for individual claims. There is no overdispersion problem such as for claim frequency, but it may of course be found that the residuals seem to depend on μ in a different way than expected from the gamma variance function. Actuaries might doubt the gamma distribution but agree about a variance function, e.g. as represented by the gamma or by some other Tweedie distribution. This could be combined with a reference to the distribution-free quasi-likelihood methods.

Uncertainty quantification, in particular construction of confidence intervals based on standard errors, is slightly more complicated. In the generalized linear models for claim frequency or claim severity, the observed or Fisher information provides not only approximate confidence intervals via standard errors for components of $\boldsymbol{\beta}$, but the uncertainty can be propagated to intervals for mean values in two steps: First for the linear predictor in the tariff cell by linear error propagation, and second by inverting the log function. The same holds with quasi-likelihood estimates, albeit on more shaky grounds.

For the pure premium, however, a difference is that the standard estimate in a tariff cell is obtained by multiplying the corresponding fitted claim frequency and fitted claim severity, so we combine two models, where one is conditional on the data of the other. On log-scale the pure premium for a tariff cell is a sum of two terms, which with some abuse of notation (since each of them is a linear predictor, linear in the $\boldsymbol{\beta}$-parameters) we will call $\hat{\boldsymbol{\beta}}_F(\boldsymbol{n})$ and $\hat{\boldsymbol{\beta}}_S(\boldsymbol{y}, \boldsymbol{n})$. Here we have indicated that $\hat{\boldsymbol{\beta}}_F$ depends on the array $\boldsymbol{n}$ of sample sizes and $\hat{\boldsymbol{\beta}}_S$ depends on both $\boldsymbol{n}$ and the severity data $\boldsymbol{y}$. Combination of the (inverse) information matrices yields the sum

$$\mathrm{var}(\hat{\boldsymbol{\beta}}_F(\boldsymbol{n})) + \mathrm{var}(\hat{\boldsymbol{\beta}}_S(\boldsymbol{y}, \boldsymbol{n}) \mid \boldsymbol{n}), \tag{9.31}$$

whereas the marginal variance is

$$\mathrm{var}(\hat{\boldsymbol{\beta}}_F(\boldsymbol{n}) + \hat{\boldsymbol{\beta}}_S(\boldsymbol{y}, \boldsymbol{n})) = \mathrm{var}(\hat{\boldsymbol{\beta}}_F(\boldsymbol{n})) + E(\mathrm{var}(\hat{\boldsymbol{\beta}}_S(\boldsymbol{y}, \boldsymbol{n}) \mid \boldsymbol{n})) \tag{9.32}$$

Which formula should we aim at? The difference is that the latter formula has the expected value over $\boldsymbol{n}$ in its second term. We cannot argue that one is more precise than the other. However, it can be argued and also shown quantitatively (Sundberg, 2003, Cor. 2) that (9.31) is *more relevant* than its expected value (9.32), in that it pays attention to the actual amount of information in the outcome of $\boldsymbol{n}$. Thus, we not only can but should use the conditional information about $\boldsymbol{\beta}_S$. The argument is similar to the argument for conditioning on S-ancillary statistics, but in the present case $\boldsymbol{n}$ is an S-ancillary precision index only for the inference about $\boldsymbol{\beta}_S$. We could also refer to the general preference of observed information to expected information, when they differ.

The categories of a rating factor often have a natural ordering that should be reflected in the estimated parameters. More generally, there is need for smoothing. Empty tariff cells and cells with very few data is another, related problem. None of them will be taken up here, but the reader is referred to the literature, e.g. Ohlsson and Johansson (2010).

10

Graphical Models for Conditional Independence Structures

This chapter is about models for multivariate distributions, specified by conditional independences. We here use a graph to describe a multivariate model structure. Data could for example represent presence/absence of various symptoms for some psychiatric patients, or student achievements in a number of subject areas, marked on a discrete or continuous scale. The patients or students are regarded as forming a random sample from a population of interest. Subject matter questions concern associations between presence of different symptoms or achievements in the different subject areas, respectively.

Note that the chapter is not about models for random graphs. That topic is treated in Chapter 11.

The components of the multivariate vector are random variables (variates) that will be represented by *nodes* (dots or circles) in a graph. Dependences between pairs of variates are indicated by *edges* (lines). An absent edge between two nodes implies an assumed conditional independence between the two variates, given the other variates. Other (marginal) dependences are indicated by less direct connections than the direct edge. In this way the graph with its absent edges specifies the model and motivates the name 'graphical' model. The graph is called an *independence graph* (or in some literature *dependence graph*).

Conditional independence structures for multivariate distributions are naturally and conveniently modelled as exponential families, in which a conditional independence (absent edge) means that a canonical parameter in the joint distribution is specified as being zero. In most of the chapter we restrict attention to the pure cases when the variates are either all categorical, represented by log-linear models for contingency tables, or all continuous, corresponding to the 'covariance selection' models for the multivariate normal, introduced in Example 2.11. Towards the end of the chapter we briefly look at the more complicated mixed case, in which some variates are categorical and others are of continuous type.

The presentation will be introductory and selective. For more comprehensive treatments, see the books by Whittaker (1990), Edwards (1995) and Lauritzen (1996). Much of the theory development and applications nowadays concern models with directed graphs, using arrows to represent causality, but the treatment here will be confined to the basic and simpler case of undirected graphs, representing (two-sided) associations, only.

10.1 Graphs for Conditional Independence

Figure 10.1 shows two simple graphs for a set of three random variables (nodes), $\{A, B, C\}$. For ease of notation we use the same symbols for nodes and random variables. In the upper graph the lack of link between the node C and the node pair $\{A, B\}$ tells that the variable C is modelled as totally independent of both A and B, while dependence is allowed between A and B. We write $\{A, B\} \perp\!\!\!\perp C$. In the lower diagram all three nodes are linked, but the link between A and C is only indirect, via B. The interpretation is that variates A and C are regarded as conditionally independent, given B, expressed by formula $A \perp\!\!\!\perp C \,|\, B$. We may say that B separates the left part A from the right part C, and generally, to clarify the conditional independence structure from a graph, we look for subsets of nodes separating one part of the graph from another.

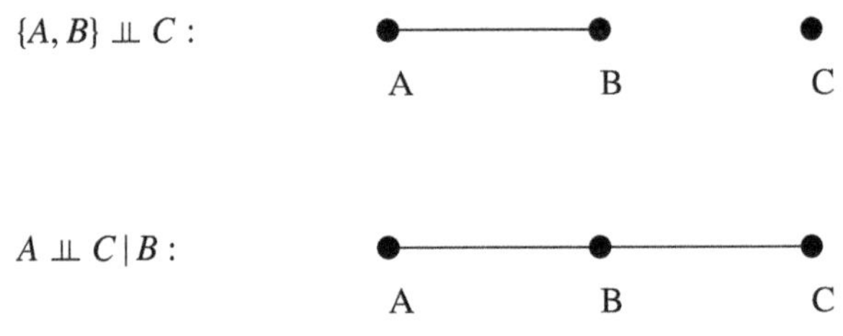

Figure 10.1 Two independence graphs with three nodes.

Example 10.1 ***A conditional independence graph with five nodes***
Figure 10.2 shows a slightly more complicated example, with five nodes. The graph specifies for example that U and $\{X, Y, Z\}$ are modelled as conditionally independent given W, that is, $U \perp\!\!\!\perp \{X, Y, Z\} \,|\, W$. Additionally, $\{U, W\} \perp\!\!\!\perp Y \,|\, \{X, Z\}$, since all paths from U or W to Y go through nodes X or Z, that is, together they separate U and W from Y.

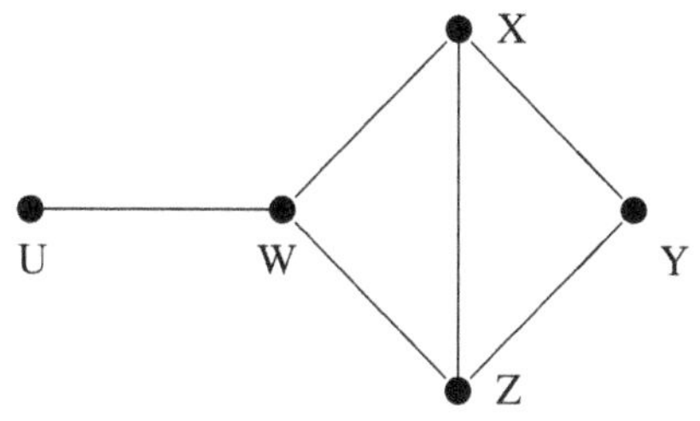

Figure 10.2 Independence graph with five nodes.

△

A conditional independence structure implies that the multivariate density can be factorized according to the independence graph. How this works more generally is explained by the theory of *Markov random fields*. This theory generalizes the Markov processes in time, for which future and past are conditionally independent, given the present. In the more general context here, the conditional independences are between two sets of variates given some other variates. In particular the *Hammersley–Clifford theorem* (Lauritzen, 1996, Sec. 3.2) tells how a conditional independence graph with its separating subsets implies a *factorization* of the joint density, and vice versa. This holds under weak regularity conditions, in particular under the assumption of a strictly positive density. The idea to connect this Markov property with log-linear models for contingency tables comes from Darroch et al. (1980). They coined the name *graphical models* for this new model class.

Each of the factors in the factorization contains only the components of a so-called *clique* of the graph. A clique is a subgraph that is both *complete* (all its node pairs directly linked) and *maximal* (not part of a larger complete subgraph). As an example, the graph in Figure 10.2 has three cliques: the subgraphs formed by $c_1 = \{U, W\}$, $c_2 = \{W, X, Z\}$, and $c_3 = \{X, Y, Z\}$. A subgraph such as $\{X, Z\}$ is complete but not maximal, and $\{W, X, Y, Z\}$ is not complete. Note that an isolated node would be a clique. In this terminology, and with variates/nodes more generally denoted $\{Y_1, \ldots, Y_k\}$, the factorization can be written

$$f(y_1, \ldots, y_k) = \prod_{\text{all cliques } c} \psi_c(y_c) \tag{10.1}$$

where $\psi_c(y_c)$ denotes a function of the variates in clique c.

We might be tempted to think of the factors ψ_c as representing marginal or conditional densities of subsets y_c. This is generally true for graphs which also have the property of being *decomposable*. A decomposable graph is loosely a graph that can be successively decomposed into smaller decomposable graphs or complete subgraphs using cliques as separators. For graphs that are undirected and *pure* (either all variates continuous or all discrete), a simpler but equivalent criterion is that of *triangulated*, which we will take as definition. But first we need a few obvious concepts. Consider a *cycle*, i.e. a path in the graph returning to the start node, after following consecutive edges along a sequence of distinct nodes. The *length* of the cycle is its number of edges (≥ 3). A *chord* is an edge connecting two nonconsecutive nodes of the cycle, thus shortening the length of the cycle.

Definition 10.1 ***Decomposability for undirected, pure graphs***
An independence graph is *decomposable* (or *triangulated*) if any cycle of length ≥ 4 has a chord.

The chord shortens the cycle (from the start node and back), and by repeating this procedure the cycle can be shortened to a triangle.

Whether a given, moderately large graph can be triangulated is usually easy to check. The graphs in Figure 10.1 do not even have a cycle, so they are decomposable. The graph of Figure 10.2 contains two distinct 3-cycles and one 4-cycle ($W - X - Y - Z - W$ with W chosen as start node), but the 4-cycle has the chord $X - Z$, so the graph is decomposable. If the chord $X - Z$ were deleted from the graph, however, the resulting graph would not be decomposable, see two paragraphs below.

A decomposable model may be factorized by successive conditioning and simplification, passing over the cliques one by one in suitable order, for example from c_1 to c_3 (or in opposite direction). Note that W separates U from the rest of the graph, and $\{X, Z\}$ separates $\{U, W\}$ from Y. This is utilized to obtain

$$f(u, w, x, z, y) = f_{c_1}(u \mid w)\, f_{c_2}(w \mid x, z)\, f_{c_3}(x, y, z). \tag{10.2}$$

Models with decomposable graphs are not only more tractable theoretically and computationally than nondecomposable models, they are also easier to interpret. In particular their structure may suggest a stochastic mechanism by which data could be imagined as having been generated.

The smallest graph that is not decomposable is the simple 4-cycle of Figure 10.3. It has 4 cliques, being the subgraphs formed by the pairs $\{W, X\}$,

$\{W, Z\}$, $\{X, Y\}$ and $\{Y, Z\}$. Note the absence of the chord $X - Z$ in comparison with Figure 10.2. Thus, if we try $f(w \mid \ldots)$ as a factorization start, we must condition on both x and z, which are not joint members of any clique. This makes the corresponding model more difficult to handle and to interpret than a model with a decomposable graph of similar size.

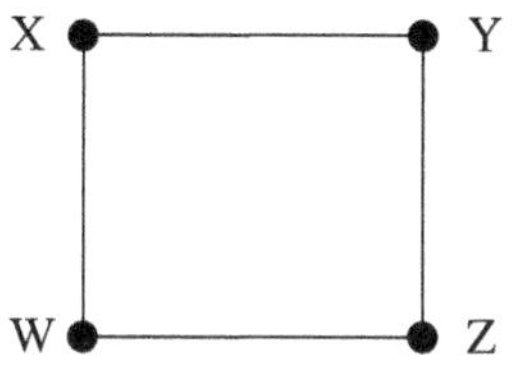

Figure 10.3 Smallest nondecomposable independence graph.

Exercise 10.1 ***Independence versus conditional independence***
Show that conditional independence of X and Z given Y does not imply marginal independence of X and Z, nor does the converse hold. △

Exercise 10.2 ***Decomposability***
First convince yourself that a complete graph (full set of edges) is decomposable. Next, suppose one edge is deleted from a complete graph. Show that the resulting graph is still decomposable. Suppose one more edge is deleted. Show that the resulting graph need not be decomposable. △

10.2 Graphical Gaussian Models

We have already met this type of model under the name covariance selection models, see Example 2.11. We have seen that reduction of the covariance structure should preferably be made in the inverse of the covariance matrix, the concentration matrix Σ^{-1}. We saw that each zero among the nondiagonal elements of Σ^{-1} could be interpreted as a pairwise conditional independence, given the other components of the vector. We can represent this type of structure by a graph with components as nodes and having edges between nodes except where a conditional independence has been specified by a zero in Σ^{-1}. The general theory states that it is sufficient to condition on a separating set of components in the graph.

Note that for the multivariate Gaussian models, marginal and conditional distributions are also Gaussian. Thus, for a decomposable model, all factors of a factorization such as (10.2) represent Gaussian densities. More precisely, f_{c_1} and f_{c_2} are conditional Gaussian densities and f_{c_3} is an unrestricted trivariate marginal Gaussian density.

10.2.1 ML Estimation

Preparing for ML estimation in $N_k(\boldsymbol{\mu}, \Sigma)$, we first note that the MLE of $\boldsymbol{\mu}$ is the sample mean vector, directly obtained as part of the general likelihood equation system $\boldsymbol{\mu}_t(\boldsymbol{\theta}) = \boldsymbol{t}$. Furthermore, the conditionality principle of Section 3.5 motivates that we need only the sample covariance matrix, S, for the inference about Σ. Later in Chapter 3, see Example 3.12, the likelihood equations for covariance selection models were introduced.

From the general exponential family theory (Proposition 3.13) we know that a unique solution exists at least if S is nonsingular (and when Σ is positive definite, S is nonsingular with probability 1 at least if $n > k$),. For a sample from a saturated multivariate normal (i.e. without restrictions), the MLE of Σ is simply $\widehat{\Sigma} = S$. For each prescribed zero in Σ^{-1}, the canonical statistic is reduced by deletion of the corresponding component of S and of the corresponding equation of the equation system. Thus, $\widehat{\Sigma} = S$ is preserved, except in the zero positions of Σ^{-1}, corresponding to the conditional independences. What are then the covariances of $\widehat{\Sigma}$ in the latter positions, and how can we find them?

We will derive these covariances by successive decompositions. In this procedure we will make use of an expression for the covariance between two disjoint subsets of variates, denoted A and B, when a third subset C *separates* A from B in the corresponding independence graph, that is, all paths between A and B pass through C, or in other words, the corresponding variate sets A and B are conditionally independent, given C. As an example, $\{X, Z\}$ separates $\{U, W\}$ from Y in Figure 10.2. We exclude the trivial case when C is empty ($C = \emptyset$), when we have total independence between A and B, $\Sigma_{AB} = 0$.

Proposition 10.2 ***Covariance lemma***
Suppose $C \neq \emptyset$ separates A from B in the model graph. Then $\mathrm{cov}(A, B) = \Sigma_{AB}$ *can be expressed as*

$$\Sigma_{AB} = \Sigma_{AC}\,\Sigma_{CC}^{-1}\,\Sigma_{CB}. \tag{10.3}$$

Proof

$$\Sigma_{AB} = E\{\text{cov}(A,\, B\,|\,C)\} + \text{cov}\{E(A\,|\,C),\, E(B\,|\,C)\}$$
$$= 0 + \text{cov}\{\Sigma_{AC}\Sigma_{CC}{}^{-1}C,\, \Sigma_{BC}\Sigma_{CC}{}^{-1}C\} = \Sigma_{AC}\,\Sigma_{CC}{}^{-1}\,\Sigma_{CB}.$$

Here the conditional covariance in the first term vanishes since C separates A from B in the graph. In the second term we utilize that in the Gaussian case, conditional mean values are expressed as linear regressions on C. □

We apply this result when $c = \{A,\, C\}$ is a clique such that C alone (i.e. not any node in A) shares nodes with other cliques (note, C is allowed to be empty). When the graph is decomposable (with more than a single clique), it is always possible to find such a clique c. Note that Σ_{AC} and the variance Σ_{CC} are parts of Σ_{cc}, which is estimated by the corresponding S_{cc}. Thus, we have an expression for $\widehat{\Sigma}_{AB}$ in terms of $\widehat{\Sigma}_{CB}$, which means that we have reduced the problem by eliminating the A-part of the graph and one of the cliques, $c = \{A,\, C\}$. We repeat this procedure on another clique, and so on, until the resulting $\{C,\, B\}$ is the last clique, for which we know $\widehat{\Sigma}_{CB} = S_{CB}$. We will exemplify by applying the procedure on the decomposable graph of Figure 10.2, with factorization (10.2).

Consider the first factor of (10.2), corresponding to the clique $c_1 = \{U,\, W\}$, where W separates U from the rest of the graph, i.e. from $\{X, Y, Z\}$. For the MLE $\widehat{\Sigma}$ we need the (estimated) marginal covariance vector between U and $\{X, Y, Z\}$. We first apply Proposition 10.2 with $A = U$, $C = W$, and $B = \{X, Y, Z\}$, to obtain

$$\widehat{\Sigma}_{U\{X,Y,Z\}} = S_{UW}\, S_{WW}{}^{-1}\, \widehat{\Sigma}_{W\{X,Y,Z\}}\,.$$

It remains to handle the last factor, $\widehat{\Sigma}_{W\{X,Y,Z\}}$, of three covariances. Two of them, $\widehat{\Sigma}_{W\{X,Z\}}$ are part of $\widehat{\Sigma}_{c_2c_2} = S_{c_2c_2}$. The last one, $\widehat{\Sigma}_{WY}$, is found by one more application of Proposition 10.2, now with $C = \{X, Z\}$ separating $A = W$ and $B = Y$. The result is

$$\widehat{\Sigma}_{WY} = S_{W\{X,Z\}}\, S_{\{X,Z\}\{X,Z\}}{}^{-1} S_{\{X,Z\}Y}\,.$$

In the nondecomposable case, an explicit solution is not possible. The procedure just described does not lead the whole way to a last clique, cf. Exercise 10.4. We need an iterative solution method. Several such methods exist, which successively run through the set of cliques again and again, and several **R** implementations, see Højsgaard et al. (2012).

10.2.2 Testing for Conditional Independence in Decomposable Models

We treat here model reduction by elimination of one edge, e, from the independence graph of a multivariate normal distribution model, or equivalently insertion of one (additional) off-diagonal zero in Σ^{-1}, or reduction of the minimal sufficient statistic by a particular sample covariance, s_{xy} between nodes/variables X and Y, say. As before, we regard the mean value as already eliminated from the discussion by conditioning or marginalizing, so s_{xy} is an explicit component of the canonical statistic.

The graphs are assumed decomposable, both with and without the edge e. This guarantees that the conditional independence hypothesis can be restated and simplified, by making use of the following properties:

- The edge e between X and Y belongs to precisely one clique, c_e say;
- If Z is defined by writing $c_e = \{X, Y, Z\}$, then, under H_0, i.e. with the edge e deleted, Z separates X from Y. The set Z is allowed to be empty.

For each of these statements, the argument is that if they were not true, there must be, under H_0, a cycle of length ≥ 4 edges that could not be shortened to a triangle, i.e. the graph under H_0 would not be decomposable. The consequence of the properties is that the hypothesis and its test can be confined to the clique c_e, i.e. to the distribution for $\{X, Y, Z\}$, neglecting all other variates and their structure.

Consider again Figure 10.2 of Example 10.1. One example is when e is the edge between X and Y. This edge belongs only to clique $c_3 = \{X, Y, Z\}$, and node Z separates X from Y when the direct edge e has been deleted. On the other hand, it would not have been an example to let e be the edge between X and Z, because this edge belongs to two cliques. This goes back to the fact that the graph under this H_0 is not decomposable.

We first consider the exact test, according to the principle of Section 5.1, but the derivation has much in common with that for the deviance given by Lauritzen (1996, Sec. 5.3.3). We will show that the test is a t-test (one-sided or symmetrically two-sided). The procedure and the derivation extends Exercise 5.2 for correlation testing in the bivariate case.

For the exact test of H_0, we need the conditional distribution of $v = s_{xy}$, given the canonical statistic u under H_0, that is, given all sample variances and covariances for $\{X, Y, Z\}$, except s_{xy}. The first step is to substitute for s_{xy} the sample *partial correlation coefficient* $r_{xy\cdot z}$, in other words the correlation between x and y adjusted for the other variates z in c_e. This can be done because $r_{xy\cdot z}$ is a linear function of $s_{x,y}$, whose coefficients only depend on statistics in u. More specifically, $r_{xy \cdot z}$ is obtained from the bivariate conditional sample covariance matrix of $\{X, Y\}$, which is $((S^{-1})_{\{x,y\}})^{-1}$ (see Example 2.10):

$$r_{xy \cdot z} = -(S^{-1})_{xy} / \sqrt{(S^{-1})_{xx}(S^{-1})_{yy}}\,. \tag{10.4}$$

The verification of (10.4) is left as Exercise 10.6, but its precise form is not needed for the argument. Being a correlation coefficient, $r_{xy \cdot z}$ has a parameter-free distribution under H_0, and Basu's theorem (see Proposition 3.24) then tells that $r_{x,y \cdot z}$ must be independent of u. Thus, we only need the marginal distribution of $r_{xy \cdot z}$. This is a classic result by Fisher from 1924, see e.g. Cramér (1946, Sec. 29.13). More precisely, the monotone and odd function $\sqrt{n-f}\; r / \sqrt{1-r^2}$ of $r = r_{xy \cdot z}$ is exactly $t(n-f)$-distributed, where f is the number of nodes in the clique c_e. Thus we have shown the following result:

Proposition 10.3 ***Exact test for conditional independence***
The exact test for a conditional independence, deleting one edge when both models are decomposable, is equivalent to a test for partial correlation within the (unique) clique from which the edge is deleted. This exact test can be executed as a t-test (degrees of freedom $n - f$; f being the number of nodes in the clique).

Among large sample tests (i.e. large n), the score test statistic W_u is explicitly based on v, or equivalently on the partial correlation coefficient, and it follows that the score test represents the normal approximation of the exact two-sided t-test. However, the deviance and Wald test statistics are also functions of the partial correlation coefficient, such that they are asymptotically equivalent to the score test.

10.2.3 Models under Symmetry Assumptions

For keeping the number of model parameters low, imposing conditional independences is not the only way. One possibility is to supplement some conditional independences with a type of assumption that some dependencies are 'equally strong', in some quantitative sense. Such models are treated in Højsgaard and Lauritzen (2008). The guidance in model formulation comes from various types of symmetry assumptions.

A simple possibility is to assume equalities in the concentration matrix Σ^{-1}. The resulting models remain regular exponential families. Whether the assumption is reasonable or not can be tested by the methods of Chapter 5, because the hypothesis is linear in the canonical parameters; it can be expressed in terms of differences of canonical parameters being zero. A problem with these models, however, can be their lack of invariance to scale changes, because Σ^{-1} is scale dependent. This might introduce some arbitrariness in the model assumptions.

An alternative is to make equality assumptions on partial correlations instead of the elements of Σ^{-1}, because the former are scale invariant. Expressed in the elements of Σ^{-1}, the partial correlations are not linear functions of the elements of Σ^{-1}, however, so the resulting models will typically be curved exponential families. In particular this implies that we cannot rely on the property of a unique root to the likelihood equations. An example of multiple roots is mentioned in Højsgaard and Lauritzen (2008), who also discuss procedures for the numerical solution of the equations.

Exercise 10.3 ***Zeros in*** Σ^{-1}
Suppose Σ^{-1} for $\{Y_1, \ldots, Y_4\}$ has zeros as indicated, with asterisk values arbitrary. Find the corresponding independence graph representation and use this to conclude that Σ^{-1} for $\{Y_1, Y_2, Y_3\}$, $\{Y_1, Y_2, Y_4\}$, or any other triple (in preserved order), must be of the 3×3 form to the right:

$$\Sigma^{-1}(Y_1, Y_2, Y_3, Y_4) = \begin{pmatrix} * & * & 0 & 0 \\ * & * & * & 0 \\ 0 & * & * & * \\ 0 & 0 & * & * \end{pmatrix}; \quad \Sigma^{-1}(Y_1, Y_2, Y_4) = \begin{pmatrix} * & * & 0 \\ * & * & * \\ 0 & * & * \end{pmatrix}$$

△

Exercise 10.4 ***Nondecomposable example***
Consider the nondecomposable graph in Figure 10.3. Try to apply Proposition 10.2 on one of the absent chord covariances, and show that it does not help, because (10.3) requires the other absent chord covariance. △

Exercise 10.5 ***SUR model***
Consider the seemingly unrelated regressions model (7.2), with the two x-variables regarded as stochastic, normally distributed and allowed to be correlated. Represent the model by a conditional independence graph. △

Exercise 10.6 ***Partial correlation formula***
Verify formula (10.4) by explicitly inverting a suitable 2×2 matrix. △

10.3 Graphical Models for Contingency Tables

Independence in 2×2 contingency tables was treated in some detail in Section 5.2 and Example 5.3. Here we extend this situation in two directions:

- More than two 'factors' (classification criteria)
- More than two index levels (values) per factor

The temptation in data analysis to restrict attention to two-dimensional tables by summing over all but two indices in a higher-dimensional table must be avoided, since it opens up for the dangerous, and not uncommon Yule–Simpson effect (Simpson's paradox), by which a qualitative relationship valid in each of some groups can change and even be reversed if data are aggregated over groups, see e.g. Lauritzen (1996, Ch. 4) for illustrations. With more than two factors, we should instead look for conditional independences, rather than only unconditional (marginal) independence. The extension to more than two index levels was briefly introduced in Exercise 5.5 and Example 5.4. Now we consider the two extensions jointly.

Section 5.2 showed an important equivalence in several aspects between different sampling schemes, in particular the multinomial and the Poisson. The theory of log-linear models holds also with Poisson variates, but for the terminology of 'independence' to be natural, we restrict the discussion to multinomial sampling. Note then as a first step that for any model representing multinomial sampling of n individuals, the data can be reduced by the sufficiency principle from the set of individual outcomes to the table of cell counts, which is our starting point.

A three-way cross-classification is rich enough to serve as example for an introduction of contingency table models with conditional independence. Let $\{A, B, C\}$ represent the classification criteria and $\pi_{ijk} > 0$ be the corresponding multinomial probability for cell (i, j, k). The saturated model has only the constraint $\sum \pi_{ijk} = 1$. The cell counts are $\{n_{ijk}\}$, and in the saturated model, n_{ijk} is the MLE of the mean value parameter $n\,\pi_{ijk}$. Complete mutual independence means that $\pi_{ijk} = \alpha_i\beta_j\gamma_k$. This and subsequent factorizations are log-linear in terms of parameters $\log\alpha_i$ etc. Less restrictive is $(A, B) \perp\!\!\!\perp C$, or $\pi_{ijk} = (\alpha\beta)_{ij}\gamma_k$. The next step is conditional independence, $A \perp\!\!\!\perp C \mid B$. We can always write $\pi_{ijk} = \pi_{ik|j}\,\pi_j$ (where π_j denotes the marginal probability over i and k). Conditional independence means that $\pi_{ik|j}$ factorizes, and we get

$$\pi_{ijk} = \pi_{i|j}\,\pi_{k|j}\,\pi_j = (\alpha\beta)_{ij}(\beta\gamma)_{jk}\beta_j\,. \tag{10.5}$$

This says that separately for every class j of B, classifications A and C are mutually independent. These last two models can be described by the independence graphs shown in Figure 10.1.

The saturated multinomial model can be written

$$\log \pi_{ijk} = \theta_0 + \theta_i^A + \theta_j^B + \theta_k^C + \theta_{ij}^{AB} + \theta_{ik}^{AC} + \theta_{jk}^{BC} + \theta_{ijk}^{ABC}, \tag{10.6}$$

with ANOVA-type constraints for uniqueness of representation. Specifically, for any term we let the sum over any of its indices be zero, e.g. $\sum_i \theta_i^A = 0$ and $\sum_i \theta_{ij}^{AB} = \sum_j \theta_{ij}^{AB} = 0$. The restricted models for π_{ijk} just discussed, for example (10.5), are specified by deleting 'interaction' terms in the saturated model. For complete independence, delete the last four but keep the first four terms of (10.6). For the two conditional independence models of Figure 10.1, delete only two or three terms, for example θ^{AC} and θ^{ABC} to represent $A \perp\!\!\!\perp C \mid B$, (10.5). Note that the θ^{ABC} term disappears together with the lower order term θ^{AC} as soon as the conditional independence given B is assumed. This is clear from the factorization of π_{ijk} and reflects a property of conditional independence models necessarily being *hierarchical* log-linear models: If a term (e.g. θ^{AC}) is deleted from the model, all higher-order terms including the same factors (θ^{ABC} in this case) must also be deleted. However, not even all hierarchical models are graphical models. If only the last term (θ^{ABC}) is deleted from model (10.6), the resulting model remains hierarchical, but cannot be interpreted in terms of conditional independences. To characterize a graphical model we must refer to its clique sets and include in the model formula precisely those terms which correspond to the clique sets and to all subsets of cliques. Thus, if $\{A, B, C\}$ is a clique or a subset of a larger clique, θ^{ABC} must be allowed in the model. This type of discussion has no analogue in the Gaussian case.

Generally, allowing > 3 factors, a graphical model can be specified by an expression of type (10.6) for the table of probabilities, including all terms representing cliques and subsets of cliques, but no other terms. However, a representation such as (10.5) is simpler to work with than representations based on (10.6). For a decomposable model with an arbitrary number of factors, we then run through the cliques in suitable order, successively conditioning on separating sets s and simplifying, as in (10.2). We will make use of such representations. Let i be a multi-index for the cells of the table and i_c (or i_s) the corresponding index for the marginal table of the subset c (or s) of factors, obtained by summing over the other factors. Then,

$$\pi(i) = \frac{\prod_c \pi(i_c)}{\prod_s \pi(i_s)^{\nu_s}}, \qquad (10.7)$$

where the numerator product is over all cliques c of the graph, and the denominator product is over all successive separating sets s, the latter with their multiplicities ν_s, if they appear more than once. The denominator is necessarily independent of the factorization order, as long as that ordering is allowed.

As an illustration, for the 5-dimensional graph of Figure 10.2, denoting indices as (u, w, x, y, z), we obtain

$$\pi(u, w, x, y, z) = \pi(u|w)\,\pi(w|x, z)\,\pi(x, y, z) = \frac{\pi(u, w)\,\pi(w, x, z)\,\pi(x, y, z)}{\pi(w)\,\pi(x, z)}.$$

10.3.1 ML Estimation, in Particular for Decomposable Models

For any of the probability representations of the preceding section we conclude that the canonical statistic is formed by the marginal tables n_c of all cliques, with their saturated multinomial distributions. Generally, this is true for any hierarchical model. Thus, the likelihood equations (expressed in the mean value parameterization, and with multi-index) are

$$n\,\hat{\pi}(i_c) = n_c(i_c).$$

For simplicity, we assume all $n_c > 0$. This is sufficient to specify the complete table of probabilities uniquely. For decomposable models the estimator of the complete table is explicit, by using the representation (10.7) with estimates inserted. The result is

$$n\,\hat{\pi}(i) = \frac{\prod_c n_c(i_c)}{\prod_s n_s(i_s)^{\nu_s}}, \qquad (10.8)$$

Note the formal similarity with (10.3). There is one difference, however. Total independence between two sets of factors corresponds to an empty separator s, which must be counted in (10.8) in the form of $n_s = n$, but is not needed in (10.7), because additional factors $\pi(i_s) = 1$ do not make any change there.

As illustrations, we first consider the simple conditional independence model in the lower graph of Figure 10.1 and in (10.5). The separator is B, with margin $n_{.j.}$, and we get the MLE

$$n\,\hat{\pi}_{ijk} = \frac{n_{ij.}\,n_{.jk}}{n_{.j.}}, \tag{10.9}$$

where dot for an index means summation over that index. In the upper graph of Figure 10.1, we have total independence between $\{A, B\}$ and C. With the corresponding empty separator we get

$$n\,\hat{\pi}_{ijk} = \frac{n_{ij.}\,n_{..k}}{n}.$$

An explicit formula for the distribution of the ML estimator (10.8) can be written down, see Lauritzen (1996, Th. 4.22).

10.3.2 Testing for Conditional Independence in Decomposable Models

In analogy with Section 10.2.2, we discuss model reduction by elimination of one particular edge e from an independence graph. We here confine the discussion to the deviance test statistic. As an example, consider the hypothesis model $A \perp\!\!\!\perp C \mid B$, already treated in (10.9), that corresponds to deletion of one edge from the saturated model. Using (10.9), we obtain the deviance expression

$$D = 2\sum_{ijk} n_{ijk}\{\log(n_{ijk}/n) - \log(\hat{\pi}_{ijk})\} = 2\sum_{ijk} n_{ijk}\log\left(\frac{n_{ijk}\,n_{.j.}}{n_{ij.}\,n_{.jk}}\right). \tag{10.10}$$

Lauritzen (1996, Prop. 4.30) demonstrates that in essence the form of the deviance in (10.10) holds much more generally. As before, we assume the two models are decomposable, both with and without edge e. This has the consequence that e belongs only to a single clique (see Section 10.2.2), that we denote c_e (the full model in (10.10)).

- The deviance test statistic D and its distribution is the same in the marginal table for the clique c_e as in the whole table;
- If index j in (10.10) is interpreted as a multi-index for all indices of c_e except the two corresponding to e, the conditional two-way (i, k)-tables for given different j are mutually independent, their contributions to D are distributed as deviances for independence in two-way tables, and they add (sum over j).

The exact distribution of the deviance test statistic is quite complicated and depends on the rest of the table, but it can be simulated when needed, see Lauritzen (1996, Sec. 4.4.3) for details. If the cell numbers in the c_e-table are large, we can use the large sample χ^2 distribution for D.

Exercise 10.7 ***MLE in Figure 10.2***
Write down MLE formulas for a probability table corresponding to Figure 10.2. △

10.4 Models for Mixed Discrete and Continuous Variates

The extension of conditional independence models to the case when some component variates are continuous and some are discrete was developed by Steffen Lauritzen and Nanny Wermuth in the 1980s (Lauritzen and Wermuth, 1989). In such mixed models, we want to continue working with normal and multinomial distributions, respectively. We are then led to introduce the so-called Conditionally Gaussian (*CG*) distributions. This will be done first, before we discuss independence structures.

10.4.1 CG Distributions

Consider a random vector $\{I, \boldsymbol{Y}\}$, where I has a discrete probability distribution (perhaps on the cells of a contingency table, but that is irrelevant so far) and $\boldsymbol{Y}$ has a multivariate continuous distribution.

Definition 10.4 ***CG distributions***
The random vector $\{I, \boldsymbol{Y}\}$ is said to have a *CG distribution* if

- I has a marginal probability distribution on a finite set of 'cells', with positive probability π_i for each cell i.
- For each possible outcome $I = i$, $\boldsymbol{Y}$ has a conditional multivariate Gaussian distribution $\mathrm{N}(\boldsymbol{\mu}_i, \Sigma_i)$.

When we have a sample of size n from a CG distribution, we get a multinomial marginal distribution for the cell frequencies n_i, and conditionally on a given i a sample of size n_i from a Gaussian distribution. Note however that the marginal distribution of $\boldsymbol{Y}$ is typically not Gaussian, but a mixture of Gaussian distributions. This is a necessary sacrifice for ensuring conditional normality.

The assumption of multinomial cell frequencies need not necessarily be satisfied for a CG type of model to be useful. As mentioned before we sometimes generate the multinomial by conditioning on a Poisson total, but in other cases we want to condition on given partitionings of the index set. Edwards (1995, Ch. 4) has two examples of drug trials where the drug allocation on individuals is by randomization into fixed size treatment groups. This implies not only that π_i is known by design for each i (or component of vector i), but even a conditioning on given cell frequencies. In one of his examples, an additional categorical variate is the individual's sex, whose frequencies are also controlled by design.

The CG distribution can be specified by the so-called *moment characteristics* $\{\pi_i, \boldsymbol{\mu}_i, \Sigma_i\}$. A general simplification of the model is to assume that Σ_i in fact does not depend on i. The model is then called *homogeneous*. We will not specifically discuss homogeneous models here, however.

A comprehensive account of properties of CG distributions is given by Lauritzen (1996, Sec. 6.1.1). Here we will rather restrict to a minimum. First we note that except for the obvious constraint $\sum \pi_i = 1$, the distribution is a regular exponential family, whose canonical parameters are easily found from the joint density $f(i, \boldsymbol{y})$:

$$\begin{aligned} f(i, \boldsymbol{y}) &\propto \pi_i \, \{\det \Sigma_i\}^{-1/2} \exp\{-(\boldsymbol{y} - \boldsymbol{\mu}_i)^T \Sigma_i^{-1} (\boldsymbol{y} - \boldsymbol{\mu}_i)/2\} \\ &= \exp\{g_i + h_i^T \boldsymbol{y} - \boldsymbol{y}^T \Sigma_i^{-1} \boldsymbol{y}/2\}, \end{aligned} \tag{10.11}$$

which implicitly defines g_i and h_i. To express them in terms of π_i, $\boldsymbol{\mu}_i$ and Σ_i is left as an exercise. It follows from (10.11) that for a saturated model, without unnecessary constraints on the parameters, the canonical statistic consists of all cell frequencies n_i and for each cell its sample mean value vector and sample covariance matrix. A convenient alternative parameterization is the mixed one with the mean value parameters π_i for the cell frequencies instead of the corresponding complicated canonical parameters g_i. Note that in the saturated case, n_i is S-ancillary (see Section 3.5) for the inference about $\boldsymbol{\mu}_i$ and Σ_i, so inference about the probabilities π_i should be made in the marginal multinomial distribution of $\{n_i\}$, and about $\boldsymbol{\mu}_i$ and Σ_i in the conditional distribution given n_i.

Maximum likelihood estimation in the saturated mixed model is trivial (marginally multinomial frequencies and conditionally Gaussian samples), provided an important condition is satisfied: For each i, the conditional sample size given i must be large enough, so all sample covariance matrices are strictly positive definite. This could be an annoying condition, when n_i are multinomially random n_i. The corresponding condition is less restrictive for a homogeneous mixed model, but another possibility to reduce the parameter dimension of the covariance matrix is to try a graphical mixed model with conditional independences, which is the next topic.

10.4.2 Graphical Mixed Interaction Models

In this section we will restrict attention to CG distributions with two categorical variates A and B and two continuous variates Y and Z. This is enough for a flavour of generality while keeping notation simpler and more concrete than in the general case. Figure 10.4 shows a selection of independence graphs. In the mixed case it is important to distinguish between continuous and discrete variates, and in this section we adhere to the conventional use of filled *d*ots for *d*iscrete and unfilled *c*ircles for *c*ontinuous variates.

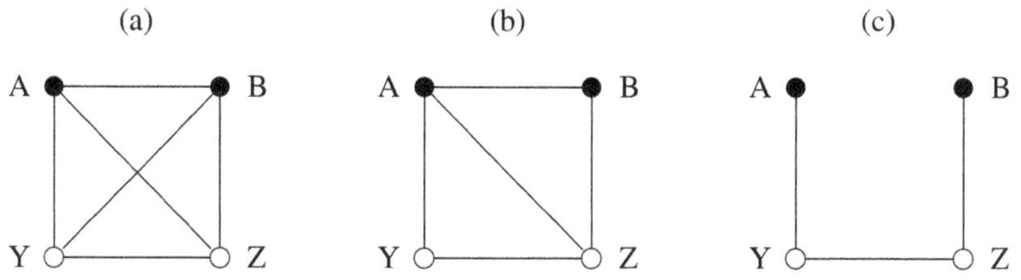

Figure 10.4 CG(2,2) conditional independence structures.
(a) Saturated; (b) $B \perp\!\!\!\perp Y \mid \{A, Z\}$; (c) Nondecomposable.

We start by reconsidering the saturated model seen in Figure 10.4(a). In analogy with the three-dimensional (10.6), a saturated model for a two-dimensional table of probabilities may generally be expressed by

$$\log \pi_{ij} = \theta_0 + \theta_i^A + \theta_j^B + \theta_{ij}^{AB}, \tag{10.12}$$

with ANOVA-type constraints as in (10.6) for uniqueness of representation. The last term is an interaction type of term, representing nonadditivity. Independence between A and B in this contingency table is equivalent to absence of this interaction term. In the same way as for $\log \pi_{ij}$ in (10.12) we may decompose the terms appearing in the exponent of (10.11), to express how they depend on (i, j). The resulting expression for the saturated model, represented by Figure 10.4(a), can be written as follows, where the letter λ is used in the decomposition of g, η in h and ψ in Σ^{-1}:

$$\begin{aligned}
g_{ij}^{AB} &= \lambda_0 + \lambda_i^A + \lambda_j^B + \lambda_{ij}^{AB}, \\
h_{ij}^{Y} &= \eta_0^Y + \eta_i^{AY} + \eta_j^{BY} + \eta_{ij}^{ABY}, \\
h_{ij}^{Z} &= \eta_0^Z + \eta_i^{AZ} + \eta_j^{BZ} + \eta_{ij}^{ABZ}, \\
\{\Sigma_{ij}^{-1}\}_{YY}/2 &= \psi_0^Y + \psi_i^{AY} + \psi_j^{BY} + \psi_{ij}^{ABY}, \\
\{\Sigma_{ij}^{-1}\}_{ZZ}/2 &= \psi_0^Z + \psi_i^{AZ} + \psi_j^{BZ} + \psi_{ij}^{ABZ}, \\
\{\Sigma_{ij}^{-1}\}_{YZ} &= \psi_0^{YZ} + \psi_i^{AYZ} + \psi_j^{BYZ} + \psi_{ij}^{ABYZ}.
\end{aligned} \tag{10.13}$$

In a model of homogeneous type, i.e. with Σ independent of $\{i, j\}$, interest is confined to the first three lines of (10.13).

Figure 10.4(b) represents a model with B and Y conditionally independent, given A and Z. By the Hammersley–Clifford theorem, which is applicable also to mixed models (because the joint density is strictly positive), there is a one-to-one relation between models specified by conditional independences and densities obtained by deleting suitable terms in the representation (10.13). In the present example this works as follows. To achieve the conditional independence between B and Y, all terms involving both of B and Y jointly are deleted from (10.13). This means deletion of $\eta_j^{BY} + \eta_{ij}^{ABY}$ from h_{ij}^Y, of $\psi_j^{BY} + \psi_{ij}^{ABY}$ from $\{\Sigma_{ij}^{-1}\}_{YY}/2$, and of $\psi_j^{BYZ} + \psi_{ij}^{ABYZ}$ from $\{\Sigma_{ij}^{-1}\}_{YZ}$. Note that the discrete part of the graph, with nodes $\{A, B\}$, forms a clique in that subgraph, represented by the maximal expression for the term g. Equivalently expressed, the elements of g_{ij}^{AB} are arbitrary, (except for λ_0 that serves as a norming constant). Free parameters $\{g_{ij}^{AB}\}$ correspond to free positive probabilities $\{\pi_{ij}\}$ (except for unit sum), so it does not matter if we parameterize by g_{ij} or by π_{ij}.

Figure 10.4(c) might look simple, but this is an illusion. That the Gaussian part is bivariate is of minor importance; most problems arise already with a univariate Gaussian variate. The root of the problems is that the discrete nodes/variables A and B are separated by continuous type node(s). In particular this destroys the decomposability from the pure case – triangulated is not enough in the mixed case. A mostly computational consequence is that there is no explicit solution to the likelihood equations. Furthermore, as in any CG model, $\{A, B\}$ is introduced by its marginal multinomial distribution, and we might be tempted to think it factorizes, i.e. that A and B are marginally independent. However, conditional independence between factors A and B given the continuous variates is expressed as $\lambda_{ij}^{AB} = 0$ in (10.13), and this does not carry over to marginal independence between factors A and B, expressed as $\theta_{ij}^{AB} = 0$ in (10.12), or vice versa. At least in situations when the discrete variates have an explanatory role, the advice is perhaps to avoid this type of structure by allowing arbitrary π_{ij}. This solves the problem completely in the case of a single continuous variate, but only connecting A with B in Figure 10.4(c) creates a 4-cycle, so this graph remains nondecomposable.

In a graphical model with arbitrary multinomial probabilities (so excluding for example Figure 10.4(c)), ML estimation is carried out as follows:

(1) The multinomial probabilities π_{ij}, representing the categorical variates, are estimated by equating them to the frequencies of the canonical statistic. Either we regard this as estimation in the marginal model for the discrete part of data, or we stay in the joint model using the mixed parameterization.

(2) The Gaussian part of the CG model is fitted by equating sample means and sample sums of squares and products to their expected values, selecting those statistics that correspond to all terms of type η and ψ present in the model. In principle this can be done explicitly for decomposable models, but unless the model is quite simple it is recommended to use a computer package for the model fitting (including the categorical variates).

Exercise 10.8 ***Comparison of CG(1,1) with other models***
(a) Compare the CG(1,1) with the standard one-way ANOVA model.
(b) Consider a CG(1,1) with a binary categorical variate, and compare with the model for a mixture of two normal distributions, Example 8.4. △

11

Exponential Family Models for Graphs of Social Networks

For modelling and studying social networks, the concept of an exponential random graph model has become an important tool, with interesting theoretical achievements in the decades around year 2000. Graphs can be directed or nondirected. We will restrict the discussion to nondirected graphs, just to keep the discussion simpler. For the same reason, we also leave aside dynamic graph models. Before we look at three different stages seen in the history of social network models, some concepts and notations must be introduced.

11.1 Social Networks

A social network will be represented by a graph, with a number of nodes, and edges between some of the nodes. The nodes represent the *social actors* under study, typically being individuals belonging to some group with social relations or interactions, but they could be corporations or other organizations. The relations could be 'mutual friendship', 'sharing information' or 'collaborating' (nondirected, mutual relations), or 'nominating as a friend', 'sending information to' or 'infecting' (directed), and they are represented by the presence of an edge between the corresponding pair of nodes. Thus, with n nodes, then $\binom{n}{2} = n(n-1)/2$ nondirected edges between distinct pairs are possible, and twice as many directed edges. We use an *indicator* y_{ij} to tell if there is a (nondirected) edge between nodes i and j ($i \neq j$), that is, $y_{ij} = 1$ when an edge is present, else $y_{ij} = 0$.

The matrix Y of y_{ij}-values, with the diagonal excluded, and with symmetry for a nondirected graph, is called the *adjacency matrix*, or sociomatrix. For a nondirected graph we only need the triangular subset $\{y_{ij}\}_{i<j}$.

Typically, we have useful additional information about the social actors, in the form of a vector $\boldsymbol{x}_i$ of auxiliary or explanatory variates (attributes), such as sex, race, age, education, school class membership. The matrix over all individuals of this vector is denoted X. We are interested in a probability description of the network, but regard X as fixed. Thus, if X is regarded as random in some sense, inference will be taken to be conditional on X, as in usual regression.

The most basic characteristics of the graph are the number of nodes and the number of edges (or the average number of edges per node), but it is less obvious how other characteristics should be chosen to yield an adequate representation of a social network. For example, the degree of a node is its number of edges to other nodes, and the *degree distribution* for a graph is a popular study object in probability theory, but has typically been regarded as less informative in the social networks context.

11.2 The First Model Stage: Bernoulli Graphs

In a nondirected Bernoulli graph model, all $\binom{n}{2}$ *dyads* (node pairs, potential edges) are assumed mutually independent. For a nondirected *homogeneous* Bernoulli graph, that is, a graph structure with exchangeable nodes, and in particular with the same probability π_0 for all edges (also called an Erdös–Rényi graph), the probability for any specified graph with an observed number $r(Y) = \sum_{i<j} y_{ij}$ of edges is the same as for a specified sequence of Bernoulli trials with success probability π_0 and the same length as there are dyads, that is,

$$Pr(\{y_{ij}\}; \pi_0) = \pi_0^{r(Y)} (1 - \pi_0)^{\binom{n}{2} - r(Y)} = e^{\,\mathrm{logit}(\pi_0)\, r(Y)}\, (1 - \pi_0)^{\binom{n}{2}}\,.$$

Thus, the canonical statistic is simply $r(Y)$, and the canonical parameter is $\theta = \mathrm{logit}(\pi_0)$, in this context called the *density parameter*. The last factor represents the norming constant,

$$C(\theta) = (1 - \pi_0(\theta))^{-\binom{n}{2}}\,.$$

We can introduce more structure in the model without losing the Bernoulli character. For example, if the nodes are divided in two blocks (e.g. by sex) with exchangeability within blocks, we can let π_0 be replaced by π_{11}, π_{22} and π_{12} for within each block and between blocks, respectively, and correspondingly use $r_{11}(Y)$, $r_{22}(Y)$ and $r_{12}(Y)$ as canonical statistics, where r_{11} and r_{22} are the within block number of edges, and r_{12} is the number between blocks, $r_{11} + r_{22} + r_{12} = r$. As an exponential family this is

$$Pr(Y; \theta_{11}, \theta_{22}, \theta_{12}) = \frac{1}{C(\boldsymbol{\theta})}\, e^{\theta_{11} r_{11}(Y) + \theta_{22} r_{22}(Y) + \theta_{12} r_{12}(Y)}\,, \qquad (11.1)$$

with $\theta_{11} = \mathrm{logit}(\pi_{11})$, etc, as components of a vector $\boldsymbol{\theta}$, and

$$C(\boldsymbol{\theta}) = (1 - \pi_{11}(\theta_{11}))^{-n_1(n_1-1)/2} (1 - \pi_{22}(\theta_{22}))^{-n_2(n_2-1)/2} (1 - \pi_{12}(\theta_{12}))^{-n_1 n_2}\,.$$

This can be seen as a logistic regression type model, if we let $x = 0$ and $x = 1$ characterize the two blocks and write $r_{11} = \sum_{i<j} y_{ij} x_i x_j$, $r_{22} = \sum_{i<j} y_{ij}(1 - x_i)(1 - x_j)$, and $r_{12} = \sum_{i<j} y_{ij}(x_i(1 - x_j) + (1 - x_i)x_j)$.

Other explanatory variates could also be introduced, without destroying the Bernoulli property. At the extreme end of the scale we have the saturated statistical model, with one parameter per dyad,

$$Pr(Y;\ \{\pi_{ij}\}) = e^{\sum_{i<j} \mathrm{logit}(\pi_{ij})\, y_{ij}} \prod_{i<j}(1 - \pi_{ij})\,.$$

Not surprisingly, however, it was realized that models with mutually independent edges were far too simple to fit real social networks. In particular, Bernoulli graphs are unable to model *transitivity*, defined as the tendency for actors to form triangles ('Friends of my friends are my friends'). The next stage was entered when Ove Frank and David Strauss in a seminal paper (Frank and Strauss, 1986) introduced social networks models with Markov dependence, inspired by developments in spatial statistics.

11.3 Markov Random Graphs

Frank and Strauss (1986) utilized the fact that a Markov graph, in a sense consistent with its nearest neighbours definition in spatial models, must have a special type of exponential family representation of its probability distribution (the Hammersley–Clifford theorem, already met a few times in Chapter 10). This Markov property means that two dyads (node pairs) are conditionally independent, given the rest of the graph, unless they have a node in common. We do not go deeper into the motivating theory here, since we will go outside this model class in the next section. Instead we refer to the Frank & Strauss paper for details. The canonical statistic of such a model involves only neighbour dependence in the form of counts of edges, triangles and stars. A *star* is a set of edges radiating from one and the same node. A k-star is a star formed by k edges, radiating from the same node. Each edge from a node taken by itself is a 1-star. If the degree of a node is k, the maximal star in that node is a k-star, but there are also a number k of $(k-1)$-stars, etc, down to k 1-stars. See Figure 11.1 for an illustration of 2-stars and 3-stars.

More precisely, the probability of observing an edge pattern (adjacency matrix) Y under a homogeneous Markov graph structure on n nodes is

$$Pr(Y;\ \boldsymbol{\theta}) \propto e^{\rho\, r\, +\, \sum_{k=2}^{n-1} \sigma_k s_k\, +\, \tau\, t} \tag{11.2}$$

where $\boldsymbol{\theta} = (\rho, \sigma_2, ..., \sigma_{n-1}, \tau)$ and the canonical statistic consists of

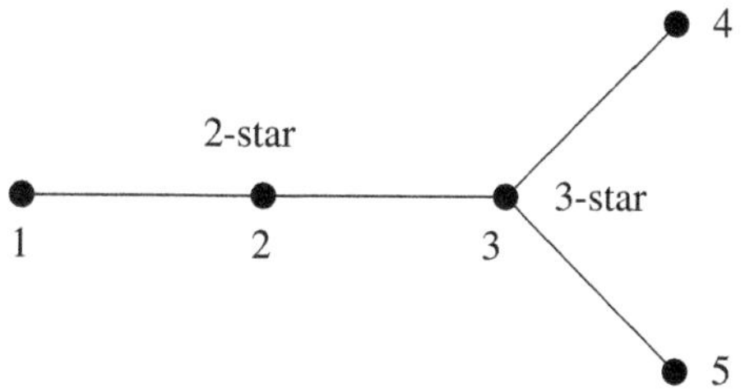

Figure 11.1 Graph with five nodes, four edges, four 2-stars, and one 3-star. One 2-star is centered at node 2 and the 3-star radiates from node 3, where the other three 2-stars are also centered.

- $r = s_1/2$ = total number of edges $= \sum_{i<j} y_{ij}$
- s_k = total number of k-stars, $k = 2, \ldots, n-1$,
  $s_k = \sum_i \sum_{j_1 < \ldots < j_k} y_{ij_1} y_{ij_2} \cdots y_{ij_k}$
- t = total number of triangles $= \sum_{i<j<k} y_{ij} y_{ik} y_{jk}$

The whole set of k-stars $\{s_k\}$, including r, is rich enough to be mathematically equivalent to the degree distribution. Details are omitted.

A special case that has been of particular interest is the so called *triad model*, with only three parameters, obtained by setting $\sigma_k = 0$ for $k > 2$. In principle this model is capable of incorporating both clustering and triangle-forming (transitivity) effects:

$$Pr(Y; \boldsymbol{\theta}) \propto e^{\rho\, r(y) + \sigma\, s(y) + \tau\, t(y)}, \tag{11.3}$$

where $s = s_2$, in previous notations. Model (11.3) is homogeneous, but as with Bernoulli graph models, we need not demand homogeneity, but can allow block structure, cf. (11.1). Details are omitted.

The norming constant for the triad model (11.3) can be written

$$C(\boldsymbol{\theta}) = \sum_{\text{all } Y} e^{\rho\, r(y) + \sigma\, s(y) + \tau\, t(y)} = \sum_{r,s,t} f(r, s, t)\, e^{\rho\, r + \sigma\, s + \tau\, t}, \tag{11.4}$$

where the first sum is over all $2^{\binom{n}{2}}$ possible adjacency matrices, and the second sum is over all possible combinations of r, s and t, and the structure function $f(r, s, t)$ is the number of adjacency matrices with specified (r, s, t).

The inference about the model parameters is more difficult than for Bernoulli models, because the norming constant $C(\boldsymbol{\theta})$ is analytically intractable, being a sum over all the $2^{\binom{n}{2}}$ possible adjacency matrices of the exponential factor (11.2) or (11.3). The number of values of the canonical statistic is much smaller, but yet large, and the structure function has no explicit form, either. As a consequence, we do not have an explicit form of the likelihood equations, nor of the information matrix. Bayesian inference also runs into problems of the same dignity. Among approximate methods devised to solve the estimation problem, the following pseudo-likelihood method has dominated practice, together with numerical simulations of graphs for different parameter values by help of MCMC-type procedures, see Section 11.3.2 for descriptions.

11.3.1 Pseudo-Likelihood Estimation

The method starts by specifying a Besag-inspired *pseudo-likelihood* (more generally regarded as a *composite likelihood*). This is formed by first calculating, for each dyad $\{i, j\}$, the conditional probability for the observed edge or no-edge y_{ij} between these nodes, given the rest of the graph, this complement denoted Y_{ij}^c. Next, these conditional probabilities are multiplied over all dyads $\{i, j\}$ as if they corresponded to mutually independent outcomes (a deliberate misspecification!):

$$L_{pseudo}(\boldsymbol{\theta}; Y) = \prod_{i<j} Pr(y_{ij} \mid Y_{ij}^c). \tag{11.5}$$

The Markov property implies that each complementary adjacency matrix Y_{ij}^c can be reduced to the set of nodes having edges in common with i or j, but this characterization is not of importance for the argument.

Factors in the product (11.5) have the form

$$Pr(y_{ij} \mid Y_{ij}^c) = \frac{Pr(y_{ij}, Y_{ij}^c)}{Pr(Y_{ij}^c)} = \frac{Pr(y_{ij}, Y_{ij}^c)}{Pr(y_{ij} = 0, Y_{ij}^c) + Pr(y_{ij} = 1, Y_{ij}^c)}$$

Again, as was the case for the Bernoulli graph models, this model is equivalent to a logistic regression. This will be demonstrated for the triad model

(11.3), being a typical example. Going over to logits, (11.3) yields

$$\text{logit}(Pr(y_{ij} = 1 \mid Y^c_{ij})) = \log \frac{Pr(y_{ij} = 1,\ Y^c_{ij})}{Pr(y_{ij} = 0,\ Y^c_{ij})} = \rho\,(\Delta r)_{ij} + \sigma\,(\Delta s)_{ij} + \tau\,(\Delta t)_{ij}\,,$$

where $(\Delta r)_{ij}$, $(\Delta s)_{ij}$ and $(\Delta t)_{ij}$ are the changes in r, s and t when an edge is added between nodes i and j in a graph with $y_{ij} = 0$, and with Y^c_{ij} fixed. In particular, necessarily $\Delta r = 1$, whereas $\Delta s \geq 0$ and $\Delta t \geq 0$ will depend on the rest of the graph, Y^c_{ij}.

Thus, to find the pseudo-ML estimate of (ρ, σ, τ) we can use an ordinary statistical package for logistic regression. We have $\binom{n}{2}$ observations, with y_{ij} as binary response and regressors $x_1 = \Delta r = 1$ (so ρ is a regression intercept), $x_2 = \Delta s \geq 0$, $x_3 = \Delta t \geq 0$. For dyad (i, j), the values of these regressors are found by considering the effect of adding an edge if the observed response was $y_{ij} = 0$, or by deleting this edge from the observed graph if $y_{ij} = 1$ was observed. For a concrete illustration, see Section 11.4, in particular Table 11.2. For general conditions under which logistic regression yields a unique root, see the references given in Example 3.5.

11.3.2 MCMC Likelihood Estimation

We start by the general exponential family formula

$$\log L(\theta) - \log L(\theta_0) = (\theta - \theta_0)^T\, t(y_{obs}) - \log\{C(\theta)/C(\theta_0)\}, \qquad (11.6)$$

where t is the canonical statistic of the model. The idea now, due to Geyer and Thompson (1992), is to let a starting value θ_0 be successively updated by a simulation procedure. The pseudo-MLE is a standard choice of starting value (when it exists). The last term of (11.6) can be written

$$\log\{C(\theta)/C(\theta_0)\} = \log E_{\theta_0}\left\{e^{(\theta-\theta_0)^T t}\right\},$$

since

$$C(\theta)/C(\theta_0) = \sum_{\text{sample space for } y} e^{(\theta-\theta_0)^T t(y)} e^{(\theta_0)^T t(y)} / C(\theta_0).$$

Now $y_1,\ldots,y_m$ are simulated from the θ_0-distribution and we refer to the law of large numbers when approximating $E_{\theta_0}\left\{e^{(\theta-\theta_0)^T t}\right\}$ by the average

$$\frac{1}{m}\sum_{i=1}^{m} e^{(\theta-\theta_0)^T t(y_i)}.$$

The simulation under θ_0 is carried out by some MCMC method, because $C(\theta_0)$ has no simple form so the distribution cannot be handled explicitly.

A risk for failure, however, is that the MCMC sequence gets stuck in degenerate or near degenerate graphs, see towards the end of Section 11.3.3.

11.3.3 Properties of Likelihood Estimates

When the MCMC-ML or pseudo-ML estimates have been found, there are still problems, some of which will be illustrated in the example of Section 11.4. The estimates of the parameter components can be strongly positively correlated. This is essentially because adding edges will not only increase the total number of edges, r, but usually also the total number of k-stars and triangles. Even conditionally on the observed number of edges r, the two statistics s and t will be correlated.

The ML theory for exponential families, from Chapter 3, tells that the MLE of θ is approximately distributed with

$$\mathrm{var}(\hat{\theta}_{ML}) \approx I(\theta)^{-1} = \mathrm{var}_\theta(r, s, t)^{-1}, \tag{11.7}$$

and at least provisionally let us assume that the pseudo-ML estimator has a similar correlation matrix. Thus, when the correlations are strong between r, s, and t, the parameter estimates are not only strongly correlated, but have large variances (inversion of near-singular matrix). This implies that confidence intervals for the individual parameters will be wide.

One way to partially avoid this particular problem could be to use a mixed parameterization. A subvector of θ together with the complementary subvector of $E_\theta(t)$ will have approximately uncorrelated MLEs.

Another problem is the interpretation of an (estimated) parameter value. What do the components of θ tell about the tendency of actors to form clusters and to form triangles (transitivity)? Handcock (2003) advocates the mean value parameterization or a mixed parameterization (see more in Section 11.3.4), as being simpler than the canonical in this respect. However, this does not mean that the interpretation becomes easy. A difference in the mean value vector can also be quite difficult to interpret in terms of social behaviour of the actors. The problem remains, that with these models the parameter vector in any parameterization seems difficult to interpret and might be difficult to connect with social science research questions posed for real networks.

A frequent and related problem with Markov graph models is a phenomenon called *near degeneracy* (Handcock, 2003). A specified model (for simulation) may turn out to have very low probability for all but one or a few possible graph configurations, which are degenerate (empty or complete) or near degenerate. Observed such graphs are not of interest to model, so the corresponding parameter values lack interest. Also, when parameters are changed, the model may change in a steep way from one type of degeneracy to the other. To get a flavour of how the complete graph can dominate, let us see how the probability in (11.3) changes from a complete graph to a graph with exactly one absent edge. Deleting one edge will reduce s and t by $2(n-2)$ and $(n-2)$, respectively. Thus any graph with one edge absent will have a probability smaller than that of the complete graph by a factor

$$e^{-\rho - (n-2)(2\sigma + \tau)}$$

Assuming n large, $\rho \geq 0$, and $2\sigma + \tau$ positive and not close to zero, this factor will be quite small. Deleting one more edge, the probability will be reduced by a similar factor again (slightly depending on what edge is deleted), etc. In other words, under these circumstances the probability will shrink quickly with each edge deleted from the complete graph, at least so long as the resulting graph is almost complete. This is a serious problem for MCMC type methods, because they can fail by getting stuck in the complete (or empty) graph, even though the actual graph observed is far from both these extremes.

The near degeneracy problem is further elucidated by Chatterjee and Diaconis (2013), who use large deviation approximations of the norming constant $\log C(\theta)$ in large sample situations (large n). They note that the Markov type graph is in the limit typically indistinguishable from an Erdös–Rényi type graph, and they also give conditions for and prove the near-degeneracy phenomenon, coupled with the occurrence of an underlying phase transition. To illustrate, they mostly use the (ρ, τ) model (edges and triangles), where the near-degeneracy appears whenever ρ is negative and numerically large enough. The exact small sample illustration here in Section 11.4 is a lightweight supplement to the heavyweight large n study by Chatterjee and Diaconis (2013).

11.3.4 Conditional Inference

The mixed parameterization, touched on in Section 11.3.3, comes natural if the conditionality principle in Section 3.5 is applied: If we are specifically interested in one component of θ, say τ, the inference should be conditional on (r, s). This yields a one-dimensional likelihood, which is easily expressed by restriction to the observed (r, s) as

$$L_c(\tau;\, t_{obs} \,|\, r = r_{obs}, s = s_{obs}) = \frac{f(r_{obs}, s_{obs}, t_{obs})\, e^{\tau t_{obs}}}{\sum_t f(r_{obs}, s_{obs}, t)\, e^{\tau t}}\,.$$

The structure function f in the numerator is a constant factor that can be omitted, but the one in the denominator causes the same type of numerical problems as in the joint model. For very small graphs, as in the next example, Section 11.4, the likelihood can be handled explicitly, but otherwise MCMC-ML is again a natural choice to try.

11.4 Illustrative Toy Example, $n = 5$

For a graph with five nodes we can make by hand a complete characterization of all possible graphs. Even if it is quite small it works well to illustrate the general discussion of Section 11.3 and its subsections. Table 11.1 contains this complete characterization of the $2^{10} = 1024$ graphs.

Table 11.1 *Frequencies of different graphs with five nodes*

r	s	t	freq	r	s	t	freq	r	s	t	freq
0	0	0	1	4	5	1	60	6	10	2	75
1	0	0	10	4	6	0	5	6	11	2	60
2	0	0	15	5	5	0	12	6	12	4	5
2	1	0	30	5	6	0	60	7	13	2	30
3	1	0	30	5	6	1	60	7	14	3	60
3	2	0	60	5	7	1	60	7	15	4	30
3	3	0	20	5	8	1	30	8	18	4	15
3	3	1	10	5	8	2	30	8	19	5	30
4	3	0	60	6	9	0	10	9	24	7	10
4	3	1	10	6	9	1	60	10	30	10	1
4	4	0	75								

Figure 11.1 showed one such graph, with $r = 4$, $s = s_2 = 4$, $t = 0$. Figure 11.2 depicts the support sets of all possible values of (r, s) and of (r, t), both sets encircled by their convex hulls. The family (11.3) for (r, s, t) (or (r, s) or (r, t)) is regular (because the support set is finite, see Proposition 3.7), so Proposition 3.13 tells that the MLE exists precisely when the observed canonical statistic is one of the support points in the interior of this convex hull. Remarkable then to see the support sets restricted to relatively narrow strips on and near the lower boundary of the convex hull! On the other hand, the possible mean values of (r, s) and (r, t), respectively, fill these open convex hulls (since the likelihood equations $\boldsymbol{t} = E_\theta(\boldsymbol{t})$ have a solution there), so the support points are concentrated in a small and curved

part of the mean value space $\boldsymbol{\mu}(\boldsymbol{\Theta})$. Furthermore, note that the points on the lower boundary do not correspond to (finite) maximum likelihood estimates in the canonical parameter.

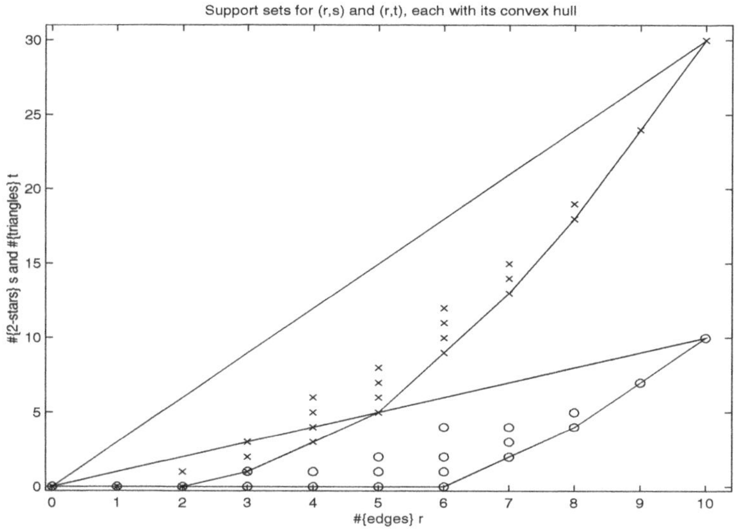

Figure 11.2 For $n = 5$, support sets for (r, s) (crosses) and for (r, t) (circles), and their convex hulls, representing possible mean value parameter points.

Figures 11.3, 11.4 and 11.5 represent the simple model having only the numbers r and s of edges and 2-stars in its canonical statistic,

$$Pr(\{y_{ij}\}; \theta) \propto e^{\rho r(y) + \sigma s(y)}, \tag{11.8}$$

that is (11.3) with $\tau = 0$. As remarked by Handcock (2003), this is an analogue both of the Ising model, Example 2.16, and of the Strauss model for point processes, Section 13.1.2.

Figure 11.3 shows the image under the mean value function of a rectangle in the canonical space $\mathbb{R}^2$ for (ρ, σ). The rectangle is centered in the origin, which corresponds to a Bernoulli model with edge probability $\pi_0 = 1/2$. The width of σ is ± 0.3, whereas the width of ρ is ± 1, to reflect a higher sensitivity to changes in σ. Note first how distorted the canonical rectangle is in the mean value space, being turned into a thin long curved strip, no longer convex. It should then not come as a surprise that ML estimation can run into problems. Considering the support set of (r, s)-values, note that almost all possible outcomes are either within the mean value

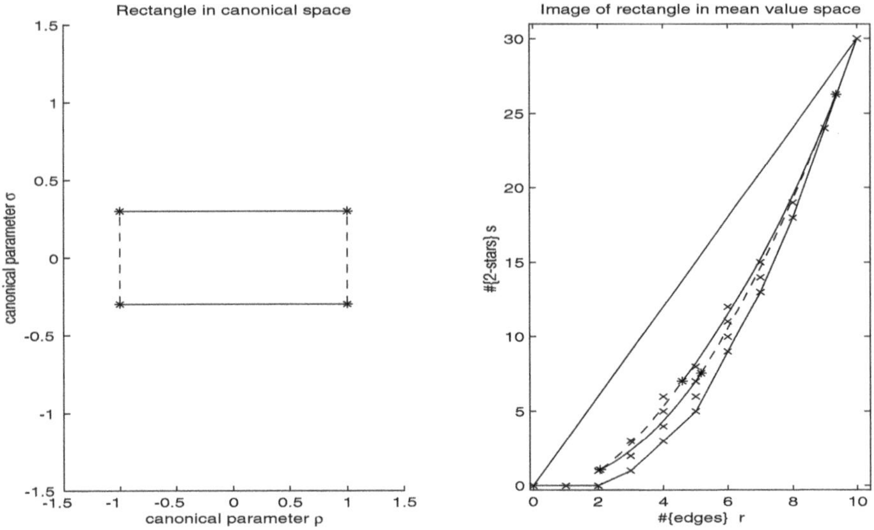

Figure 11.3 For $n = 5$ and the canonical statistic (r, s), the left part shows a rectangle in the canonical parameter space, centered at the origin, and the right part its image as a long but thin, curved strip in the mean value parameter space. The support set is also shown, as crosses, cf. Figure 11.2

space image of our rectangle or at the boundary of the mean value space, corresponding to infinite canonical parameter values.

The information matrix for (ρ, σ). that is, the variance-covariance matrix of (r, s), has a correlation of magnitude 0.9 in most of the rectangle, but as we move towards its upper right corner the already strong correlation increases to 0.996. These values then also represent the correlation between $\hat{\mu}_r = r$ and $\hat{\mu}_s = s$, and with a crude approximation also between the MLEs of ρ and σ. With an analogous crude approximation for a mixed parameterization the MLE components are uncorrelated. However, these approximations when (ρ, σ) is in the upper right corner are not at all reliable. As seen (with some difficulty) in the right part of Figure 11.3, the corresponding mean value point has $E(r) > 9$, so it is in the region where all outcomes are on the boundary.

The small n-value in this case allows $C(\boldsymbol{\theta})$ to be calculated by enumeration, formula (11.4). Figure 11.4 shows a surface plot for $\log C(\boldsymbol{\theta})$. We see a flat region and a region of constant slope, with a sharp transition from one to the other (in particular for $\rho < 0$). The mean values are the par-

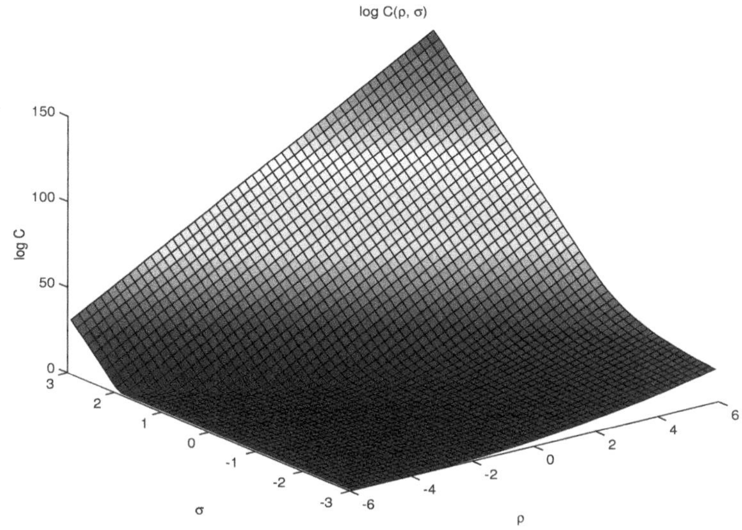

Figure 11.4 Surface plot of $\log C(\rho, \sigma)$ over the rectangle $|\rho| < 6, |\sigma| < 3$,

tial derivatives of $\log C(\theta)$, so in the flat region the mean values of r and s are close to zero (empty graph). In the steep slope region the gradient components are essentially the maximal r and s values, corresponding to a completely filled graph, that is, $r = \binom{n}{2} = 10$ and $s = \binom{n}{3}\binom{3}{1} = 30$. Thus, the transition from slope to flatness occurs quite dramatically near the line $\sigma = -\rho/(n-2) = -\rho/3$ (see also Park and Newman (2004) and Chatterjee and Diaconis (2013) for corresponding large n results). This feature of $\log C(\boldsymbol{\theta})$ has a great influence on the log-likelihood and the maximum likelihood estimation, that we will look further into.

The upper part of Figure 11.5 illustrates the log-likelihood surface contours in $\boldsymbol{\theta} = (\rho, \sigma)$ for two possible outcomes with quite different r-values, $(r, s) = (3, 3)$ and $(r, s) = (6, 12)$. Note that the MLE is not very sensitive to the statistics but the likelihood slopes are quite different. The form of $\log C(\boldsymbol{\theta})$ explains the forms seen. Since the log-likelihood is

$$\log L(\rho, \sigma) = \rho\, r + \sigma\, s - \log C(\rho, \sigma),$$

it will be essentially $\rho\, r + \sigma\, s$ to the lower left of $\sigma = -\rho/3$ and $\log L \approx \rho\,(r - 10) + \sigma\,(s - 30)$ $(r \leq 10,\ s \leq 30)$ to the upper right of the same line. Thus the MLE will be found near this line whatever r and s is observed, as long as the graph is not essentially empty or complete.

Model degeneration is a problem that is even more pronounced for larger graphs, but it can be seen already in this example. For (ρ, σ) in the rectangle of Figure 11.3, the upper right corner yields a probability of 0.86 for the graph to be full or have only one edge absent. This is not a dramatically high probability. However, going twice as far out in that direction, to $(\rho, \sigma) = (+2, +0.6)$, the corresponding probability is 0.999. Such parameter values will not be of interest in modelling, but can be a serious problem for MCMC-ML techniques.

We return to Figure 11.5 with its two outcomes $(r, s) = (3, 3)$ and $(r, s) = (6, 12)$. In both cases the number of 2-stars is the maximal s for the given r-value. Under a Bernoulli model with density parameter $\rho = 0$ ($\pi_0 = 1/2$), the outcome $(r, s) = (6, 12)$, should be rare, having respectively the joint and conditional probabilities

$$\Pr\{(r, s) = (6, 12)\} = 0.5\%,$$
$$\Pr\{(s = 12 \mid r = 6)\} = 2.4\%.$$

The corresponding probabilities for the outcome $(3, 3)$ are much larger,

$$\Pr\{(r, s) = (3, 3)\} = 3.0\%,$$
$$\Pr\{(s = 3 \mid r = 3)\} = 25\%.$$

The likelihood surface for $(r, s) = (3, 3)$ is remarkably flat in any lower left direction from the MLE and steep in the opposite direction. The lower left part of the contour plot corresponds to mean value points located towards the origin, in particular with small $E(s)$, whereas the upper right corresponds to points near the maximum mean value point, $(10, 30)$.

In these diagrams, the solid innermost oval curves show the parameter regions with deviance < 6.0 (as compared with the MLE). This corresponds to an asymptotic 95% confidence under a $\chi^2(2)$ distribution, but since the graph is small, the actual degree of confidence may differ substantially. While keeping this in mind, we will talk about it as a 95% confidence region. Note how the confidence regions tend to fill almost the whole mean value space, much more densely near its lower boundary, but leaving a more or less visible strip near the boundary.

The ridge-like form of the likelihood contours in canonical space indicates that the observed information matrix is not far from singular, The lines of circle points in the sparse middle region of the mean value space correspond to lines in the canonical space parallel with and close to the ridge (close to the major axis of the confidence 'ellipse'). The sparsity of

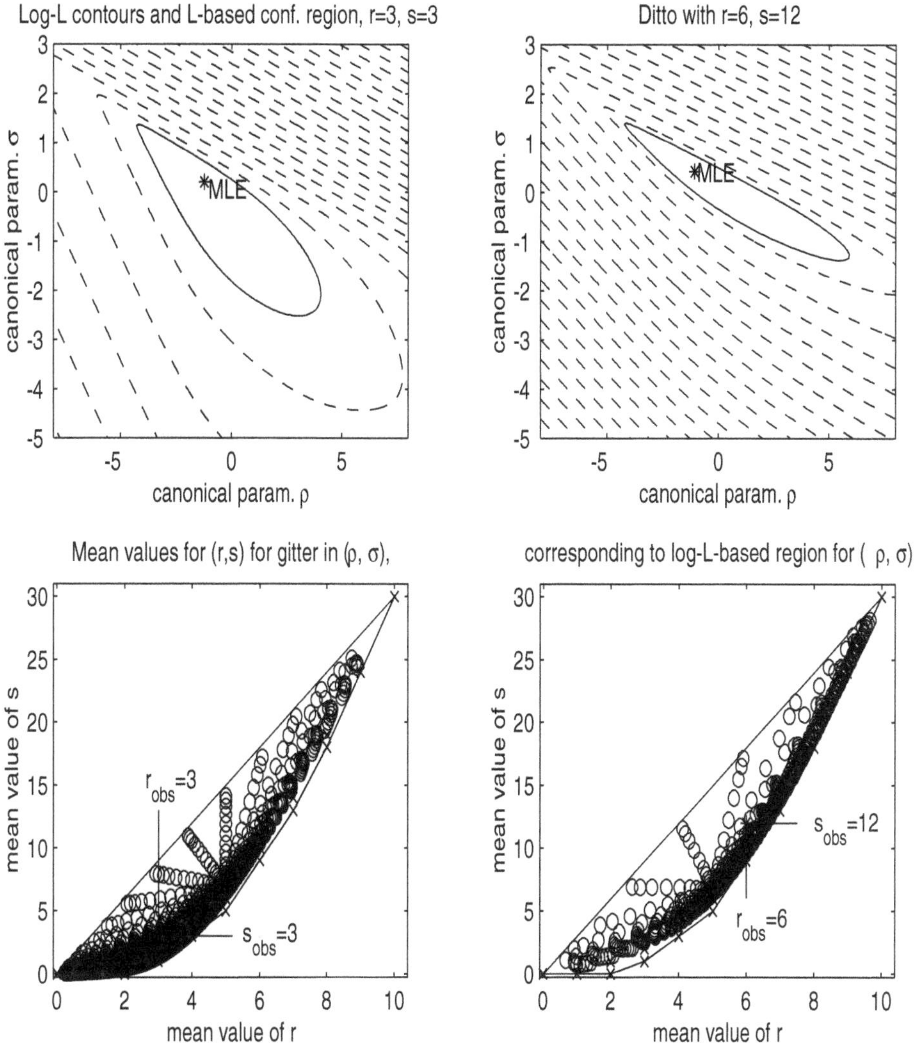

Figure 11.5 Likelihood contours and confidence regions in model (11.8), with $n = 5$ and canonical statistic (r, s);
Left: Outcome $r = 3$, $s = 3$.
Right: Outcome $r = 6$, $s = 12$.
Top: Likelihood surface contour curves in (ρ, σ), with MLE and standard approximate 95% likelihood-based confidence regions (solid curve).
Bottom: Images in mean value space (cf. Figure 11.3) of a dense gitter representation of the confidence regions in (ρ, σ). The convex hull of the set of all such images (circles) approximates the confidence region well. In both cases, this region covers almost the whole convex hull of the support set.

such lines in the diagrams shows that the mean value point reacts strongly to changes in the canonical parameter orthogonally to the ridge.

Pseudo-likelihood estimation: This method was described in and around formula (11.5). Consider $(r, s) = (3, 3)$ and estimation of (ρ, σ). There are 10 dyads to consider, 3 of which have $y = 1$, since $r = 3$. The value $s = 3$ appears for example when a 3-star is formed at one node, which leaves one node isolated. Table 11.2 shows how s will change by adding an edge at any dyad of this graph, already observed edges corresponding to $y = 1$. Thus we have a logistic regression with an intercept ρ, and with Δs as regressor with three different values. Fitting this logistic regression yields the pseudo-MLE $(\tilde{\rho}, \tilde{\sigma}) = (-2.38, 0.82)$. However, the data of Table 11.2 are quite bad for estimation, and do not allow any precision, because only the middle value of the regressor Δs yields response values $y = 1$. As a consequence, the standard errors of the coefficient estimates are relatively high, $s.e.(\tilde{\rho}) = 2.4$, $s.e.(\tilde{\sigma}) = 1.2$. The pseudo-MLE is also quite different from the MLE, which is $(\hat{\rho}, \hat{\sigma}) = (-1.22, 0.20)$. However, the log-likelihood in the pseudo-MLE point is only about one unit lower than in the proper MLE (i.e. two units in deviance).

Table 11.2 *For each of the 10 dyads, Δr and Δs values followed by the observed y-value, assuming $r = s = 3$ with a 3-star and an isolated node:*

$\sharp$\{dyads\}	Δr	Δs	y	Comment
3	+1	+1	0	connecting the isolated node to a 1-star
3	+1	+2	0	forming a triangle among the 1-stars
3	+1	+2	1	adding one of the actually existing edges
1	+1	+3	0	making a 4-star of the 3-star

If $(r, s) = (6, 12)$, the situation is even worse for the pseudo-likelihood method. The regressor Δs has only two different values, one of them yielding $y = 0$ and the other yielding $y = 1$. This is a degenerate situation, and there is no finite pseudo-ML estimate. One practical implication is that the pseudo-likelihood estimate cannot possibly be used as a starting point for MCMC type procedures. This is the default starting point in `package: ergm`, see next section.

Conditional inference: In Section 3.5 generally, and Section 11.3.4 specifically, conditional inference was recommended for a canonical parameter, in particular in conjunction with a mixed parameterization. In the present little toy example, how does conditional inference work for σ, given r?

Suppose again that we have observed $(r, s) = (6, 12)$. The conditional family for s given $r = 6$ is a regular exponential family with canonical statistic s, canonical parameter σ, with the whole real line as parameter space (see Proposition 3.7). The support set is $\{9, 10, 11, 12\}$. Since $s = 12$ is maximal, and thus in particular located on the boundary of the convex hull of this support set, it will not correspond to a finite ML estimate of σ (referring to Proposition 3.13). An analogous conclusion holds of course for $s = 9$, but with a different sign for infinity, whereas $s = 10$ or $s = 11$ would yield finite likelihood maxima. Returning to $s = 12$, the conditional likelihood function is a monotone function tending to 1 as $\sigma \to \infty$, and it can be used to compare different σ-values. For example, $2 \log L$ is almost 8 units lower for $\sigma = 0$ than for large σ-values ($\sigma > 5$, say). Thus, a conditional likelihood-based interval would be completely on the positive side, whereas $\sigma = 0$ was more centrally positioned in the joint likelihood-based region in Figure 11.5. Personally, I would rather trust the conditional result. However, the example is too small to be at all convincing.

11.5 Beyond Markov Models: General ERGM Type

Even though the concept of Markov graph models was a great step forward, applications soon showed that they were still insufficient for describing social networks in any realistic generality, and they had intrinsic problems connected with the near degeneracy phenomenon. The simple triad model was studied extensively, and adding higher order k-star numbers to the sufficient statistic did not help. These models were all found inadequate both for describing transitivity (triangle formation) in reality, and for describing 'core–periphery' structures, having some particularly attractive actors in the core. It was found that the Markov graph models could be extended to much more general exponential families without further complications. The class *ERGM* of Exponential Random Graph Models was coined (an essentially parallel name is p^* models). Simulation and estimation techniques were the same as needed already for the Markov graphs, that is, pseudo-likelihood estimation and MCMC type methods.

In the first decade of the new millennium, software for handling network data has been developed that allows very large networks to be comprehensively analysed. Two examples in the literature are an adolescent friendship network of $n = 1681$ actors (Goodreau, 2007) and a school pupils network of $n = 1461$ (Goodreau et al., 2008). They were analysed using the `ergm` program package, a toolbox in the `statnet` library of **R**-based packages for handling network models, network data, and other relational data sets.

The `ergm` package provides tools for model parameter estimation, simulation of networks under a specified model, and various goodness of fit statistics. The parameter estimation assumes a full or curved exponential family. The simulation uses MCMC procedures. To check the goodness of fit of a model, the observed network can be compared with suitable statistics for simulated networks, for example by comparing observed and simulated degree distributions.

`RSIENA` is another **R**-package (SIENA = Simulation Investigation for Empirical Network Analysis). Its focus is on longitudinal and dynamical network studies (outside the scope here).

One of the model extensions that is incorporated in the packages is to curved exponential families. As mentioned in the previous section, a Markov graph model with all possible k-stars in its sufficient statistic is not a tractable model (high dimension and degeneracy problems). Snijders et al. (2006) were the first to propose that, in particular for modelling transitivity, all s_k for $k \geq 2$ be replaced by a single statistic, an *alternating k-star statistic* with geometrically decreasing weights,

$$u(Y;\lambda) = \sum_{k=2}^{n-1} (-1)^k \frac{s_k}{\lambda^{k-2}},$$

for some $\lambda > 1$. However, because we cannot say in advance what value this λ should have, the term 'statistic' is somewhat misleading; as long as λ is an unknown parameter, the dimension of the sufficient statistic is not reduced. The model is a curved subfamily of model (11.2): All $n-2$ k-star counts ($k \geq 2$) are still needed, but the corresponding parameters are only two: a σ for the degree of k-star influence and the geometric factor $1/\lambda$.

Jointly with the number r of edges ($r = s_1$), the set of all *k-star counts* is a one-to-one linear function of the set of *degree counts*. Snijders et al. (2006) show that when $u(Y;\lambda)$ is expressed in terms of r and the degree counts, $u(Y;\lambda)$ is a non-alternating sum of the latter, whose terms also have geometrically decreasing coefficients. This contributes to a more intuitive understanding of the alternating k-star statistic. Hunter (2007) and Chatterjee and Diaconis (2013, last section) provide further discussion on this topic. Snijders et al. (2006) went on to formulate other curved families for modelling transitivity and related properties, going outside the traditional star and triangle counts.

A recent book on ERGMs, edited by Lusher et al. (2013), targeted towards social scientist interested in social networks, ends with a chapter by Pattison and Snijders on the progress made so far and 'Next steps'. They first list four features of ERGMs that should be attractive to social scientists:

- ERGMs conceptualize social networks as outcomes of social processes;
- ERGMs are able to reproduce important observed network characteristics;
- ERGMs are flexible, applicable to many different types of networks or relational data;
- ERGMs can be used to assess dynamic, interactive local processes, when longitudinal data are available.

When talking of future problem areas and demands for development, Pattison and Snijders mention for example model specification avoiding near-degeneracy, methods for assessing homogeneity, methods for more rigorous model comparisons, models with latent variables, models with hierarchical dependence, and models with categorical or ordinal variables extending binary y_{ij}.

12

Rasch Models for Item Response and Related Model Types

In the years around 1960 the Danish statistician Georg Rasch introduced path-breaking models for item response in psychological and educational testing, later to be called Rasch models. A standard situation is the following. Consider a set of n subjects (persons), all taking a test consisting of k questions or other test items. The test is intended to show the person's ability in some specific respect (a 'latent trait'). For simplicity, we assume that the response is dichotomous, 1/0 representing right/wrong or success/failure. Data may look like in Table 12.1, that will be repeatedly used for illustration purposes, but a realistic example is likely to have some more items and many more subjects. In most applications the subjects are humans, but otherwise they might be lab animals given tasks to perform.

A book edited by Fischer and Molenaar (1995) contains many scattered application examples. A quite different type of application is found in the Encyclopedia of Biostatistics entry on Rasch models, by Carriquiry and Fienberg (1998): 263 persons were followed over four influenza seasons. Susceptibility varied between persons, and the four influenza types had different infectivity. The data matrix told which persons had or had not got the different influenzas (corresponding to items exposed to), see further Example 12.1 in Section 12.3.

Table 12.1 *A toy data table for an item response study with 4 items and 9 subjects, The table has been ordered by row and column sums.*

Item\Person	1	2	3	4	5	6	7	8	9	10	11	12	Row sum
Item 1	0	0	0	0	0	0	1	1	0	1	1	1	5
Item 2	1	0	0	0	1	1	0	0	1	1	1	1	7
Item 3	0	1	1	1	1	1	1	1	1	0	0	1	9
Item 4	0	1	1	1	0	1	1	1	1	1	1	1	10
Column sum	1	2	2	2	2	3	3	3	3	3	3	4	Total = 31

12.1 The Joint Model

Let data be represented as a $k \times n$ table (matrix) $Y = \{y_{ij}\}$ of zeros and ones, as in Table 12.1. Rasch's model for such data specifies that the entries are mutually independent Bernoulli variates with probabilities $Pr(y_{ij} = 1) = \alpha_i\beta_j/(1 + \alpha_i\beta_j)$, where α_i is a parameter representing item difficulty (or rather simplicity) and β_j is a person ability parameter. By raising the numerator $\alpha_i\beta_j$ to y_{ij} we can as usual represent both outcomes by the same formula. The probability for the whole table Y then takes the form

$$f(Y;\ \{\alpha_i\}, \{\beta_j\}) = \prod_{i=1}^{k}\prod_{j=1}^{n} \frac{(\alpha_i\beta_j)^{y_{ij}}}{1 + \alpha_i\beta_j} = e^{\sum_i \psi_i y_{i.} + \sum_j \lambda_j y_{.j}} \frac{1}{\prod_i \prod_j (1 + e^{\psi_i + \lambda_j})}, \quad (12.1)$$

where $\psi_i = \log\alpha_i$, $\lambda_j = \log\beta_j$, and $y_{i.}$ and $y_{.j}$ are the row and column sums. This is clearly an exponential family with row and column sums forming its canonical statistic, and with ψ_i and λ_j as canonical parameters, a row plus column effect model on the logit scale. In standard statistical analysis of data, all such parameters are regarded as unknown. The representation is not minimal, however. There is a lack of uniqueness in the parameterization that is eliminated by for example setting $\alpha_1 = 1$ $(\psi_1 = 0)$.

One of Rasch's ideas was that the marginals of the table should be sufficient for the parameters. This is clearly the case with (12.1). Alternative sets of assumptions leading to the Rasch model are discussed in Fischer and Molenaar (1995, Ch. 2). Following Martin-Löf (1970), we will here demonstrate that the model can also be seen as a consequence of Boltzmann's law. We then assume that row and column sums represent all statistical information in a table, in the sense discussed in Chapter 6. Specifically, this is equivalent to saying that all tables having the same row and column sums are equally probable. Boltzmann's law then yields the probability for a specific table Y as

$$f(Y) \propto e^{\sum_i \psi_i y_{i.} + \sum_j \lambda_j y_{.j}}$$

Thus, we get the numerator of (12.1). The denominator of (12.1) is the norming constant. Hence we see that Boltzmann's law can be used to motivate the Rasch model (12.1).

As usual, the likelihood equations are $y_{i.} = E(y_{i.})$ and $y_{.j} = E(y_{.j})$, more specifically (see Exercise 12.1)

$$y_{i.} = \sum_j \frac{\alpha_i \beta_j}{1 + \alpha_i \beta_j} \qquad i = 1, \ldots, k \tag{12.2}$$

$$y_{.j} = \sum_i \frac{\alpha_i \beta_j}{1 + \alpha_i \beta_j} \qquad j = 1, \ldots, n. \tag{12.3}$$

The equations are too many in several respects. First we must reduce the parameter dimension by 1 to obtain a unique parameterization. If we set $\alpha_1 = 1$ ($\psi_1 = 0$) we can delete the first equation.. Second, note that if $k < n$ there are at most $k + 1$ *different* equations of type (12.3), corresponding to the $k+1$ possible values $0, \ldots, k$ for $y_{.j}$. This corresponds to an aggregation of persons by just counting the number of them, n_s, who attained the sum s ($\sum n_s = n$). So for example, in Table 12.1 we have $n_0 = 0$, $n_1 = 1$, $n_2 = 4$, $n_3 = 6$, $n_4 = 1$. Thus the n equations can be replaced by $k + 1$ equations, and with some abuse of notation the equation system can be rewritten as

$$y_{i.} = \alpha_i \sum_s n_s \frac{\beta_s}{1 + \alpha_i \beta_s} \qquad i = 1, \ldots, k - 1 \tag{12.4}$$

$$s = \beta_s \sum_i \frac{\alpha_i}{1 + \alpha_i \beta_s} \qquad s = 0, \ldots, k, \tag{12.5}$$

where β_s now represents the common value estimated for all persons with column sum s.

Third, some estimates are degenerate and trivial, and the corresponding subjects should be identified and deleted. In Table 12.1, Person 12 was too good for the items used and will get a $+\infty$ estimate of β. That value should preferably be inserted in (12.5) and the equation for $k = 4$ be deleted.

There is no explicit solution to the likelihood equations. A slow but safe iterative method is the following intuitively reasonable updating procedure:

$$\alpha_i^{(\nu+1)} = y_{i.} \Big/ \sum_s n_s \frac{\beta_s^{(\nu)}}{1 + \alpha_i^{(\nu)} \beta_s^{(\nu)}} \qquad i = 2, \ldots, k \tag{12.6}$$

$$\beta_s^{(\nu+1)} = s \Big/ \sum_i \frac{\alpha_i^{(\nu)}}{1 + \alpha_i^{(\nu)} \beta_s^{(\nu)}} \qquad s = 0, \ldots, k, \tag{12.7}$$

With the not very good starting value zero for all ψ_i and λ_j (Person 12 excluded), the algorithm immediately found its way towards the MLE solution

$\alpha_1 = 1$; $\hat{\alpha}_2 = 2.36$; $\hat{\alpha}_3 = 6.03$; $\hat{\alpha}_4 = 11.05$; or equivalently
$\psi_1 = 0$; $\hat{\psi}_2 = 0.86$; $\hat{\psi}_3 = 1.80$; $\hat{\psi}_4 = 2.40$;
$\beta_0 = 0$; $\hat{\beta}_1 = 0.077$; $\hat{\beta}_2 = 0.280$; $\hat{\beta}_3 = 1.036$; ($\hat{\beta}_4 = +\infty$).

Categories $s = 0$ and $s = k$ represent those who fail completely or have complete success. The corresponding estimates are $\beta_0 = 0$ and $\beta_k = \infty$, which are on the boundary of the parameter space, $\pm\infty$ in the canonical parameter space. For n increasing, this type of event becomes more probable. Analogously, $\alpha = 0$ and $\alpha = +\infty$ for tasks on which all persons fail or all succeed, but for fixed k this probability decreases as n increases.

Thus, there are some problems connected with ML estimation in this model. The worst feature, however, is that the item parameters α_i are not consistently estimated as $n \to \infty$. This is highly undesirable, because the items are often calibrated by exposing them to a large number of persons, and then we want the estimation error in α_i to be small. The situation has the β_js as *incidental parameters*, each added person bringing its own personal ability parameter. Like in the classic Neyman and Scott (1948) example (concerning ML estimation of σ^2 in a one-way ANOVA with many small samples, each with its own incidental mean value, Exercise 3.16), the MLE of the item parameters approach wrong values due to the presence of the incidental parameters β_j. For a proof of this, see Exercise 12.2.

Exercise 12.1 ***Likelihood equations for Rasch model***
Check the expected values in the likelihood equations (12.2) and (12.3) for the joint model by
(a) using the explicit distribution of y_{ij},
(b) differentiating the log of the norming constant. △

Exercise 12.2 ***Inconsistency of joint model MLE***
Assume $k = 2$ (two items) and $n \to \infty$. Consider the ratio α_1/α_2, which is invariant under normalization.
(a) Show that equations (12.5) simplify to $\hat{\alpha}_1/\hat{\alpha}_2 = (n_{11}/n_{21})^2$, where n_{i1} denotes the number of persons with $y_{ij} = 1$ for item i and only this item.
(b) For simplicity, assume all individuals have the same ability ($\beta_j = \beta_0$), and show lack of consistency of the item parameters, more precisely that $E(n_{i1}) = n\,\alpha_i\beta_0\,/\,(1+\alpha_1\beta_0)\,(1+\alpha_2\beta_0)$, and thus that $\hat{\alpha}_1/\hat{\alpha}_2 \to (\alpha_1/\alpha_2)^2$ in probability as $n \to \infty$ (Martin-Löf, 1970). △

12.2 The Conditional Model

In typical applications of item response models n is much larger that k. However, we saw that a large n did not help to yield good joint model ML estimates of the k item parameters. If the calibration of the items is not satisfactory, this has consequences for the future use of these items. A standard method for dealing with problems involving incidental parameters (the β_js) is, if possible, to condition on a sufficient statistic for them. In this case this is also the message of the conditionality principle of Section 3.5: For statistical inference about the canonical item parameters α_i, we should condition on the set of column sums $\{y_{.j}\}$, corresponding to the canonical parameters β_j. Thus, we now study this conditional model.

By Proposition 3.16, and in particular formula (3.22), the conditional distribution of the table is proportional to $\exp\{\sum_i \psi_i y_{i.}\}$, and the norming constant is a function of ψ and of u, that is, of the column sums $y_{.j}$. The norming constant can be derived column-wise, since the columns are mutually independent, and formally it is just to sum over all possible outcomes. Hence, the norming constant for column j, given $y_{.j}$, is

$$C_j(\psi_1, \ldots, \psi_k) = \sum_{\substack{y_{1j},\ldots,y_{kj} \\ \text{given } y_{.j}}} \prod_i \alpha_i^{y_{ij}} = \sum_{\substack{y_{1j},\ldots,y_{kj} \\ \text{given } y_{.j}}} e^{\sum_i \psi_i y_{ij}}, \tag{12.8}$$

where the sum is taken over all possible columns with the given column sum. In mathematics, this function of $\boldsymbol{\alpha}$ is known as an *elementary symmetric polynomial* of degree $y_{.j}$ and (for $y_{.j} = s$) often denoted $\gamma_s(\alpha_1, \ldots, \alpha_k)$. The first three of them are $\gamma_0 = 1$, $\gamma_1 = \sum \alpha_i$, $\gamma_2 = \sum_{i_1<i_2} \alpha_{i_1}\alpha_{i_2}$. Thus, the conditional density for the table Y can be written

$$f(Y \mid \{y_{.j}\}; \psi_1, \ldots, \psi_k) = \frac{e^{\sum_i \psi_i y_{i.}}}{\prod_j \gamma_{y_{.j}}(e^{\psi_1}, \ldots, e^{\psi_k})} = \frac{\prod_i \alpha_i^{y_{i.}}}{\prod_s \gamma_s(\alpha_1, \ldots, \alpha_k)^{n_s}}, \tag{12.9}$$

where in the second expression we have utilized that there are n_s columns with sum s. As in the joint model, a parameter restriction must be imposed for uniqueness, for example $\psi_1 = 0$, or $\sum_i \psi_i = 0$.

12.2.1 Likelihood Equations

We now derive the likelihood equations. For insertion in $E(y_{i.} \mid \{y_{.j}\})$ we need an expression for $E(y_{ij} \mid y_{.j} = s) = Pr(y_{ij} = 1 \mid y_{.j} = s)$. To get this conditional probability we repeat the summation in (12.8), but now under the additional constraint $y_{ij} = 1$. We can extract a factor α_i and a new elementary symmetric polynomial remains, now of reduced degree $s - 1$. More precisely,

$$E(y_{ij} \mid y_{.j} = s) = \frac{\alpha_i \gamma_{s-1}(\alpha_1, \ldots, \alpha_k;\ \text{except}\, \alpha_i)}{\gamma_s(\alpha_1, \ldots, \alpha_k)} \tag{12.10}$$

Thus, the likelihood equations, $i = 1, \ldots, k$, can be written

$$
\begin{aligned}
y_{i.} &= \alpha_i \sum_j \frac{\gamma_{y_{.j}-1}(\alpha_1, \ldots, \alpha_k;\ \text{except}\ \alpha_i)}{\gamma_{y_{.j}}(\alpha_1, \ldots, \alpha_k)} \\
&= \alpha_i \sum_s n_s \frac{\gamma_{s-1}(\alpha_1, \ldots, \alpha_k;\ \text{except}\ \alpha_i)}{\gamma_s(\alpha_1, \ldots, \alpha_k)} .
\end{aligned}
\tag{12.11}
$$

Computationally, the conditional equations are more difficult than those of the joint model. However, there is an extensive and well-developed theory for symmetric functions, including efficient tools for their computation. An intuitively natural, simple, but slow algorithm (Martin-Löf, 1970) would use the current α-values in a computation of the complicated second factor (the ratio), and then update α_i by dividing $y_{i.}$ with that ratio. As remarked by Tjur (1982), this is equivalent to the so-called *iterative proportional scaling* algorithm for log-linear models. More standard, however, is to use Newton–Raphson and similar methods, see Fischer and Molenaar (1995, Ch. 3) for a more extensive discussion. A quite different possibility will be introduced in Section 12.4.

Several computer packages are freely available, for example `eRm` in **R**, for statistical analyses in the conditional Rasch model and in a number of generalizations of it. For comparison with the MLE of the joint model, here are the item parameters calculated by the `eRm` program:
$\psi_1 = 0$; $\hat{\psi}_2 = 0.59$; $\hat{\psi}_3 = 1.29$; $\hat{\psi}_4 = 1.77$;
It is clear that these estimates spread much less than those of the joint model. This is probably a reflection of the fact that the joint model MLE exaggerates the differences between the ψ-values, cf. Exercise 12.2.

12.2.2 Information Matrix

The Fisher information matrix for row sums given column sums, $I(\psi) = \text{var}(y_{1.}, \ldots, y_{k.} \,|\, y_{.1}, \ldots, y_{.n}; \psi)$, is derived by analogy with the expected value in Section 12.2.1, either by differentiating the log norming constant (Exercise 12.3), or as follows. First note that since the columns are mutually independent,

$$
\text{var}(y_{1.}, \ldots, y_{k.} \,|\, y_{.1}, \ldots, y_{.n}; \psi) = \sum_s n_s \,\text{var}(y_{1j}, \ldots, y_{kj} \,|\, y_{.j} = s; \psi),
$$

for an arbitrary column number j with column sum s. The conditional variances and covariances are next obtained as

$$\mathrm{var}(y_{ij}\,|\,y_{.j}=s;\,\psi) = E(y_{ij}\,|\,y_{.j}=s)\left(1 - E(y_{ij}\,|\,y_{.j}=s)\right), \qquad (12.12)$$
$$\mathrm{cov}(y_{i_1j}, y_{i_2j}\,|\,y_{.j}=s;\,\psi) = E(y_{i_1j}y_{i_2j}\,|\,y_{.j}=s) - E(y_{i_1j}\,|\,y_{.j}=s)\,E(y_{i_2j}\,|\,y_{.j}=s).$$

The variance here is obtained from the conditional Bernoulli for a single y_{ij}, in combination with (12.10), For the covariance, we additionally need the product moment:

$$\begin{aligned} E(y_{i_1j}\,y_{i_2j}\,|\,y_{.j}=s) &= Pr\{y_{i_1j} = y_{i_2j} = 1\,|\,y_{.j}=s;\,\psi\} \\ &= \alpha_{i_1}\,\alpha_{i_2}\,\frac{\gamma_{s-2}(\alpha_1,\ldots,\alpha_k;\ \text{except}\,\alpha_{i_1},\alpha_{i_2})}{\gamma_s(\alpha_1,\ldots,\alpha_k)}, \end{aligned} \qquad (12.13)$$

where the last formula is obtained in the same way as (12.10). Formulas (12.10) and (12.13) will reappear in Proposition 12.1.

Exercise 12.3 ***Likelihood equations for the conditional Rasch model***
(a) Derive the expected values of the row sums, $E(y_{i.})$, in the conditional Rasch model by differentiating the log of the norming constant.
(b) Derive the variance-covariance matrix by differentiating once more. △

12.3 Testing the Conditional Rasch Model Fit

We here construct a score test of the conditional Rasch model, analogous with, but somewhat more complicated than the test for additivity in a two-way ANOVA model and the χ^2 test for homogeneity or independence in a two-dimensional contingency table. An exact test for model reduction is practically impossible in this case. A log-likelihood ratio test will be briefly discussed at the end of the section. The score test was proposed and derived by Martin-Löf (1970). Later an alternative formulation and an alternative proof were introduced by Glas (1988), see also Fischer and Molenaar (1995, Ch. 5). The new test derivation was due to a misunderstanding of Martin-Löf's proof, due to an impression that the proof relied on a Poisson assumption. This 'Poisson trick' was only a technical device, however, used by Martin-Löf to simplify the derivation, see the proof of Proposition 12.1, and cf. Section 5.6.

The wider model, in which we will test the conditional Rasch model, allows the item difficulties to depend on s. The natural canonical statistic will then be the table $\{n_{is}\}_{i,s}$, where n_{is} is the number of persons j with $y_{ij} = 1$ among those with raw score $y_{.j} = s$. Such a table is illustrated in Table 12.2. The first and last columns ($s = 0$ and $s = k$) should be deleted, since they carry no information about the model. Under the Rasch model we can reduce the table $\{n_{is}\}_{i,s}$ to its vector of row sums $y_{i.}$. Conditioning on the raw scores n_s implies a linear restriction for each column, so the test corresponds to a reduction of dimension from $(k-1)^2$ to $k-1$, that is, by $(k-2)(k-1)$, which will be the number of degrees of freedom of the test.

Table 12.2 *Aggregation of Table 12.1 (with 4 items and 12 subjects) to a* 4×5 *table* $\{n_{is}\}$ *of the number of persons with* $y_{ij} = 1$ *and* $y_{.j} = s$.

Item\Raw sum s	0	1	2	3	4	Row sum = $y_{i.}$
Item 1	0	0	0	4	1	5
Item 2	0	1	1	4	1	7
Item 3	0	0	4	4	1	9
Item 4	0	0	3	6	1	10
n_s	0	1	4	6	1	Total $\sum s\, n_s = 31$

The $k-1$ nontrivial linear restrictions $\sum_i n_{is} = s n_s$ complicate the derivation. If we allow them, we have to deal with singular covariance matrices. If we instead work with a reduced set of linearly independent data, we lose symmetry and find matrix inversion difficult. We will instead use the technical trick introduced in Section 5.6, implying that we regard sample sizes as outcomes of Poisson variates, instead of as fixed quantities. By doing so we eliminate the restrictions and simplify the derivation.

In the wider model just introduced, we consider the model reduction

$$\mathrm{H}_0 : \quad \{n_{is}\}_{i,s} \to (\{n_s\}_s, \{y_{i.}\}_i)$$

or parametrically expressed,

$$\mathrm{H}_0 : \quad \text{For all } i,\ \alpha_{is} = \alpha_i \ \text{(i.e. independence of } s\text{)},$$

where α_{is} are item parameters allowed to depend on s.

Whether H_0 is assumed to hold or not, the model specifies that the raw scores n_s are mutually independent and marginally Poisson distributed, $\mathrm{Po}(\lambda_s)$, where the two parameter sets $\{\lambda_s\}$ and $\{\alpha_{is}\}$ (or $\{\alpha_i\}$) are variation independent. This implies that the set of raw scores $\{n_s\}$ is S-ancillary for $\{\alpha_{is}\}$ (or $\{\alpha_i\}$). As a consequence, ML estimates of α_{is} or α_i are the same in the conditional and the Poisson models, and the large sample test statistics, including the score test statistic, are the same. Thus we can derive the form of the test statistic within the Poisson model.

Equivalently to assuming all n_s independent and Poisson distributed as a start, we can regard n as Poisson and the raw scores n_s as multinomially distributed, given their sum n, and multinomial probabilities proportional to λ_s, in combination with a total sample size that is regarded as an outcome of a $\text{Po}(\sum \lambda_s)$. In the latter interpretation it is clear that we have an example of a single sample, and the Poisson trick works in the form given in Proposition 5.6 of Section 5.6. To motivate the multinomial for $\{n_s\}$, we may either start from random ability parameters with a completely unknown (latent) distribution and form the 'extended random model' (Tjur, 1982), or simpler just motivate it as an allowed construction that is a convenient supplement to the conditional model. In this way, Martin-Löf (1970) proved Proposition 12.1.

We need one more notation. Let $\boldsymbol{n}_s$, $s = 1, \ldots, k-1$ denote the columns of the $\{n_{is}\}$ table, that is, $\boldsymbol{n}_s$ is a vector with the k elements n_{is}, $i = 1, \ldots, k$, and total $s\, n_s$.

Proposition 12.1 ***Large sample test of fit for conditional Rasch models*** *The following score test statistic W_u for test of fit for the conditional Rasch model is asymptotically χ^2 distributed with $(k-2)(k-1)$ degrees of freedom as the total sample size n increases:*

$$W_u = \sum_{s=1}^{k-1} (\boldsymbol{n}_s - E(\boldsymbol{n}_s; n_s, \hat{\alpha}))^T \operatorname{var}(\boldsymbol{n}_s; n_s, \hat{\alpha})^{-1} (\boldsymbol{n}_s - E(\boldsymbol{n}_s; n_s, \hat{\alpha})) . \quad (12.14)$$

Here $E(\boldsymbol{n}_s; n_s, \hat{\alpha})$ and the diagonal of $\operatorname{var}(\boldsymbol{n}_s; n_s, \hat{\alpha})$ *both have elements*

$$n_s\, \hat{\alpha}_i \, \frac{\gamma_{s-1}(\alpha_1, \ldots, \alpha_k, \text{except}\, \alpha_i)}{\gamma_s(\alpha_1, \ldots, \alpha_k)} ,$$

which differs only by the factor n_s from formula (12.10). The nondiagonal elements of $\operatorname{var}(\boldsymbol{n}_s; n_s, \hat{\alpha})$ *are*

$$n_s\, \hat{\alpha}_{i_1}\, \hat{\alpha}_{i_2} \, \frac{\gamma_{s-2}(\alpha_1, \ldots, \alpha_k, \text{except}\, \alpha_{i_1}, \alpha_{i_2})}{\gamma_s(\alpha_1, \ldots, \alpha_k)} ,$$

which differs only by the factor n_s from formula (12.13).
The convergence is uniform as $n \to \infty$, provided all n_s/n stay away from 0 and $\psi_{i_1} - \psi_{i_2}$ remain bounded.

Proof We will derive the form of the test statistic, but not discuss its uniform convergence property.

We make use of the Poisson trick, Proposition 5.6 from Section 5.6. It allows us to derive the test statistic under the assumption that the total n is Poisson distributed, and hence that all n_s are mutually independent Poisson variates, This makes the columns independent. This yields W_u as a sum over the columns of the $\{n_{is}\}_{i,s}$ table. So next we can restrict the attention to a single column, corresponding to a single vector $\boldsymbol{n}_s$. Since $n_{is} = \sum_{y_{.j}=s} y_{ij}$

is the sum over the n_s persons having raw score s, whose outcomes are mutually independent, $E(n_{is} \mid n_s) = n_s E(y_{ij} \mid y_{.j} = s)$ and $\mathrm{var}(n_{is} \mid n_s) = n_s \mathrm{var}(y_{ij} \mid y_{.j} = s)$, and correspondingly for the conditional covariances.

In the next step we go to the unconditional moments and utilize the Poisson properties $E(n_s) = \mathrm{var}(n_s) = \lambda_s$:

$$
\begin{aligned}
E(n_{is}) &= E(E(n_{is} \mid n_s)) = E(n_s)\, E(y_{ij} \mid y_{.j} = s) = \lambda_s\, E(y_{ij} \mid y_{.j} = s),\\
\mathrm{var}(n_{is}) &= E(\mathrm{var}(n_{is} \mid n_s)) + \mathrm{var}(E(n_{is} \mid n_s))\\
&= \lambda_s \left\{\mathrm{var}(y_{ij} \mid y_{.j} = s) + E(y_{ij} \mid y_{.j} = s)^2\right\},\\
\mathrm{cov}(n_{i_1 s}, n_{i_2 s}) &= \lambda_s \left\{\mathrm{cov}(y_{i_1 j}, y_{i_2 j} \mid y_{.j} = s) + E(y_{i_1 j} \mid y_{.j} = s) E(y_{i_2 j} \mid y_{.j} = s)\right\}.
\end{aligned}
$$

Now finally, we note that the MLE of λ_s is n_s and that the conditional moments of y_{ij} were calculated in the preceding section. For example, using (12.12) and (12.10), we obtain

$$
\begin{aligned}
\mathrm{var}(y_{ij} \mid y_{.j} = s) + E(y_{ij} \mid y_{.j} = s)^2 &= E(y_{ij} \mid y_{.j} = s)\\
&= \frac{\alpha_i \gamma_{s-1}(\alpha_1, \ldots, \alpha_k, \text{except}\, \alpha_i)}{\gamma_s(\alpha_1, \ldots, \alpha_k)},
\end{aligned}
$$

and analogously for the covariance. This concludes the derivation of the test statistic. □

Example 12.1 ***Two model test examples***
We apply the test on two data sets. For the data in Table 12.2 the three contributions to the score test statistic, for $s = 1$, $s = 2$ and $s = 3$, are 5.83, 2.68 and 2.57, respectively (a little **R** or MATLAB calculation). Note that the uninformative last column ($s = 4$) is deleted. The total is $W_u = 11.1$, which corresponds to a p-value just below 10% in the asymptotic $\chi^2(6)$ distribution. Considering the small sample size, this p-value must be taken with a pinch of salt. Anyhow, it is not surprising that $s = 1$ yields the largest contribution, considering that the single subject with $s = 1$ had $y = 1$ for the relatively difficult item 2 but not for the much simpler items 3 and 4.

We also try the test on the influenza data mentioned in the chapter introduction. An $y_{ij} = 1$ means that individual j got influenza i. The 4×5 table $\{n_{is}\}$ of the number of persons with $y_{ij} = 1$ and $y_{.j} = s$. is shown in Table 12.3.

The MLE and the score test statistic in this case uses 123 persons, since those with $s = 0$ are uninformative. The three contributions to the score test statistic, for $s = 1$, $s = 2$ and $s = 3$, are 8.50, 9.94 and 5.52, with a total of $W_u = 24.0$, which is highly significant ($df = 6$). It is clear from the n_{is} table what has caused the high value of W_u. Influenza type 4 has about

Table 12.3 *The n_{is} table for the 4 influenza types ('item' i) and 263 persons, divided according to the number s of influenza types contracted by a person. Conditional MLE are found in the rightmost column.*

Item\Raw sum s	0	1	2	3	4	$y_{i.}$	$\hat{\alpha}_i$	$\hat{\psi}_i \pm s.e.$
Item 1	0	20	23	5	0	48	1	0
Item 2	0	17	19	6	0	42	0.84	-0.17 ± 0.24
Item 3	0	16	17	5	0	38	0.74	-0.29 ± 0.24
Item 4	0	31	7	2	0	40	0.79	-0.23 ± 0.24
n_s	140	84	33	6	0	Total $\sum s\, n_s = 168$		

the same row sum and $\hat{\alpha}$-value as the others, but with disproportionate n_{4s}-values; too high for $s = 1$ and too low for $s = 2$ and $s = 3$. Carriquiry and Fienberg (1998) make a different model test, see Example 12.2, but draw the same conclusion that the simple Rasch model does not fit well. They go on to try more complex models allowing interactions between the influenza types. △

The likelihood ratio test is an alternative. It requires ML estimates of the item parameters also under the alternative, that is, of α_{is} separately for each raw score value s. The corresponding likelihood equations are the same as (12.11), except that the summation over s on both sides is not carried out.

Other model tests that might be of interest concern questions such as

- If the individuals form two subgroups, do these groups fit the same item parameter values (i.e. same α_is)?
- If the individuals form two subgroups, do these groups have the same ability distribution?
- If a set of items forms two different subsets, do these require separate Rasch models, or can they be combined in a single model (have the same Rasch scale)?

Likelihood ratio tests for these questions were constructed by Martin-Löf (1970). The first two are relatively straight forward to derive. For the last question, which is more complicated, and for many other model tests, see also Fischer and Molenaar (1995, Ch. 5).

It should be kept in mind that likelihood ratio tests and analogous tests are justified as large sample tests, and thus their p-values are (perhaps) not trustworthy unless there are reasonably many individuals in each class of the basic model.

12.4 Rasch Model Conditional Analysis by Log-Linear Models

For each individual j we observe a k-vector $(y_{1j}, \ldots, y_{kj})$ of zeros and ones. Thus, data can alternatively be arranged as a k-dimensional contingency table of size 2^k. Individuals in the same cell of this table cannot be distinguished, and in particular they have the same s. For the conditional analysis the 2^k table represents a reduction that goes less far than the reduction to the n_{is} table considered in Section 12.3 and exemplified in Table 12.1 and Table 12.2. Tjur (1982) observed and exploited that this contingency table can be furnished with a log-linear model, in the following way.

We first use the Poisson trick of imagining the total number of observations n as being arbitrary Poisson, and $\{n_s\}$ as conditionally multinomial, as in the proof of Proposition 12.1. Equivalently we start by assuming all $\{n_s\}$ independent and Poisson, with arbitrary mean values $E(n_s) = \lambda_s$. As a consequence the contingency table becomes a Poisson table with independence between cells (this is not true for the corresponding $\{n_{is}\}$). Let any of these independent Poisson variates be denoted n_A, where A is a subset of $\{1, \ldots, k\}$ representing the positions of the s ones in the sequence of zeros and ones. From the basics of the Rasch model it follows that the Poisson parameters satisfy

$$E(n_A) \propto \prod_{i \in A} \alpha_i\,,$$

where the omitted proportionality constant depends on s. If it were not for this omitted factor, we would have a standard multiplicative Poisson model. The proportionality constant will not be the same as the $\lambda_s = E(n_s)$ parameter introduced in the proof of Proposition 12.1, but it will be proportional to λ_s, so like λ_s it will be a free parameter in the model. Thus we should introduce one more factor in the model, a partitioning with respect to s. With abuse of notation, using the arbitrary λ_s in this partially new meaning, we have the Poisson model with

$$E(n_A) = \lambda_s \prod_{i \in A} \alpha_i\,, \tag{12.15}$$

where s is the number of ones in A. This is a log-linear Poisson model that may be fitted using a package for generalized linear models (GLMs). Not all of the output will be meaningful for the Rasch model, but the estimates of the ψ_is or equivalently α_is are the conditional model MLEs, and their standard errors are valid in the conditional Rasch model.

Such packages also yield, as a standard, the deviance for testing model fit. However, this will not be the same deviance as in Proposition 12.1. The model reduction would be from a saturated Poisson model, with k^2 parameters, to the model characterized by (12.15), which has $(k+1)+(k-1) = 2k$

parameters ($k+1$ s-values and k α-values with one linear restriction). The degrees of freedom for the test, $k^2 - 2k = k(k-2)$ exceeds by $k-2$ the previous $(k-1)(k-2)$. The alternative model of Proposition 12.1 assumes a Rasch model to hold for each s, but with item parameters depending on s, which is a more restrictive model than the saturated model for the 2^k table.

Example 12.2 ***The influenza example again***
We apply the preceding theory to the influenza data of Example 12.1. The first thing to notice is that the n_{is} data given in Table 12.3 are not detailed enough. Here is the table we need now:

Table 12.4 *The 2^4 contingency table for the 4 influenza types.*

Item \ s	0	1	1	1	1	2	2	2	2	2	2	3	3	3	3	4
Item 1	0	1	0	0	0	1	1	1	0	0	0	1	1	1	0	1
Item 2	0	0	1	0	0	1	0	0	1	1	0	1	1	0	1	1
Item 3	0	0	0	1	0	0	1	0	1	0	1	1	0	1	1	1
Item 4	0	0	0	0	1	0	0	1	0	1	1	0	1	1	1	1
n_A	140	20	17	16	31	12	9	2	5	2	3	4	1	0	1	0

The log-linear model fitted by `glm` in **R** yields the same α-values as in Table 12.3, and the same standard errors, as expected. The deviance for the model test is 25.8, with $df = 8$. This was the test statistic reported by Carriquiry and Fienberg (1998). Its value corresponds quite well to the previous score test value, 24.0 with $df = 6$. The increase by 1.8 units matches almost perfectly the $k - 2 = 2$ extra degrees of freedom. △

12.5 Rasch Models for Polytomous Response

With more than two responses possible, for example in multiple-choice questions, extensions of the dichotomous models were formulated by Rasch and extensively studied by E.B. Andersen. They are all notationally and numerically much more complicated than the dichotomous models.

With polytomous responses, the probability that subject j responds to item i by selecting response category h can be analogously modelled as

$$\frac{e^{\psi_{hi} + \lambda_{hj}}}{\sum_h e^{\psi_{hi}+\lambda_{hj}}}$$

For uniqueness of representation, we may set $\psi_{1i} = \lambda_{1j} = 0$ for all i and j, together with $\psi_{h1} = 0$ for all h.

Some reduced models have achieved special interest, for example the curved family with $\lambda_{hj} = \phi_h \lambda_j$, but also conditional models that are free from the subject parameters and allow consistent estimation of the item parameters. For detailed discussions of polytomous Rasch models, see Fischer and Molenaar (1995, Ch. 15–18).

12.6 Factor Analysis Models for Binary Data

Here we will look at some *latent variable* models and methods for binary data (Bartholomew et al., 2011, Ch. 4), which may be regarded as extensions of the Rasch model. Basically, think of subjects (persons) exposed to k items, yielding 0–1 responses (wrong–correct). The person ability parameter is regarded as a latent variable, varying randomly between persons.

As before, the binary response y_{ij} by person j on item i , $i = 1, \ldots, k$, is thought of as a Bernoulli variable, with probability π_{ij} depending on item difficulty and person ability. To represent person ability, we introduce a latent (unobserved) random variable z, with outcome z_j for person j. We make the following basic assumptions, specifying what is called a *latent trait model*:

- For person j, given its z_j-value, responses y_{ij} are mutually independent with *item response function* $\pi_{ij} = \pi_i(z_j) = Pr\{y_{ij} = 1 \mid z_j\}$;
- The functions $\pi_i(z)$ depend on z by

$$g(\pi_i(z)) = \alpha_i + \beta_i z \tag{12.16}$$

  for a specified function g and some item-dependent coefficients $\{\alpha_i, \beta_i\}$;
- Persons $j = 1, ..., n$ form a random sample $\{z_j\}$ from some continuous distribution for the 'trait' z.

For given z-values, this specifies a set of k generalized linear models (see Chapter 9), one for each item, with a common link function g, mapping the unit interval on the real line. We next assume that the link function is chosen to be the canonical link, that is, the logit function. Then the generalized linear model will be a full exponential family. We further assume a normal distribution for the latent z. Without restriction this can be taken to be the standardized normal, since we can play with the coefficients $\{\alpha_i, \beta_i\}$ in (12.16). Thus, for $\pi_{ij} = \pi_i(z_j)$ we have the model specification

$$\text{logit}\, \pi_i(z_j) = \alpha_i + \beta_i z_j, \quad z_j \text{ iid N}(0, 1), \quad i = 1, \ldots, k; \; j = 1, \ldots, n. \tag{12.17}$$

This model (12.17) is called the *logit/probit model* or the *logit/normit model* or, more briefly, the logit model. The probit or normit part of the name is due to the use of the standard normal distribution for z (which relates to a uniform variable on (0, 1) by the probit function Φ^{-1} transformation). Alternative models are constructed by using a different link function and/or a different distribution for z in (12.17). It can also be extended by allowing more than one latent variable in (12.17), each normally distributed and with additive effects via their respective β coefficients.

Before we consider parameter estimation in model (12.17), we show that the model generalizes a random component version of the Rasch model. In the Rasch model the probability $\pi_{ij} = Pr\{y_{ij} = 1\}$ for item i, individual j, satisfies

$$\operatorname{logit} \pi_{ij} = \alpha_i + \beta_j\,. \tag{12.18}$$

In the logit/probit model (12.17), β_i has a different meaning, but let its β_i be the same for all i, $\beta_i = \beta_0$, say. If we now identify β_j in the Rasch model with $\beta_0 z_j$ in the specialized logit/probit model, we see that the only difference is that the person parameter β_j in the Rasch model has been made random, $N(0, \beta_0^2)$ when z_j is $N(0, 1)$. In contrast to the Rasch model, however, the general logit/probit model also allows the logit scale for the difference in difficulty between items to vary between individuals (an interaction type of effect). In educational testing the β_j parameters are therefore called the *discrimination parameters*.

We now consider MLE computation for the parameters of (12.17). The model is an example of incomplete data from a full exponential family, so we can use the EM algorithm (Section 8.3). This is generally advocated for latent trait models by Bartholomew et al. (2011). The procedure is analogous to the EM version for the classic Gaussian factor analysis model, described and discussed in Section 8.8.

Here are the details. If we had also known the z_j-values we would have had a full exponential family for $\{\{y\}_{ij}, \{z_j\}\}$, essentially a set of k univariate logistic regressions. Note, namely, that the probability density of $\boldsymbol{x} = \{\{y\}_{ij}, \{z_j\}\}$ can be factorized as

$$f(\boldsymbol{x}) = \prod_i \left(\prod_j f(y_{ij} \,|\, z_j; \{\alpha_i, \beta_i\}) \prod_j f(z_j) \right). \tag{12.19}$$

Since z_j is $N(0, 1)$, the product of factors $f(z_j)$ is free of parameters, as indicated, so only the products of $f(y_{ij} \,|\, z_j)$ contribute to the Fisher score function and the observed information. More precisely, for each row i this product corresponds to a set of logistic regression data (see Example 3.5), with regressor z and with canonical statistic consisting of $y_{i.} = \sum_{j=1}^{n} y_{ij}$

(the row sums) and $\sum_{j=1}^{n} y_{ij} z_j$. To carry out these regressions forms the M-step. Unfortunately, logistic regression does not allow explicit maximization, but there are many packages available for logistic regression. For the E-step we form the conditional expected value of the canonical statistics, given the observed data $\{y_{ij}\}$. That is, we need $E(y_{i.} \mid \{y_{ij}\}) = y_{i.}$ and $E(\sum_j y_{ij} z_j \mid \{y_{ij}\}) = \sum_j y_{ij} E(z_j \mid \{y_{ij}\})$. The formula for $E(z_j \mid \{y_{ij}\})$ does not allow a simple expression, but must be numerically calculated according to

$$E(z \mid \{y_i\}) = \frac{\int z \prod_i \pi_i(z)^{y_i}(1 - \pi_i(z))^{1-y_i} \phi(z)\, dz}{\int \prod_i \pi_i(z)^{y_i}(1 - \pi_i(z))^{1-y_i} \phi(z)\, dz},$$

where index j has been omitted, and $\phi(z)$ is the standard normal density.

Exercise 12.4 ***Transformations in the logit/probit model***
In the logit/probit model (12.17), we would like the model type to be preserved if we change sign of z and use $z' = -z$ in place of z, or if we denote 'success' in the Bernoull trials as 'failure', that is, we change $\pi(z)$ to $1 - \pi(z)$. Confirm that this is the case, and identify how the parameters α_i and β_i in (12.17) are transformed when these changes are made. △

Exercise 12.5 ***Alternative parameters in the logit/probit model***
In the logit/probit model (12.17), alternative parameters that could be used are the probabilities $\pi_i(0)$, corresponding to the median (=mean) latent value $z = 0$. Show that $\pi_i(0) = 1/(1 + e^{-\alpha_i})$. △

12.7 Models for Rank Data

An item response study leads to estimated rankings of the items and of the participating individuals. When the ranking is the primary goal, a different type of study design may be preferable. Suppose $n \geq 1$ assessors are asked to rank a set of items (competitors).

One such design is the pairwise comparison study, discussed in Section 6.5, in which all items are compared pairwise. A standard model here is the Bradley–Terry model (6.15): Item i 'wins' over j with probability $\alpha_i/(\alpha_i + \alpha_j)$, for a set of item parameters $\alpha_i = e^{\theta_i} > 0$, and a more or less complete ranking by the α_is is achieved by comparing the number of wins.

Another type of design is to ask assessors to provide a complete ranking of all items (without ties), yielding *rank data*. We may imagine the assessors determining their rankings through pairwise comparisons of Bradley–Terry type, but where we condition on the event that the paired comparisons are consistent with a unique ranking. We are then led to the *Bradley–Terry/Mallows model* (Mallows, 1957) for complete rank data, stating that the probability for an assessor giving the ranks $\{y_1, \ldots, y_k\}$, a permutation of $\{1, \ldots, k\}$, is

$$f(y_1, \ldots, y_k; \alpha_1, \ldots, \alpha_k) \propto \prod_{i=1}^{k} \alpha_i^{k-y_i} \propto \prod_{i=1}^{k} \alpha_i^{-y_i} = e^{\sum_i \theta_i y_i}, \qquad (12.20)$$

where any $k - 1$ of the $\theta_i = \log \alpha_i$ can be used as canonical parameters in the $(k - 1)$-dimensional exponential family (constraint $\overline{y} = (k + 1)/2$). The norming constant has no more explicit form than the sum over all $k!$ outcomes of $\{y_1, \ldots, y_k\}$. With a homogeneous group of n assessors ranking the items independently, we regard the resulting data $\{y_{ij}\}$ as a sample of size n. The canonical statistics then have the components $t_i = \sum_j y_{ij}$.

A different approach to the modelling of rank data is represented by Martin-Löf (1970), who referred to Boltzmann's law, see Section 6.2, to motivate that if the set of rank sums t_i over assessors summarizes the relevant information in data, it should form the canonical statistic in a particular exponential family. This again yields the Bradley–Terry/Mallows model (12.20). There is no closed form ML estimate in the model. Martin-Löf (1970) proposed approximate solutions and tests of the model and in the model. One hypothesis model of particular interest is the uniform, when all α_i are equal.

A related approach is represented by Diaconis (1988, Ch. 9), who advocated exponential families with $f(y; \theta) \propto \exp^{\theta\, t(y)}$, where $\boldsymbol{t}(\boldsymbol{y})$ is a statistic of suitable dimension, selected to be the sufficient statistic. Many such models are distance models, in which $\boldsymbol{t}(\boldsymbol{y})$ is the distance of $\boldsymbol{y}$ from an ideal 'modal' ranking. This includes the so-called *Mallows' θ-model*, that assumes $\theta_i \propto i$, that is, $\theta_i = \theta\, i$ in (12.20). This model thus has a single parameter, a measure of dispersion. The canonical statistic $t = \sum i\, y_i$ is an affine function of Spearman's rank correlation coefficient, and the model is therefore also called *Spearman's model.*

Spearman's model brings some annoyance to the concept of exact tests (Section 5.1 and Section 6.3), because the null distribution for t is not unimodal, but for very small n seemingly irregular. Therefore it provides a kind of counter-example to the strict formulation of the exact test principle, that the p-value be calculated as the sum over all probabilities at most as high as $\Pr(t_{obs})$. A tail p-value calculated by summing over t-values at least as extreme as t_{obs} would appear more reasonable. On the other hand, for larger n, with $n \geq 10$ to be on the safe side, there is no such problem in practice.

The books by Fligner and Verducci (1993) and Marden (1995) show the state of the art in the middle 90'es. We have not touched here on various aspects of incomplete or partial data. Such data are often due to design, when only partial rankings are requested (e.g. assessors rank their top three). Also worth mentioning are generalized linear models as tools for modelling influence of covariates. In more recent years several **R** packages have appeared for the analysis of rank data.

13

Models for Processes in Space or Time

13.1 Models for Spatial Point Processes

We first discuss some use of exponential family models for spatial point process data. Starting from the simplest Poisson process of complete spatial randomness, we will turn to some examples of more complicated models, in particular the Strauss model for point patterns with interaction. We do not aim at complete stringency, and for a proper measure-theoretic treatment the interested reader is referred to the literature of the field, see e.g. Møller and Waagepetersen (2007), which was the source of inspiration for this section. A particularly detailed account of spatial modelling and inference, with much useful practical advice, is provided by the book Baddeley et al. (2016). Baddeley and Turner are also responsible for the powerful open source **R**-package `spatstat`, used not only in the book mentioned, but also for the illustrations in the present section.

13.1.1 Poisson Process Models

A homogeneous (i.e. constant intensity) Poisson process on a bounded subset S of $\mathbb{R}^2$, e.g. the unit square[1], is most easily defined by combining a total Poisson number N of points, Po(λ), with a conditional uniform spatial distribution and mutual locational independence between the N points. The density for the point pattern data $\boldsymbol{y}_n = \{y_1, ..., y_n\}$ generated in this way is

$$f_{\boldsymbol{y}_N}(n, \boldsymbol{y}_n; \lambda) = f_N(n; \lambda) f_{\boldsymbol{y}_N|N=n}(\boldsymbol{y}_n; \lambda) = \frac{\lambda^n}{n!} e^{-\lambda} \operatorname{Area}(S)^{-n}. \qquad (13.1)$$

The uniform conditional density for each point is the constant $1/\operatorname{Area}(S)$.

For inference about λ (or the intensity per unit area, $\lambda/\operatorname{Area}(S)$), n is sufficient and nothing essential is new in comparison with the simple Poisson model of Example 2.3. The difference is in the conditional distribution of

[1] or more generally a subset of $\mathbb{R}^k$, but we use terminology appropriate in $\mathbb{R}^2$

data given n, and model diagnostics should make use of the spatial character of data and their assumed uniform conditional density on S. An example of a model check is the quadrat counting test, where S is partitioned into equal area subsets, and the number of points are counted and compared in a χ^2 test. Other tests could be for dependence on covariates, see below.

In a spatially inhomogeneous Poisson process, the (conditional) spatial density for a randomly selected point, $y = y_i$ say, is allowed to vary with features of its position in S. Thus we let this density be proportional to a function $\rho(y)$ over S. Now its conditional spatial density, replacing $1/\text{Area}(S)$, is $\rho(y_i) / \int_S \rho(y)\,\mathrm{d}y$. Convenient and typical parametric models used for ρ are log-linear,

$$\rho(y) = e^{\boldsymbol{\beta}^T \boldsymbol{u}(y)}, \tag{13.2}$$

where $\boldsymbol{u}(y)$ is some specified (possibly vector-valued) function of the position. For example, $\boldsymbol{u}$ could incorporate the latitude coordinate, and the topographic altitude of the position (when y includes covariate information only implicit in the position in S). If $\dim \boldsymbol{\beta} = 1$, we may simply write $\rho(y) = \gamma^{u(y)}$, with $\gamma = e^{\beta}$. The likelihood for $(\lambda, \boldsymbol{\beta})$ becomes

$$L(\lambda, \boldsymbol{\beta}) = \frac{\lambda^n}{n!} e^{-\lambda} \frac{e^{\boldsymbol{\beta}^T \sum_1^n \boldsymbol{u}(y_i)}}{\left(\int_S e^{\boldsymbol{\beta}^T \boldsymbol{u}(y)}\,\mathrm{d}y\right)^n}. \tag{13.3}$$

This is an exponential family with canonical statistic $\boldsymbol{t} = \{n, \sum_1^n \boldsymbol{u}(y_i)\}$, and canonical parameter $\boldsymbol{\theta} = (\alpha, \boldsymbol{\beta})$, with $e^{\alpha} = \lambda / \int_S e^{\boldsymbol{\beta}^T \boldsymbol{u}(y)}\,\mathrm{d}y$. The marginal distribution for n is Poisson with mean value λ, so the simpler parameterization by $(\lambda, \boldsymbol{\beta})$ is a *mixed* parameterization (Section 3.3.3), by one mean value and one canonical parameter component. The parameter components are variation independent already by construction, and they are information orthogonal by Proposition 3.20. This is then of course also true for the pair (λ, γ), but not for $(\alpha, \boldsymbol{\beta})$ or (α, γ).

Computationally, ML estimation is conveniently carried out by utilizing the formal equivalence with the Poisson log-linear models of Chapter 9. In the latter models, we would think of the $\boldsymbol{u}$-value as externally selected or generated, whereas in the spatial model it forms an intrinsic property of the random point y. This method is for example used as default in the **R**-package `spatstat`.

Example 13.1 ***A spatially inhomogeneous Poisson process***
Figure 13.1 illustrates two realizations of Poisson processes with the same number of points, $n = 71$ (chosen to match Figure 13.2). The left plot represents the complete spatial randomness of a homogeneous Poisson and

the right plot a Poisson of type (13.3), with $u(y)$ being the longitudinal coordinate of a point y. In principle, simulation of a Poisson process is easy, because it can be done in two steps, first a Poisson number of points and next one and the same distribution over S for each point. For Figure 13.1, this was carried out by the function `rpoint` of `package: spatstat`.

When the inhomogeneous model with covariate u was fitted to the two data sets of Figure 13.1 by the `spatstat` default method, the ML estimates $\hat{\beta} = 0.34$ and $\hat{\beta} = 0.86$ were obtained, with standard errors 0.41 and 0.42, respectively. An approximate z-test shows that the $\hat{\beta}$ of the inhomogeneous model is just barely significant on the 5% level. What would you have guessed? △

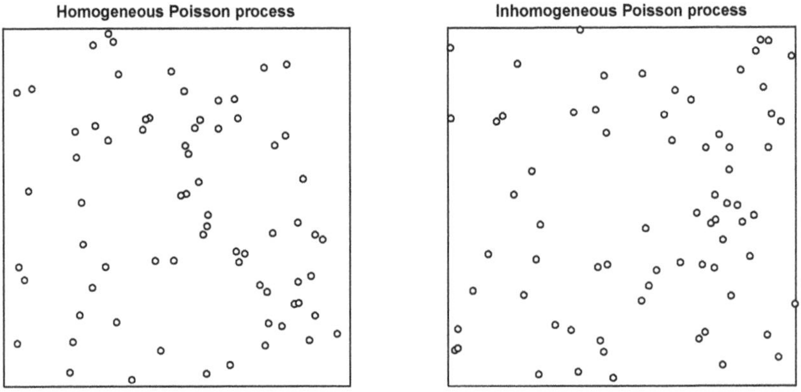

Figure 13.1 Simulated Poisson processes in the unit square, with $n = 71$.
Left: Homogeneous Poisson.
Right: With the longitudinal coordinate u of y as covariate, $\beta = 1$, i.e. $\rho(u) \propto e^u$; MLE $\hat{\beta} = 0.86$, $s.e.(\hat{\beta}) = 0.42$

An alternative, approximate estimation method is to divide the area in a large number of small gridboxes, preferably so small that no gridbox contains more than one point (and most contain none), the crucial condition being that the covariate is essentially constant within gridbox. Each gridbox has then a binary outcome and can be regarded as a Bernoulli trial, and the data can be analysed using a binomial logistic generalized linear model. This kind of technique is popular in Geographic Information Systems (GIS).

Inference about the pattern parameter $\boldsymbol{\beta}$ is usually more important than inference about the mean point density over S, λ. By the conditionality principle, Proposition 3.21, of Section 3.5, the inference about $\boldsymbol{\beta}$ should be conditional, given $N = n$. The conditional model implies an ordinary iid fixed sample size exponential family inference for the data $\boldsymbol{y}_n = (y_1, ..., y_n)$. The factorization (13.3) in its λ and $\boldsymbol{\beta}$ factors shows that we are in the even stronger situation of a *cut*, separating the two parameter components completely. In other words, the statistic n is S-ancillary for inference about $\boldsymbol{\beta}$, similarly to the Poisson Example 3.2, that contains the essential ingredients also of the present example. In particular, the MLEs of $\boldsymbol{\beta}$ and λ are the same in the joint model as in the conditional and marginal models, respectively (and in particular $\hat{\lambda} = n$). Furthermore, the cut in two factors guarantees that the conditional observed information for $\boldsymbol{\beta}$ is the same as the joint model observed information for $(\lambda, \boldsymbol{\beta})$. Thus, for large sample inference based on the observed information, the joint and conditional models will yield identical standard errors.

13.1.2 Strauss model for pairwise interaction

The Poisson models assume that realized points do not interact, and therefore such models are often too simple to fit point patterns in real life. As examples involving repulsion, think of a model for tree positions in a forest, when we might want to express that trees compete for space and therefore are rarely found tight together, or similarly for animals who possess and defend individual territories. Opposite to repulsion is attraction or clustering. Good models for attraction have been more difficult to define, however, so most statistical models for interaction are models for repulsion. Common to all models for interaction is that they are analytically intractable, and for exponential families both the norming constant and the structure function lack explicit form.

The simplest parametric model for pairwise interaction, also being the simplest non-Poisson Markov spatial point process, is the Strauss model (Strauss, 1975), in which the constant Poisson intensity is modified by an interaction factor γ for close neighbour pairs. The model is represented by a density of type

$$f(n, \boldsymbol{y}_n) = \frac{\lambda^n}{n!} \frac{\gamma^{m(\boldsymbol{y}_n)}}{C(\lambda, \gamma)}, \qquad (13.4)$$

where $m(\mathbf{y}_n)$ is the number of *close neighbour pairs*, defined as point pairs being at mutual distance less than a fixed interaction range r. Thus, each such pair contributes a factor γ.

Example 13.2 ***Strauss model for Swedish pines data***
A data set frequently used to represent a Strauss model is shown in Figure 13.2. It originates from L. Strand and goes under the name Swedish pines data. It shows the positions of 71 pine tree saplings in an area of about $100\,\text{m}^2$ and it has been estimated to have $r \approx 1$ m and $\gamma \approx 0.3$. △

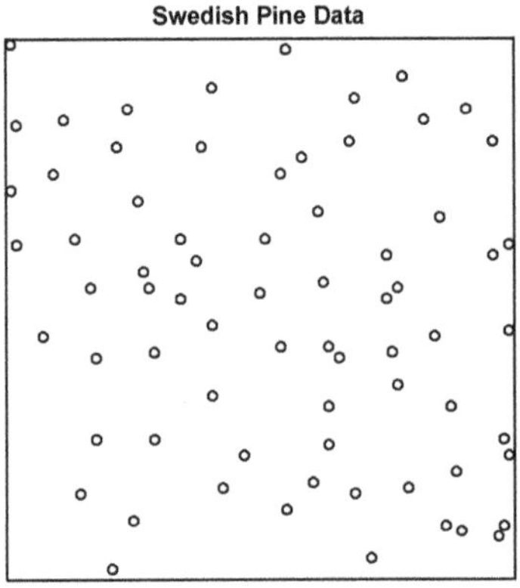

Figure 13.2 Swedish pines data, used as example of a Strauss process, with parameters $r \approx 1$ m, $\gamma \approx 0.3$.
Observation window $10 \times 10\,\text{m}^2$.

The norming constant $C(\lambda, \gamma)$ in (13.4) that makes $f(n, \mathbf{y}_n)$ a probability density cannot be given in a tractable form, so we will just write $C(\lambda, \gamma)$. Anyhow, it is clear that the model is an exponential family with canonical statistic $(n, m(\mathbf{y}_n))$ and corresponding canonical parameter $\theta = (\theta_1, \theta_2) = (\log \lambda, \log \gamma)$.

The canonical parameter space $\mathbf{\Theta}$ is less obvious. For $\gamma = 1$, i.e. $\theta_2 = 0$, we recover the homogeneous Poisson model. For $0 < \gamma < 1$ ($\theta_2 < 0$) the model represents different degrees of repulsion, and the limiting value $\gamma = 0$ ($\theta_2 \to -\infty$) makes close neighbours impossible, a so-called *hard core model*. The original idea by Strauss was to represent attraction (clustering) by $\gamma > 1$ ($\theta_2 > 0$). However, it was pointed out by Kelly and Ripley (1976) that the model is not summable/integrable jointly over n and $\boldsymbol{y}_n$ when $\gamma > 1$. A technical reason is that the exponent of γ will be of magnitude n^2, not n, and the implication is that a Strauss model with $\gamma > 1$ would have an infinity of points in S. Hence, the canonical (i.e. maximal) parameter space $\boldsymbol{\Theta}$ is the restricted set $\theta_2 \leq 0$ $(\gamma \leq 1)$.

Note that $\boldsymbol{\Theta}$ is not open, so the family is *not regular*. The boundary $\theta_2 = 0$ is formed by simple homogeneous Poisson models, for different intensity values θ_1. Furthermore, for such Poisson models, $E_{\theta_1}(N)$ and $E_{\theta_1}(m(\boldsymbol{y}_N))$ are both finite. This implies that the gradient of $\log C(\theta_1, \theta_2)$ has a finite limit as $\theta_2 \to -0$, and hence that the family is *not even steep*. One of the consequences is that there is a positive probability for observed pairs (n, m) which do not match their expected values, whatever choice of parameter values. By intuition, this happens when $m(\boldsymbol{y}_n)$ is larger than expected under a homogeneous Poisson model with $E_{\theta_1}(N) = n$. The likelihood must then be maximized over θ_1 along the boundary of $\boldsymbol{\Theta}$, for given $\theta_2 = 0$, which means we are back in the simple homogeneous Poisson model.

Two techniques for fitting the Strauss model are analogous to techniques used when fitting random graph models for social networks. MCMC-ML is based on simulation by MCMC, and the other method (used in **spatstat**) maximizes a pseudo-likelihood instead of the proper likelihood. The pseudo-likelihood method avoids the calculation of the complicated normalizing constant in (13.4) by replacing the likelihood with a product of pointwise conditional likelihoods for each point y_i, given the set of surrounding points, To some degree, this avoids the complicated norming constant. Due to the restricted range of interaction, only points within distance r need be considered in the conditioning.

By conditioning on n we can eliminate the parameter θ_1 (i.e. λ), and obtain a conditional distribution family for $\boldsymbol{y}_n$ given $N = n$, and a corresponding conditional likelihood for θ_2 (or γ) alone. It follows from the conditionality principle, Proposition 3.21 of Section 3.5, that the inference about θ_2 (or γ) should be made conditional, given n, and if θ_2 is supplemented by $E(n)$ as parameter in a mixed parameterization, n is the MLE of $E(n)$. For the Strauss process, (13.4), the conditional density for $\boldsymbol{y}_n$ is proportional to $\gamma^{m(\boldsymbol{y}_n)}$. In this case the norming constant $C_n(\gamma)$ for fixed n is conceptually simpler than the norming constant for random n, so the conditional distribution is more easily handled than the joint distribution. It is still quite complicated, however, see Exercise 13.1 to get a feeling for the likelihood for very small n.

Interestingly, the Strauss model is an example of the somewhat pathological property that the conditional model has a wider parameter space for θ_2

than the joint model. When n is fixed, any θ_2-value can be allowed. Hence the conditional model is able to represent attraction, whereas this feature is not consistent with the joint model. From the practical point of view, however, it turns out that the conditional model is an uninteresting model for clustering. The reason is that when $\theta_2 > 0$ each realization of the process tends to look much like one of the two limits $\theta_2 = 0$ or $\theta_2 \to \infty$ (Geyer and Thompson, 1995; Geyer, 1999). In the former, cluster-free model, n points are independently spread out over S, according to a uniform distribution. In the other limiting model all points are concentrated in a single clump of radius r. In a very short interval of intermediate values of θ_2, the distribution of $m(\boldsymbol{y}_n)$ in the conditional model is bimodal, with the two modes representing the two limiting models mentioned. This is an example of a phase transition similar to the phase transitions discussed for social networks models in Section 11.3. An illustration of the peculiar properties possible is provided by Geyer and Thompson (1995, Fig. 1 & 2). These figures show that if the observed value of $m(\boldsymbol{y}_n)$ is neither small or large, we can estimate the canonical parameter precisely. However, this does not help us to estimate the mean value parameter with precision, and furthermore it is quite implausible that we will see such an observation in practice. Hence the conditional model with $\theta_2 > 0$ not only lacks joint interpretation, it is also not a good model for clustering or attraction.

The factorization in marginal and conditional densities is not a cut. If it had been, the conditional and joint model ML estimates of θ_2 would have been the same, but they are not. Geyer and Møller (1994) compared them empirically by MCMC on simulated data, but did not find larger differences, however, than what they expected from the uncertainty in the MCMC method, see also Geyer (1999).

So far, we have assumed r known, but when fitting a Strauss model to data, this is typically not realistic. Thus, r is also a parameter, but of a different character. We can calculate the profile likelihood for it by maximizing the likelihood (13.4) for a number of chosen r-values. In the second stage, the profile likelihood for r can be maximized.

13.1.3 Gibbs Models

The Strauss model can of course be generalized. Ripley and Kelly (1977) modified and applied the Hammersley–Clifford theorem for Markov random fields to spatial point processes. We first only make the reasonable assumption that the density is *hereditary*, which means that if $f(\boldsymbol{y}_n) > 0$, then $f(\boldsymbol{x}) > 0$ must hold for all subsets $\boldsymbol{x}$ of $\boldsymbol{y}_n$. Ripley & Kelly now showed that $f(\boldsymbol{y}_n)$ must be of the form

$$f(\boldsymbol{y}_n) = e^{\sum_x U(\boldsymbol{x})} \tag{13.5}$$

where the sum is over all subsets $\boldsymbol{x}$ of $\boldsymbol{y}_n$. Such models are often called *Gibbs models*, with terminology from statistical physics.

Usually Gibbs models are formulated as log-linear exponential families and with the functions U chosen to depend on distances. In Poisson models $U(\boldsymbol{x}) = 0$ for all subsets $\boldsymbol{x}$ of two or more points. The Strauss model is the special case when $U(x) = \theta_1$ for single points x and $U(x_1, x_2) = \theta_2$ for point pairs with mutual distance $\leq r$, else $U(x_1, x_2) = 0$. All higher-order U-functions, for triplets etc, are zero, and this holds for most Gibbs models suggested in the literature. Finally, $U(\emptyset)$ for the empty set $\emptyset$ is always minus the log of the normalizing constant.

Ripley and Kelly (1977) further introduced a local Markov property, with interaction radius r, and showed that it is equivalent to $U(\boldsymbol{x}) = 0$ as soon as not all point pairs in $\boldsymbol{x}$ are within mutual distance r. Thus the Strauss model is Markov in this sense. The Markov property simplifies both estimation and simulation. In particular it justifies the use of pseudo-likelihood (composite likelihood) in the Strauss model.

For a discussion of other Gibbs models, in which one or several potential functions has the role of the function $m(\boldsymbol{y}_n)$, see Møller and Waagepetersen (2007, Sec. 5.3) and its references, and see `spatstat` for a rich library of pairwise interaction models. We mention just a few here. An obvious generalization of the simple Strauss model is to allow different γ-factors in different range intervals. Another suggested model is a *triplets process* that allows an additional interaction effect for each triplet of three close neighbours. A model including the highest order interaction is the *area–interaction model*, obtained if $m(\boldsymbol{y}_n)$ in (13.4) is replaced by $A(\boldsymbol{y}_n)$, being the area part of S that is within distance r from at least one of the points in $\boldsymbol{y}_n$ (i.e. the area of a union of n discs with radius r).

A problem with models allowing interaction, that we have neglected here and which is often neglected in practice due to its complications, is the *edge problem*. Think of the Strauss model, as an example. If the observation window S is part of a larger region, then what we observe in S is related to what we cannot observe outside S, within a distance r from S. In principle at least, we could regard the observed data within S as incomplete data, the data in the wider region as complete, and use the methodology of Chapter 8, including for example the EM algorithm. Simpler is to use an inner observation window S_0, such that all points in S_0 have distance $\geq r$ to the boundary of S, and condition on the outcome in the part of S outside

S_0. However, we have to pay for the simplicity by some loss of information.

It can be interesting to compare this Markovian modelling of spatial point processes, conditional on n, with Markov random graph modelling of social networks, and to note analogies. In both cases we have intractable norming constants, requiring MCMC or pseudo-likelihood methods. Both $m(\boldsymbol{y}_n)$ and the number of edges in a network can be of magnitude n^2, and partially for that reason phase transitions can occur in parts of the parameter space. The triplets process model appears to have an analogy in the combined two-star and triangle models for networks, Is it perhaps possible to formulate a general theory simultaneously covering some properties of both of these types of situations (and perhaps others)?

Exercise 13.1 ***Strauss model with very small*** n
(a) Given $n = 2$, find expressions for the two possible conditional Strauss model likelihood functions, and find the corresponding (degenerate) maximum likelihood values for γ (or θ_2).
(b) Try the same for $n = 3$ and $m(\boldsymbol{y}_n) = 1$ or $= 2$, and find how difficult the calculations are already for this small sample size. △

Exercise 13.2 ***Local Markov models***
Check that the criterion for a local Markov model (Ripley and Kelly, 1977) is satisfied by the triplets model and by the area–interaction model. △

13.2 Time Series Models

13.2.1 State Space Models

Here is a brief introduction to the topic of state space models from the viewpoint of exponential families. For further and more detailed reading, Durbin and Koopman (2001) and Fahrmeir and Tutz (2001, Ch. 8) are proposed, supplementing each other by different paths into the subject.

A univariate linear state space model can be written

$$y_t = \boldsymbol{x}_t^T \boldsymbol{\alpha}_t + \epsilon_t\,, \qquad (13.6)$$

which by intention looks like a (simple or) multiple linear regression model, except that its observations are indexed by t (for discrete time), and more seriously that the parameter vector $\boldsymbol{\alpha}_t$ has got an index t. The variation in $\boldsymbol{\alpha}_t$, the *state vector*, is intended to make the model *dynamic* in character. With freely varying regression coefficients $\boldsymbol{\alpha}_t$, however, the situation would be hopeless, with too many parameters. Instead we additionally assume a structured variation, by an underlying Markovian random process for $\boldsymbol{\alpha}_t$,

$$\boldsymbol{\alpha}_t = F_t \boldsymbol{\alpha}_{t-1} + \boldsymbol{\delta}_t \,. \tag{13.7}$$

Here the matrix F_t is called the *transition matrix*, and is assumed known, and the linear formula (13.7) is called the *transition equation* or *state equation*. In combination with (13.7), formula (13.6) for the observable $\boldsymbol{y}_t$ is called the *observation equation*. The random process $\boldsymbol{\alpha}_t$ could include trends and seasonal variation.

We have already met a similar structure, corresponding to the latent random process for $\boldsymbol{\alpha}_t$, in the class of hidden Markov models (HMM), Section 8.7, where the latent process was assumed to be a finite state space Markov chain. Right now we rather think of Gaussian variation in both $\boldsymbol{\alpha}_t$ and y_t, so with the whole of $\mathbb{R}^{\dim \alpha}$ as state space (provided F_t is full rank). More precisely, think of $\boldsymbol{\delta}_t$ and $\boldsymbol{\epsilon}_t$ as white noise, with mutual independence, in combination with a completely independent start value $\boldsymbol{\alpha}_0$ for the $\boldsymbol{\alpha}_t$ process.

The model given by (13.7) and (13.6) may be extended in several respects. The observation equation (13.6) may be a multivariate linear regression (see first part of Example 7.4), and in applications often is so. Then the observable is a vector $\boldsymbol{y}_t$, and $\boldsymbol{x}_t$ is replaced by a matrix X_t. The scalar x_t, vector $\boldsymbol{x}_t$, or matrix X_t is allowed to depend on previous observed values. The noise variances or covariance matrices for $\boldsymbol{\delta}_t$ and $\boldsymbol{\epsilon}_t$ are allowed to depend on t, as long as the form of this dependence is regarded as known. Provisionally let us think of them as known.

The main task when data follow the (linear) state space model is to make inference about the $\boldsymbol{\alpha}_t$ sequence. Here the famous so called *Kalman filter* (and Kalman smoother) solves the problem, in different forms depending on if the problem is *filtering* (about the last $\boldsymbol{\alpha}_t$), *smoothing* (about the whole sequence realized so far), or *forecasting* of future values. Further statistical inference problems concern model characteristics such as the variances of $\boldsymbol{\delta}_t$ and $\boldsymbol{\epsilon}_t$ and the initial value $\boldsymbol{\alpha}_0$. They may be wholly or partially unknown. Such parameters (sometimes called hyperparameters) can be ML-estimated from the likelihood for the observed $\boldsymbol{y}_t$ data, that is, the marginal likelihood not involving $\boldsymbol{\alpha}_t$ explicitly. If $\boldsymbol{\alpha}_t$ had been observed, the estimation problem had been simpler. This motivates the point of view of Chapter 8, to utilize that the observed $\boldsymbol{y}_t$-data can be regarded as incomplete data from an exponential family whose complete data also include $\boldsymbol{\alpha}_t$. One advantage is that the EM algorithm (Section 8.3) can be used to find the MLE numerically.

Another type of extension of the model, of particular interest in the present text, is to nonlinear, non-Gaussian observation equations replacing (13.6). We could think of any generalized linear model, for example letting y_t be Poisson distributed with a mean value relating to $\boldsymbol{\alpha}_t$. Fahrmeir and Tutz (2001) call such models *dynamic generalized linear models*. A dynamic log-linear Poisson model will then have $\log E(y_t) = \boldsymbol{x}_t^T \boldsymbol{\alpha}_t$, whereas in a dynamic logit model for Bernoulli variables, $\operatorname{logit} \pi_t = \boldsymbol{x}_t^T \boldsymbol{\alpha}_t$. The Kalman filter and smoother part of the analysis becomes more complicated, and several different methods compete. The Poisson type of model is used by Fahrmeir and Tutz (2001) on a data set of counts of phone calls. The dynamic logit model has for example been used to model rainy/dry days in Tokyo and Oxford/Cambridge boat race winners:

Example 13.3 ***A dynamic logit model in practice***
Durbin and Koopman (2001, Sec. 14.5) analyse the binary outcomes from 1829 to 2000 of the Oxford–Cambridge annual boat race by a model in which the logit $\alpha_t = \operatorname{logit} \pi_t$ of the winning probability π_t for (say) Cambridge is assumed to have varied according to a simple random walk,

$$\alpha_t = \alpha_{t-1} + \delta_t\,, \tag{13.8}$$

with $\delta_t \sim \mathrm{N}(0, \sigma_\delta^2)$. They do not fit the dynamic logit model directly, but a recursive linear approximation of it. Whether the dynamic model is reasonable can certainly be discussed, and they do not try to check statistically the fit of their model, but it is simple, and it allows the clustering of wins and losses seen in much of their data.

Fahrmeir and Tutz (2001) use the same model (13.8) for α_t to fit to a data set of dry and wet days in Tokyo. A crucial difference to the boat race data, however, is that t for this data set represents the calendar day number ($t = 1, \ldots, 365(366)$), even though there are two years of data. So y_t is assumed $\sim \mathrm{Bin}(2, \pi_t)$ distributed. Even though the model also in this case is able to show a form of clustering, here into dryer and wetter periods, the assumption that these periods occur the same dates in different years appears too naive. If the state equation explains serial correlation in y_t to a sufficient extent, is another question. △

13.2.2 Other Models for Processes in Time

The book by Küchler and Sørensen (1997) represents a quite different direction in modelling of time series data. They study parametric statistical models, more precisely exponential families, corresponding to a wide variety of probabilistic models for stochastic processes. The underlying processes treated include Markov processes in general, martingales, counting processes, diffusion type processes, and random fields. Particular themes of the book include asymptotic likelihood theory and sequential ML estimation. The asymptotic theory is relatlively simple for Lévy processes, essentially equivalent to processes having independent stationary increments, which can be made correspond to successive independent observations.

The general definition of an exponential family for a random process in Küchler and Sørensen (1997) is formally quite like the definition we have used so far, but the ingredients are more complex. The likelihood, which is defined as a Radon–Nikodym derivative and constructed in a stochastic basis corresponding to a suitable probability space, is written

$$L_t(\boldsymbol{\theta}) = a_t(\boldsymbol{\theta})\, q_t\, e^{\gamma_t(\boldsymbol{\theta})^T B_t}\,.$$

Here q_t and the canonical B_t are random processes, the latter allowed to be vector-valued. As stated by the authors, we could think of deleting the index t from γ_t, because a vast majority of families met in statistical practice satisfy this demand. We leave the topic there, and refer to their book (Küchler and Sørensen, 1997) for further reading.

14

More Modelling Exercises

14.1 Genotypes under Hardy–Weinberg Equilibrium

Let A and a denote the only possible alleles (i.e. forms) of a certain gene in the human genome. Since humans have a set of two genes (one from each parent), they must be either both A (denoted AA), both a (denoted aa) or one of each (denoted Aa). These letter combinations denote the three possible genotypes. Let the individuals of a random sample, fixed size n, from a large population be classified by genotype as AA, Aa or aa. The sample frequencies of the three genotypes are counted. Their expected values depend not only on the population frequencies of A and a, but also on the current population structure.

(a) Formulate a first, basic model for such genotype data, without any assumptions about the population frequencies of the three different types, and characterize this model as an exponential family.

(b) Express in terms of the three genotype population frequencies what are the population frequencies of gene A versus gene a.

(c) The population is said to be in *Hardy–Weinberg equilibrium* if the population proportions of the genotypes AA, Aa and aa are p^2, $2pq$ and q^2, respectively, for some p, $0 < p < 1$, and $q = 1 - p$. Characterize as an exponential family the resulting model for the same sample of data, if the population is known to be in H–W equilibrium. Here, p and q can of course be interpreted as the population proportions of the A and a alleles, respectively, and the sampling of individuals is equivalent to independent random sampling of two alleles from this population.

(d) Extend the model for genotypes frequencies under Hardy–Weinberg equilibrium to the case of three different alleles, with population proportions p, q, and r, $p + q + r = 1$. In blood group analysis the alleles may be A, B and 0 (zero), see Section 8.6.

14.2 Model for Controlled Multivariate Calibration

As a basic structure, suppose q-dimensional $\boldsymbol{y}$ $(q \times 1)$ has a multivariate regression on $\boldsymbol{x}$ $(p \times 1)$, $q \geq p \geq 1$, for $\boldsymbol{x}$ fixed and given (controlled), so

$$\boldsymbol{y} \sim \mathrm{N}(\boldsymbol{\alpha} + \boldsymbol{B}\,\boldsymbol{x},\, \Sigma),$$

where $\boldsymbol{\alpha}$ $(q \times 1)$ and $\boldsymbol{B}$ $(q \times p)$ represent arbitrary intercept and regression coefficients parameters, and the covariance matrix Σ $(q \times q)$ is also wholly unknown. Suppose we have a sample of n pairs $\{\boldsymbol{x}_i, \boldsymbol{y}_i\}$, for the calibration, and additionally the y-value $\boldsymbol{y}_0$ for a completely unknown $\boldsymbol{x}$, to be called $\boldsymbol{\xi}$, and which is the parameter of main interest.
It is natural to factorize the likelihood as

$$L(\boldsymbol{\alpha}, \boldsymbol{B}, \Sigma, \boldsymbol{\xi}) = L_1(\boldsymbol{\alpha}, \boldsymbol{B}, \Sigma;\, \boldsymbol{y}_1, \ldots, \boldsymbol{y}_n)\, L_2(\boldsymbol{\alpha}, \boldsymbol{B}, \Sigma, \xi;\, \boldsymbol{y}_0).$$

(a) When $p = q = 1$, argue that the calibration factor L_1 yields all information for estimation of the regression parameters $\{\alpha, B, \sigma\}$.
(b) When $p = q > 1$, show that we have still a full exponential family, now of parametric dimension $2q + q^2 + q(q + 1)/2$.
(c) When $q > p$, however, show that we have a curved exponential family instead of a full family, and that the argument in (a) does no longer hold (there is information in y_0 about the regression parameters). For simplicity, you are allowed to let $p = \dim \boldsymbol{x}_i = \dim \boldsymbol{\xi} = 1$.

Comment on Controlled Multivariate Calibration

It is difficult to make full use of the information in $\boldsymbol{y}_0$ in case (c), and the additional amount of estimation precision is little. In practice it may be wiser to use only L_1 for estimation of α and B also in this case, as in (a); see Brown and Sundberg (1987) and Brown (1993, Ch. 5). Furthermore, if inference about $\boldsymbol{\xi}$ is regarded as prediction of a random quantity rather than estimation of an unknown parameter, we are led to regress $\boldsymbol{x}$ on $\boldsymbol{y}$ instead of $\boldsymbol{y}$ on $\boldsymbol{x}$. When this is done separately for each component of $\boldsymbol{x}$, the situation has been transferred to 'ordinary' multiple regressions.

14.3 Refindings of Ringed Birds

Birds are caught and ringed (get a ring around a leg). A few of them are refound (dead) in subsequent years, and the number of years until they are found gives information about the survival probabilities for that species (refindings of live birds also provide such information, of course, but are not considered here). We will consider some simplistic models for such kinds of data. Your task is to construct these models from assumptions and characterize them as exponential families (canonical statistics and parameters, parameter space, etc.).

(a) Suppose n of N ringed birds are found dead within the next five years. For each such bird, $j = 1, \dots, n$, we know precisely how many years it survived, and let n_k be the number found dead after surviving precisely k years, $\sum_0^4 n_k = n$. Assume birds have some common probability π_r of *r*efinding at death, as if iid Bernoulli trials. Assume further that all birds, independently, have some probability π_s to *s*urvive each next year (so in particular, and not very realistic, calendar year and bird age are irrelevant). Under these assumptions we want a model for data that allows inference about π_s.

(b) Since we are interested only in the probability π_s for survival, show and motivate that the parameter π_r can and should be eliminated by conditioning on the random number n of refindings.

Comments on Refindings of Ringed Birds

Whether n is handled as part or data, as in (a), or conditioned on and thus treated as fixed, as in (b), the models for data are of multinomial type, but not standard saturated multinomial models. The natural models in these settings are not even full exponential models.

In (a), there are six classes (found after $0, \dots, 4$ years, or not found), and a model of order 2. A bird found year 1 has not survived the first year, and this outcome has probability $\pi_r (1-\pi_s)$, etc. (add factor π_s for each year). As canonical statistic we can take $\boldsymbol{t} = (\sum_0^4 k\, n_k,\, n)$. The canonical parameter has the components $\theta_1 = \log \pi_s$ and $\theta_2 = \log\big(\pi_r(1-\pi_s)/(1-\pi_r(1-\pi_s^5))\big)$. For the parameters to be interpretable in the model, we must have $0 < \pi_s < 1$ and $0 < \pi_r \leq 1$. This defines a region in $\mathbb{R}^2$ bounded by $\theta_1 \leq 0$ and $\theta_2 \leq \log\{(1-\pi_s)/\pi_s^5\} = \log(1-e^{\theta_1}) - 5\theta_1$. However, this region is only part of the canonical parameter space, which is $\boldsymbol{\Theta} = \mathbb{R}^2$ (follows from Proposition 3.7, because the number of possible values of $\boldsymbol{t}$ is finite).

In case (b), the conditionality principle (Proposition 3.21) asks us to condition on n for more relevant inference about π_s (or θ_1). Data then represent a sample from a truncated geometric (or 'Time to first success') distribution (Example 2.6). The probabilities p_k for survival precisely $k \geq 0$ years satisfy $p_k \propto \pi_s^k$, both without and with truncation. With truncation the probability for outcome k is

$$p_k = p^k/(1 + p + p^2 + p^3 + p^4), \qquad k = 0, \dots, 4, \tag{14.1}$$

for some p, where we naturally first think of $p = \pi_s$. The distribution (14.1) is an exponential family with canonical statistic $\sum k n_k$ and canonical parameter $\theta_1 = \log p$. Its canonical parameter space is $\Theta_1 = \mathbb{R}$ (again by Proposition 3.7), corresponding to the whole positive halfline $\mathbb{R}_+$ for p, whereas the interpretation of p as probability $p = \pi_s < 1$ is constrained to the subset $\Theta_1 = (-\infty, 0)$. One consequence is that π_s^k is a necessarily decreasing sequence, whereas the full family allows p^k to be increasing with k. Also, if $p \geq 1$, (14.1) does not remain a distribution in the limit, if we let the point of truncation increase from 5 towards infinity.

As a consequence, in the two models for bird refindings, data may turn out to be inconsistent with the probabilistic model formulated, but nevertheless correspond to a unique ML estimate in the wider, canonical parameter space. For example, when n is regarded as fixed and only the refound birds yield data, all such birds have a value between 0 and 4 for the number of years survived. From (14.1) with $p = \pi_s < 1$ we expect a mean value less than 2, but the actual sample mean may be as large as 4, see Figure 14.1. A sample mean > 2 is either bad luck or indicates serious model lack of fit.

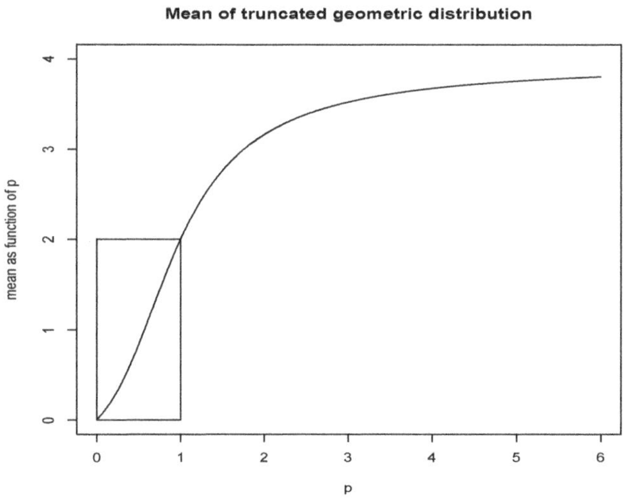

Figure 14.1 Expected number of survived years in the truncated geometric distribution (14.1), plotted as a function of p. For $p < 1$ (rectangle), p can be interpreted as the probability π_s of surviving next year, but the full model allows all $p > 0$.

14.4 Statistical Basis for Positron Emission Tomography

Background: Positron Emission Tomography (PET) is a medical imaging technique. A positron-emitting radionuclide (a tracer) is attached to some biologically active molecule which is introduced into a body. The emissions are used to see where the tracer is accumulated. If, for example, the molecule is a glucose analogue, the accumulation will reflect the metabolic activity in different regions. More precisely, positrons emitted by the tracer immediately annihilate with nearby electrons, yielding two gamma photons that fly off in random but opposite directions. Some of these photons hit gamma detectors placed around the body, and are counted, whereas others simply miss the detectors. When a pair of simultaneous hits is registered, it is known that an emission has occurred, and it is counted. The detected counts are used to form an image of the activity in the body region, via the estimated spatial distribution of tracer concentration.

For a simplified statistical treatment and modelling, regard the region of interest as consisting of J voxels (volume elements, 3-d pixels), and let the number of possible pairs of detectors be K. We further make the standard assumption in this context, that the number x_{jk} of positrons emitted from voxel j and detected by detector pair k, during an observation time interval, is a Poisson variate with expected value $c_{jk}\,\lambda_j$. Here the c_{jk} are mathematically calculated constants (calculated from knowledge of the geometry of the detector system and other factors), whereas λ_j are unknown parameters representing the intensity of emission from the amount of tracer substance in the voxels $j = 1, ..., J$. We want to make inference about $\{\lambda_j\}$. We will first unrealistically assume that for each detected pair of gamma photons, we can observe the voxel where it originated. This means that we have observed all x_{jk}. Later we will make the model more realistic.

Statistical model assumptions: The variates x_{jk} are mutually independent and Poisson distributed with expected value $c_{jk}\,\lambda_j$,

$$x_{jk} \sim \text{Po}(c_{jk}\,\lambda_j), \quad j = 1, ..., J, \quad k = 1, ..., K,$$

where $c_{jk} \geq 0$ are known constants and λ_j are unknown parameters. For tasks A1–A3 we assume that we have observed the whole data matrix $X = \{x_{jk}\}$ (in combination with the c_{jk}-values). We want to make infererence about $\{\lambda_j\}$, $j = 1, ..., J$.

Tasks A: (A1) Write down the likelihood for the parameters $\{\lambda_j\}$.
(A2) This is an exponential family. Identify the canonical statistic vector

$t(X)$, the canonical parameters, and find the expected value and variance-covariance matrix of $t(X)$.
(A3) Write down the likelihood equations and the MLE of the parameter vector $\{\lambda_j\}$. Find the distribution of this MLE, and its large-sample normal approximation.

Tasks B and C: In practice it is impossible to observe the x-values (that is, in which voxel a photon originated), as was unrealistically assumed known in task A. More realistic data are then restricted to the column sums $y_k = \sum_j x_{jk}$, for each $k = 1, ..., K$, (summation over all voxels j that could yield a hit in detector pair k). Statistical inference is still possible, but with less simplicity and less precision. In what sense do we now have an exponential family? In other words, assume the following:
Suppose that our data are $y_k = \sum_j x_{jk}$, where the x_{jk} are as defined for task A. Also assume $K > J$, so we have more observations than parameters. Show that data $\{y_k\}$ could be regarded either as
(B) following a curved exponential family, or as
(C) being incomplete data from the family in (A).
Your tasks now are to investigate each of these representations:

(B1) For the curved family representations, introduce first a K-dimensional full exponential family having the set of y_k as its canonical statistic. Tell what the corresponding canonical parameter is, expressed in the λ_j parameters. Derive the score function first for this canonical parameter of the full family for $\{y_k\}$, and from there the score function for $\{\lambda_j\}$ in the curved family representation.
(B2) Use this to write down the likelihood equations for $\{\lambda_j\}$ based on the data $\{y_k\}$.

(C1) As a preparation for task (C2), show that the conditional distribution of $\{x_{jk}\}_j$ for fixed k given the observed $y_k = \sum_j x_{jk}$ is multinomial, and specify this multinomial.
(C2) Find the likelihood equations for $\{\lambda_j\}$ based on data $\{y_k\}$, using an incomplete data interpretation of data $\{y_k\}$.

Compare the results obtained in (B2) and (C2). Do not spend time trying to solve the likelihood equations, because there is no explicit solution. The EM algorithm is a standard tool is this field. Another possible interpretation of the model is as a generalized linear model with a noncanonical link function. This means that standard programs for such models are also possible, at least in principle.

Comments on Positron Emission Tomography

A: The likelihood for data X is a quite simple Poisson likelihood, with $t = \{x_{j.}\}$, where dot denotes summation over k. For later reference, $E(t)$ has components $c_{j.}\,\lambda_j$. It is evident that the MLE is explicit.

B: With $\theta_k = \log(\sum_j c_{jk}\lambda_j)$ as canonical parameters, we get a simple, full, extended model. However taking into account that all θ_k depend nonlinearly on the λ_js, the actual model is a curved subfamily.

C1: This property has been used previously, in particular in connection with Poisson tables and contingency tables, see for example Section 5.2. The result we need is the expression for the conditional expected value, $E(x_{jk}\,|\,y_k) = y_k\, c_{jk}\lambda_j / \sum_j c_{jk}\lambda_j$.

C2: When only incomplete data $\{y_k\}$ are available, instead of the full data $t = \{x_{j.}\}$ (assumed in task *A*), the likelihood equations can be written $E(t\,|\,\{y_k\}) = E(t)$. Here the right-hand side has components $c_{j.}\lambda_j$, see task *A*, and the left-hand side has the components

$$E(x_{j.}\,|\,\{y_k\}) = \sum_k E(x_{jk}\,|\,y_k) = \sum_k y_k \frac{c_{jk}\lambda_j}{\sum_j c_{jk}\lambda_j},$$

using the result from *C1*. The likelihood equations can be simplified by deletion of the factor λ_j from both sides. Then we get the same form of the equations as in the *B* approach, but on the other hand this destroys the theoretical basis for use of the EM algorithm, so it might be better to keep these factors in the equations.

A good old basic reference in this application area is Vardi et al. (1985).

Appendix A

Statistical Concepts and Principles

A parametric statistical model for an observed or imagined set of data is mathematically represented by a family of probability models, indexed by a finite-dimensional parameter $\boldsymbol{\theta}$. The model is aimed at describing and explaining the variability in data, both systematic variation and random variability. This role for probability, in modelling, is called phenomenological.

Another role for probability is in inferential measures of uncertainty, called epistemological. Examples from frequentist inference are the degree of confidence attached to a (confidence) interval and the p-value of a significance test.

Frequentist inference is based on the *repeated sampling principle:* Measures of uncertainty are to be interpreted as long run frequencies in hypothetical repetitions under the same conditions.

The *likelihood* $L(\boldsymbol{\theta}; \text{data})$ is the probability function or probability density for the actually observed data, but regarded as a function of the parameter vector. The *maximum likelihood (ML) principle* states that we choose as parameter estimator the value maximizing the likelihood (MLE). The *likelihood ratio* (or equivalently, *deviance*) test compares the maximum likelihoods in a restrictive hypothesis model and a wider alternative model.

The *(Fisher) score function* $U(\boldsymbol{\theta})$ is the gradient vector of $\log L(\boldsymbol{\theta})$. Under some regularity conditions it satisfies $E_{\theta}(U(\boldsymbol{\theta})) = 0$, which contributes to the motivation for the *likelihood equation* system $U(\boldsymbol{\theta}) = 0$.

The *observed information* matrix $J(\boldsymbol{\theta}; \text{data})$ is minus the matrix of second derivatives of $\log L$ (written $-D^2 \log L(\theta)$). Under some regularity conditions, its expected value equals the variance matrix of U, denoted the *Fisher information* matrix.

A *statistic* $\boldsymbol{t}$ is a function of data, $\boldsymbol{t} = \boldsymbol{t}(\text{data})$. The *sampling distribution* of a statistic (e.g. of an estimator) is its probabiity distribution induced by the underlying probability distribution for data.

If the statistic $\boldsymbol{t}$ is such that the conditional distribution of data, given $\boldsymbol{t}$, is free of parameters, $\boldsymbol{t}$ is *sufficient* (for the parameter). A sufficient statistic is *minimal* if it is a function of every other sufficient statistic. The ratio of the likelihood function relative a reference parameter value is minimal sufficient in the model.

Sufficiency principle: Provided that we trust the model for data, all possible outcomes of data having the same value of a sufficient $\boldsymbol{t}$ must lead to the same conclusions about the parameter.

This is essentially equivalent to the following principle.

Weak likelihood principle: Provided that we trust the model for data, all possible outcomes of data having mutually proportional likelihood functions must lead to the same conclusions about the parameter.

The role of the complementary conditional distribution for data, given a sufficient statistic $\boldsymbol{t}$, is to be a basis for formal or informal diagnostic model checking.

Conditionality principle: In some situations a conditionality principle applies, stating that inference should be conditional on a component, $\boldsymbol{u}$ say, of the minimal sufficient statistic $\boldsymbol{t}$. These situations are essentially of one of two kinds. One type is when the marginal distribution of $\boldsymbol{u}$ depends on *nuisance parameter* components which are eliminated by conditioning. The simplest such case, that goes under the name S-ancillarity, assumes that the likelihood factorizes into one conditional factor (given $\boldsymbol{u}$) and one marginal factor (for $\boldsymbol{u}$) with completely separate parameter subvectors. The other, more tricky situation can only occur when the minimal statistic $\boldsymbol{t}$ is of higher dimension than the model parameter vector $\boldsymbol{\theta}$, with some component $\boldsymbol{u}$ of $\boldsymbol{t}$ having a (marginal) distribution that does not depend on $\boldsymbol{\theta}$ ($\boldsymbol{u}$ is then called distribution-constant or *ancillary*). Typically $\boldsymbol{u}$ then has the role of a *precision index* (e.g. a randomly generated sample size), and should be conditioned on for increased *relevance*.

A *parametric hypothesis* H_0 about the model or its parameter $\boldsymbol{\theta}$ will be assumed to specify the value of a subvector $\boldsymbol{\psi}$ (in a suitable parameterization), $H_0 : \boldsymbol{\psi} = \boldsymbol{\psi}_0$. This covers both the situation when $\boldsymbol{\psi}$ is a parameter of particular subject matter importance and we either want to reject the hypothesis or to find support for it, and the situation when we want to simplify the model by deleting parameter components (meaning they are given specified values). The latter situation includes the case when only the hypothesis model is of interest but it is embedded in a wider model in order to check model adequacy. We will treat all these cases under the name *model-reducing hypotheses*.

A *significance test* will use a test statistic, that is, a statistic selected for its ability to distinguish different values of the hypothesis parameter. Let the test statistic be denoted $g(\boldsymbol{t})$, with larger values indicating less consistency with H_0, and let the realized value of $g(t)$ be $g(t)_{obs}$. The probability under H_0 of the interval $g(t) \geq g(t)_{obs}$ is the *p-value* of the test. A small p-value indicates that we should not believe in H_0. In the Neyman–Pearson framework we specify in advance a critical such probability α, called the *size* (or the level of significance) of the test, and we decide to reject H_0 if the p-value is smaller than the desired size of the test.

Confidence regions for ψ, with confidence degree $1 - \alpha$ can be constructed via significance testing (and vice versa). Choose as confidence region the set of those ψ_0-values which are not rejected in a size α test of $H_0 : \psi = \psi_0$, given the available data, and a selected test procedure applicable for each possible ψ_0.

Exact sampling distributions for MLEs and test statistics, and the corresponding confidence degrees of confidence regions, are rarely explicit, and therefore almost only confined to some simple exponential families. Instead, much practice is based on the application of 'large sample' results of normal approximations, for which only mean vector and variance-covariance matrices are required. Large sample approximations are not valid only in the iid case of a sample of size n, but strict proofs are typically found carried out only in the iid case. The present text is no different in that respect. On the other hand, this text includes the refined approximation theory that sometimes goes under the name small sample approximation and is based on the so-called saddlepoint approximation.

An *extended Fisherian reduction* (Cox, 2006, Sec. 4.3.1) is a formulated procedure for inference that essentially follows the steps outlined here. In a very condensed form, thus: find the likelihood function, reduce first to a minimal sufficient statistic. Reduce next to a statistic for the parameter of interest alone, that allows hypothesis testing etc. This may involve conditioning on part of the minimal statistic, to eliminate nuisance parameters and/or to increase the relevance of the analysis.

The so-called Neyman–Pearson framework is more decision-oriented, with a consequentially reduced role of the likelihood. For example, statistical tests are constructed to have optimal power against specified alternatives within some class of tests, and the test outcome leads to a decision of 'rejection' or 'acceptance' of the null hypothesis.

There are many textbooks on a general statistical theory, that can be suggested to the reader. Among the sources of inspiration for the present book are Cox and Hinkley (1974), Davison (2003) and Pawitan (2001), but Bickel and Doksum (2001), Casella and Berger (2002), Lehmann and Casella (1998) and Lehmann and Romano (2005) should also be mentioned. For discussion of principles, Cox (2006) is a balanced treatise.

Appendix B

Useful Mathematics

B.1 Some Useful Matrix Results

In statistics generally, and not least in this text, we often need to invert partitioned covariance matrices and information matrices, and the partitioning allows useful expressions for the corresponding parts of the inverse. Let a positive definite matrix M be partitioned as

$$M = \begin{pmatrix} M_{11} & M_{12} \\ M_{21} & M_{22} \end{pmatrix}, \tag{B.1}$$

where M_{11} and M_{22} are square matrices, and $M_{21} = M_{12}^T$. Let the corresponding partitioning of the inverse M^{-1} be

$$M^{-1} = \begin{pmatrix} M^{11} & M^{12} \\ M^{21} & M^{22} \end{pmatrix}. \tag{B.2}$$

Because M is positive definite, its inverse exists. The same is true of M_{11}, M_{22}, M^{-1}, M^{11} and M^{22}, because they are also necessarily positive definite matrices.

Proposition B.1 ***Properties of partitioned matrices***
A. Inversion. *For a positive definite matrix M partitioned as in (B.1), the diagonal and off-diagonal blocks of M^{-1} may be expressed in terms of the blocks of M as*

$$M^{11} = (M_{11} - M_{12}M_{22}^{-1}M_{21})^{-1} \tag{B.3}$$
$$M^{12} = -M^{11}M_{12}M_{22}^{-1}, \tag{B.4}$$

and analogously for M^{21} and M^{22}.
B. Determinant. *The determinant of M can be factorized as*

$$\det M = \det\left((M^{11})^{-1}\right) \det M_{22} = \det M_{22} \,/\, \det M^{11}. \tag{B.5}$$

Proof **A.** To see that these expressions hold, we just check that they function as blocks of M^{-1}:

$$M^{11}M_{11} + M^{12}M_{21} = M^{11}M_{11} - M^{11}M_{12}M_{22}^{-1}M_{21} = M^{11}(M^{11})^{-1} = \mathrm{I}_{11},$$

and

$$M^{11}M_{12} + M^{12}M_{22} = M^{11}M_{12} - M^{11}M_{12} = 0_{12},$$

etc.

B. A direct matrix theory proof is based on the identity

$$\begin{pmatrix} \mathrm{I}_{11} & -M_{12}M_{22}^{-1} \\ 0_{21} & \mathrm{I}_{22} \end{pmatrix} M = \begin{pmatrix} M_{11} - M_{12}M_{22}^{-1}M_{21} & 0_{12} \\ M_{21} & M_{22} \end{pmatrix},$$

where I denotes an identity matrix of adequate dimension. From this equality the desired relationship follows, by calculation rules for determinants (of products, block-diagonal matrices, and inverses). □

An alternative, indirect probabilistic proof of the determinant relation was given in Example 2.10, where it was also mentioned that theoretical multivariate normal regression theory could be used to prove the other results. Note how in Example 2.10 the regression of the first component y_1 on the second component y_2 was derived, in terms of parts of the inverse covariance matrix, and how (B.4) and (B.3) helped to express this in terms of the covariance matrix itself.

B.2 Some Useful Calculus Results

B.2.1 Differentiation

The *Jacobian matrix* for a function $y = g(x)$ extends the gradient to the case when x and y are vector-valued. The Jacobian matrix has elements $\partial y_i/\partial x_j$ in position (i, j) and is denoted $\left(\frac{\partial y}{\partial x}\right)$. The Jacobian matrix corresponds to the best linear approximation of a differentiable function locally near a given point. The determinant of the Jacobian matrix, when $\dim x = \dim y$, is often called simply the *Jacobian*, Its absolute value represents how the transformation from x to y expands or shrinks local volume. In that role it appears in the determination of the density of a transformed random variable.

The *chain rule* for a composite scalar function $f(y)$, when $y = g(x)$:

$$\frac{\partial f}{\partial x_j} = \sum_i \frac{\partial f}{\partial y_i}\frac{\partial y_i}{\partial x_j},$$

or in vector/matrix form for column gradient vectors $D_x f$ and $D_y f$,

$$D_x f = \left(\frac{\partial y}{\partial x}\right)^T D_y f.$$

Local *inversion* of a function $y = g(x)$, when $\dim x = \dim y$, is possible if the Jacobian matrix is nonsingular, and the Jacobian matrix for the inverse g^{-1} is given by inversion of the Jacobian matrix for g,

$$\left(\frac{\partial x}{\partial y}\right) = \left(\frac{\partial y}{\partial x}\right)^{-1}$$

B.2.2 Factorials and the Gamma Function

The factorials $n!$ of the nonnegative integer n are special values of the gamma function,

$$n! = \Gamma(n+1)$$

Stirling's approximation formula holds not only for $n!$, but for the gamma function. We have referred to it for $n!$ in the form

$$n! \approx \sqrt{2\pi n}\, n^n e^{-n}.$$

For the gamma function a corresponding expression is

$$\Gamma(z) \approx \sqrt{2\pi/z}\, z^z e^{-z}$$

It is perhaps not immediate that these expressions are essentially the same when $z = n+1$. However, when $z = n$ the two approximations differ exactly by a factor n, and this is precisely what is wanted, since $\Gamma(n+1) = n\Gamma(n)$ (which actually holds also for noninteger z).

Bibliography

Agresti, A. (2013). *Categorical Data Analysis*. 3rd edn. Wiley. (Cited on pages 10, 51, and 56)

Albert, A., and Anderson, J.A. (1984). On the existence of maximum likelihood estimates in logistic regression models. *Biometrika*, **71**, 1–10. (Cited on page 51)

Andersen, E.B. (1977). Multiplicative Poisson models with unequal cell rates. *Scand. J. Statist.*, **4**, 153–158. (Cited on pages 21 and 57)

Baddeley, A.J., Rubak, E., and Turner, R. (2016). *Spatial Point Patterns: Methodology and Applications with R*. Chapman and Hall. (Cited on page 246)

Barndorff-Nielsen, O.E. (1978). *Information and Exponential Families in Statistical Theory*. Wiley. (Cited on pages x, 33, 34, 36, and 59)

Barndorff-Nielsen, O.E., and Cox, D.R. (1994). *Inference and Asymptotics*. Chapman and Hall. (Cited on pages 35, 47, 73, 94, 118, 124, 125, 131, 135, and 139)

Bartholomew, D.J., Knott, M., and Moustaki, I. (2011). *Latent Variable Models and Factor Analysis: A Unified Approach*. 3rd edn. Wiley. (Cited on pages 162, 241, and 242)

Basu, D. (1964). Recovery of ancillary information. *Sankhyā A*, **26**, 3–16. (Cited on page 138)

Baum, L.E., Petrie, T., Soules, G., and Weiss, N. (1970). Maximization technique occurring in the statistical analysis of probabilistic functions of Markov chains. *Ann. Math. Statist.*, **41**, 164–171.

Bernardo, J.M., and Smith, A.F.M. (1994). *Bayesian Theory*. Wiley. (Cited on page 31)

Beyer, J.E., Keiding, N., and Simonsen, W. (1976). The exact behaviour of the maximum likelihood estimator in the pure birth process and the pure death process. *Scand. J. Statist.*, **3**, 61–72. (Cited on page 123)

Bickel, P.J., and Doksum, K.A. (2001). *Mathematical Statistics: Basic Ideas and Selected Topics, Vol. 1*. 2nd edn. Prentice-Hall. (Cited on pages 61 and 267)

Box, G.E.P., and Cox, D.R. (1964). An analysis of transformations (with discussion). *J. Roy. Statist. Soc. Ser. B*, **26**, 211–252. (Cited on page 185)

Brown, L.D. (1988). *Fundamentals of Statistical Exponential Families*. IMS Lecture Notes 9. Institute of Mathematical Statistics. (Cited on page x)

Brown, P.J. (1993). *Measurement, Regression, and Calibration*. Oxford University Press. (Cited on page 259)

Brown, P.J., and Sundberg, R. (1987). Confidence and conflict in multivariate calibration. *J. Roy. Statist. Soc. Ser. B*, **49**, 46–57. (Cited on page 259)

Cappé, O., Moulines, E., and Rydén, T. (2005). *Inference in Hidden Markov Models.* Springer. (Cited on page 159)

Carriquiry, A.L., and Fienberg, S.E. (1998). Rasch models. Pages 3724–3730 of: *Encyclopedia of Biostatistics*, vol. 5. Wiley. (Cited on pages 228, 238, and 240)

Casella, G., and Berger, R.L. (2002). *Statistical Inference.* 2nd edn. Duxbury Thomson Learning. (Cited on page 267)

Ceppellini, R., Siniscalco, M., and Smith, C.A.B. (1955). The estimation of gene frequencies in a random-mating population. *Annals of Human Genetics*, **20**, 97–115. (Cited on page 153)

Chatterjee, S., and Diaconis, P. (2013). Estimating and understanding exponential random graph models. *Ann. Statist.*, **41**, 2428–2461. (Cited on pages 217, 221, and 226)

Cox, D.R. (2006). *Principles of Statistical Inference.* Cambridge University Press. (Cited on pages 56, 83, 86, 123, and 267)

Cox, D.R., and Hinkley, D.V. (1974). *Theoretical Statistics.* Chapman and Hall. (Cited on pages 32, 85, 86, 90, and 267)

Cox, D.R., and Oakes, D. (1984). *Analysis of Survival Data.* Chapman and Hall. (Cited on page 123)

Cramér, H. (1946). *Mathematical Methods of Statistics.* Princeton University Press. (Cited on pages 30, 54, 90, 111, 113, and 199)

Daniels, H.E. (1954). Saddlepoint approximations in statistics. *Ann. Math. Statist.*, **25**, 631–650. (Cited on page 70)

Darroch, J.N., Lauritzen, S.L., and Speed, T.P. (1980). Markov fields and log-linear interaction models for contingency tables. *Ann. Statist.*, **8**, 522–539. (Cited on page 193)

Davison, A.C. (2003). *Statistical Models.* Cambridge University Press. (Cited on pages 15, 32, and 267)

Dempster, A.P. (1972). Covariance selection. *Biometrics*, **28**, 157–176. (Cited on page 14)

Dempster, A.P., Laird, N.M., and Rubin, D.B. (1977). Maximum likelihood from incomplete data via the EM algorithm (with discussion). *J. Roy. Statist. Soc. Ser. B*, **39**, 1–38. (Cited on page 153)

Diaconis, P. (1988). *Group Representations in Probability and Statistics.* IMS Lecture Notes 11. Institute of Mathematical Statistics. (Cited on pages 17 and 244)

Drton, M. (2008). Multiple solutions to the likelihood equations in the Behrens–Fisher problems. *Stat. Prob. Letters*, **78**, 3288–3293. (Cited on page 132)

Drton, M., and Richardson, T.S. (2004). Multimodality of the likelihood in the bivariate seemingly unrelated regressions model. *Biometrika*, **91**, 383–392. (Cited on page 121)

Durbin, J., and Koopman, S.J. (2001). *Time Series Analysis by State Space Methods.* Oxford University Press. (Cited on pages 254 and 256)

Edwards, D. (1995). *Introduction to Graphical Modelling.* Springer. (Cited on pages 192 and 206)

Edwards, D., and Lauritzen, S.L. (2001). The TM algorithm for maximizing a conditional likelihood function. *Biometrika*, **88**, 961–972. (Cited on page 46)

Efron, B. (1975). Defining the curvature of a statistical problem (with application to second order efficiency) (with discussion). *Ann. Statist.*, **3**, 1189–1242. (Cited on pages 118, 129, 130, and 135)

Efron, B. (1978). The geometry of exponential families. *Ann. Statist.*, **6**, 362–376. (Cited on pages 39, 118, 124, 131, 133, 134, and 141)

Efron, B., and Hinkley, D.V. (1978). Assessing the accuracy of the maximum likelihood estimator: observed versus expected Fisher information (with discussion). *Biometrika*, **65**, 457–487. (Cited on page 139)

Efron, B., and Tibshirani, R. (1996). Using specially designed exponential families for density estimation. *Ann. Statist.*, **24**, 2431–2461. (Cited on pages 15, 58, and 166)

Fahrmeir, L., and Tutz, G. (2001). *Multivariate Statistical Modelling Based on Generalized Linear Models.* 2nd edn. Springer. (Cited on pages 164, 254, and 256)

Feller, W. (1968). *An Introduction to Probability Theory and its Applications, Vol. I.* 3rd edn. Wiley. (Cited on page 68)

Fischer, G.H., and Molenaar, I.W. (eds). (1995). *Rasch Models: Foundations, Recent Developments, and Applications.* Springer. (Cited on pages 228, 229, 233, 234, 238, and 241)

Fisher, R.A. (1922a). On the interpretation of χ^2 from contingency tables, and the calculation of P. *J. Roy. Statist. Soc.*, **85**, 87–94. (Cited on pages 89 and 95)

Fisher, R.A. (1922b). On the mathematical foundations of theoretical statistics. *Phil. Trans. Royal Soc. London, Ser. A*, **222**, 309–368. (Cited on pages 100 and 101)

Fisher, R.A. (1935). The logic of inductive inference (with discussion). *J. Roy. Statist. Soc.*, **98**, 39–82. (Cited on page 77)

Fisher, R.A. (1953). Dispersion on a sphere. *Proc. Roy. Soc. Ser. A*, **217**, 295–305. (Cited on page 17)

Fisher, R.A. (1956). *Statistical Methods and Scientific Inference.* Oliver and Boyd. (Cited on page 141)

Fligner, M.A., and Verducci, J.S. (eds). (1993). *Probability Models and Statistical Analyses for Ranking Data.* Springer. (Cited on page 244)

Frank, O., and Strauss, D. (1986). Markov graphs. *J. Amer. Statist. Assoc.*, **81**, 832–842. (Cited on page 212)

Geyer, C.J. (1999). Likelihood inference for spatial point processes. Pages 79–140 of: Barndorff-Nielsen, Kendall, and van Lieshout (eds), *Stochastic Geometry: Likelihood and Computation.* Chapman and Hall. (Cited on page 252)

Geyer, C.J., and Møller, J. (1994). Simulation procedures and likelihood inference for spatial point processes. *Scand. J. Statist.*, **21**, 359–373. (Cited on page 252)

Geyer, C.J., and Thompson, E.A. (1992). Constrained Monte Carlo maximum likelihood for dependent data (with discussion). *J. Roy. Statist. Soc. Ser. B*, **54**, 657–699. (Cited on page 215)

Geyer, C.J., and Thompson, E.A. (1995). Annealing Markov chain Monte Carlo with applications to ancestral inference. *J. Amer. Statist. Assoc.*, **90**, 909–920. (Cited on page 252)

Ghosh, M., Reid, N., and Fraser, D.A.S. (2010). Ancillary statistics: A review. *Statistica Sinica*, **20**, 1309–1332. (Cited on page 139)

Glas, C.A.W. (1988). The derivation of some tests for the Rasch model from the multinomial distribution. *Psychometrika*, **53**, 525–546. (Cited on page 234)

Goodreau, S.M. (2007). Advances in exponential random graph (p^*) models applied to a large social network. *Social Networks*, **29**, 231–248. (Cited on page 225)

Goodreau, S.M., Handcock, M.S., Hunter, D.R., Butts, C.T., and Morris, M. (2008). A statnet tutorial. *J. Statist. Software*, **24**, 1–26. (Cited on page 225)

Handcock, M.S. 2003. *Assessing degeneracy in statistical models of social networks*. Tech. rept. Working paper 39. Center for Statistics and the Social Sciences, University of Washington. (Cited on pages 216 and 219)

Hilbe, J.M. (2011). *Negative Binomial Regression*. 2nd edn. Cambridge University Press. (Cited on page 180)

Hinkley, D.V. (1973). Two-sample tests with unordered pairs. *J. Roy. Statist. Soc. Ser. B*, **35**, 337–346. (Cited on page 144)

Hinkley, D.V. (1977). Conditional inference about a normal mean with known coefficient of variation. *Biometrika*, **64**, 105–108. (Cited on page 138)

Højsgaard, S., and Lauritzen, S.L. (2008). Graphical Gaussian models with edge and vertex symmetries. *J. Roy. Statist. Soc. Ser. B*, **70**, 1005–1027. (Cited on pages 199 and 200)

Højsgaard, S., Edwards, D., and Lauritzen, S. (2012). *Graphical Models with R*. Springer. (Cited on page 197)

Huang, K. (1987). *Statistical Mechanics*. 2nd edn. Wiley. (Cited on page 19)

Hunter, D.R. (2007). Curved exponential family models for social networks. *Social Networks*, **29**, 216–230. (Cited on page 226)

Jacobsen, M. (1989). Existence and unicity of MLEs in discrete exponential family distributions. *Scand. J. Statist.*, **16**, 335–349. (Cited on pages 34 and 51)

Jensen, J.L. (1995). *Saddlepoint Approximations*. Oxford University Press. (Cited on pages 70, 71, 94, and 139)

Johansen, S. 1979. *Introduction to the Theory of Regular Exponential Families*. Lecture Notes, 3, Inst. of Mathem. Statistics, University of Copenhagen. (Cited on page 34)

Jørgensen, B. (1993). *The Theory of Linear Models*. Chapman and Hall. (Cited on page 11)

Jørgensen, B. (1997). *The Theory of Dispersion Models*. Chapman and Hall. (Cited on page 181)

Kagan, A.M., Linnik, Yu.V., and Rao, C.R. (1973). *Characterization Problems in Mathematical Statistics*. Wiley. (Cited on page 27)

Keiding, Niels. (1974). Estimation in the birth process. *Biometrika*, **61**, 71–80. (Cited on page 123)

Kelly, F.P., and Ripley, B.D. (1976). A note on Strauss' model for clustering. *Biometrika*, **63**, 357–360. (Cited on page 250)

Kolassa, J.E. (1997). Infinite parameter estimates in logistic regression, with application to approximate conditional inference. *Scand. J. Statist.*, **24**, 523–530. (Cited on page 51)

Küchler, U., and Sørensen, M. (1997). *Exponential Families of Stochastic Processes*. Springer. (Cited on pages 256 and 257)

Kulldorff, Martin. (2001). Prospective time periodic geographical disease surveillance using a scan statistic. *J. Roy. Statist. Soc. Ser. A*, **164**, 61–72. (Cited on page 80)

Kumon, M. (2009). On the conditions for the existence of ancillary statistics in a curved exponential family. *Statistical Methodology*, **6**, 320–335. (Cited on page 137)

Lauritzen, S.L. (1975). General exponential models for discrete observations. *Scand. J. Statist.*, **2**, 23–33. (Cited on page 33)

Lauritzen, S.L. (1996). *Graphical Models*. Oxford University Press. (Cited on pages 192, 193, 198, 201, 204, 205, and 206)

Lauritzen, S.L., and Wermuth, N. (1989). Graphical models for associations between variables, some of which are qualitative and some quantitative. *Ann. Statist.*, **17**, 35–57. (Cited on page 205)

Lehmann, E.L., and Casella, G. (1998). *Theory of Point Estimation.* 2nd edn. Springer. (Cited on pages 61 and 267)

Lehmann, E.L., and Romano, J.P. (2005). *Testing Statistical Hypotheses.* 3rd edn. Springer. (Cited on pages 59 and 267)

Leone, F.C., Nelson, L.S., and Nottingham, R.B. (1961). The folded normal distribution. *Technometrics*, **3**, 543–550. (Cited on page 144)

Little, R.J.A., and Rubin, D.B. (2002). *Statistical Analysis with Missing Data.* 2nd edn. John Wiley and Sons. (Cited on page 146)

Liu, C., Rubin, D.B., and Wu, Y.N. (1998). Parameter expansion to accelerate EM: The PX-EM algorithm. *Biometrika*, **85**, 755–770. (Cited on page 163)

Lusher, D., Koskinen, J., and Robins, G. (eds). (2013). *Exponential Random Graph Models for Social Networks.* Cambridge University Press. (Cited on page 226)

Mallows, C.L. (1957). Non-null ranking models. I. *Biometrika*, **44**, 114–130. (Cited on page 243)

Marden, J.I. (1995). *Analyzing and Modeling Rank Data.* Chapman and Hall. (Cited on page 244)

Mardia, K.V., and Jupp, P.E. (2000). *Directional Statistics.* Wiley. (Cited on pages 17 and 147)

Martin-Löf, P. 1970. *Statistiska modeller.* Lecture Notes, in Swedish, written down by R. Sundberg. Mathem. Statistics, Stockholm University, available at www.math.su.se. (Cited on pages xii, 17, 65, 66, 87, 100, 104, 105, 107, 108, 109, 113, 229, 231, 233, 234, 236, 238, and 244)

Martin-Löf, P. (1974a). Exact tests, confidence regions and estimates. Pages 121–138 of: Barndorff-Nielsen, Blæsild, and Schou (eds), *Proc. Conf. on Foundational Questions in Statistical Inference, Aarhus 1973.* Dept. Theoret. Stat., University of Aarhus. (Cited on pages 100, 101, and 108)

Martin-Löf, P. (1974b). The notion of redundancy and its use as a quantitative measure of the discrepancy between a statistical hypothesis and a set of observational data (with discussion). *Scand. J. Statist.*, **1**, 3–18. (Cited on pages 100, 108, 109, 110, and 113)

Martin-Löf, P. (1975). Reply to Sverdrup's polemical article Tests without power. *Scand. J. Statist.*, **2**, 161–165. (Cited on pages 100 and 108)

McCullagh, P., and Nelder, J.A. (1989). *Generalized Linear Models.* 2nd edn. Chapman and Hall. (Cited on pages 164, 180, 186, and 187)

Meng, X.L., and Rubin, D.B. (1993). Maximum likelihood estimation via the ECM algorithm: A general framework. *Biometrika*, **80**, 267–278. (Cited on page 154)

Meng, X.L., and van Dyk, D. (1997). The EM algorithm — an old folk-song sung to a fast new tune (with discussion). *J. Roy. Statist. Soc. Ser. B*, **59**, 511–567. (Cited on pages 153 and 154)

Møller, J., and Waagepetersen, R.P. (2007). Modern Statistics for Spatial Point Processes (with discussion). *Scand. J. Statist.*, **34**, 643–711. (Cited on pages 246 and 253)

Morris, C.N. (1982). Natural exponential families with quadratic variance functions. *Ann. Statist.*, **10**, 65–80. (Cited on page 39)

Neyman, J., and Scott, E.L. (1948). Consistent estimates based on partially consistent observations. *Econometrica*, **16**, 1–32. (Cited on pages 50 and 231)

Ohlsson, E., and Johansson, B. (2010). *Non-Life Insurance Pricing with Generalized Linear Models*. Springer. (Cited on pages 187 and 190)

Park, J., and Newman, M.E.J. (2004). Solution of the two-star model of a network. *Phys. Rev. E*, **70**, 066146, 1–5. (Cited on page 221)

Pavlides, M.G., and Perlman, M.D. (2010). On estimating the face probabilities of shaved dice with partial data. *The American Statistician*, **64**, 37–45. (Cited on page 124)

Pawitan, Y. (2001). *In All Likelihood: Statistical Modelling and Inference using Likelihood*. Oxford University Press. (Cited on pages 26, 32, 42, 59, 61, 73, and 267)

Reid, N. (1988). Saddlepoint methods and statistical inference (with discussion). *Statistical Science*, **3**, 213–238. (Cited on page 73)

Ripley, B.D., and Kelly, F.P. (1977). Markov point processes. *J. London Math. Soc.*, **15**, 188–192. (Cited on pages 252, 253, and 254)

Rubin, D.B., and Thayer, D.T. (1982). EM algorithms for ML factor analysis. *Psychometrika*, **47**, 69–76. (Cited on page 162)

Schweder, T., and Hjort, N.L. (2016). *Confidence, Likelihood, Probability. Statistical Inference with Confidence Distributions*. Cambridge University Press. (Cited on pages 76 and 86)

Skovgaard, I.M. (1990). On the density of minimum contrast estimators. *Ann. Statist.*, **18**, 779–789. (Cited on page 139)

Skovgaard, I.M. (2001). Likelihood asymptotics. *Scand. J. Statist.*, **28**, 3–32. (Cited on pages 94 and 139)

Small, C.G., Wang, J., and Yang, Z. (2000). Eliminating multiple root problems in estimation (with discussion). *Statistical Science*, **15**, 313–341. (Cited on pages 124 and 131)

Smyth, G.K., and Verbyla, A.P. (1996). A conditional approach to residual maximum likelihood estimation in generalized linear models. *J. Roy. Statist. Soc. Ser. B*, **58**, 565–572. (Cited on page 49)

Snijders, T.A.B., Pattison, P., Robbins, G.L., and Handcock, M.S. (2006). New specifications for exponential random graph models. *Sociological Methodology*, **36**, 99–163. (Cited on page 226)

Song, P.X-K., Fan, Y., and Kalbfleisch, J.D. (2005). Maximization by parts in likelihood inference. *J. Amer. Statist. Assoc.*, **100**, 1145–1158. (Cited on page 46)

Strauss, D. (1975). A model for clustering. *Biometrika*, **63**, 467–475. (Cited on page 249)

Sugiura, N., and Gupta, A.K. (1987). Maximum likelihood estimates for Behrens–Fisher problem. *J. Japan Statist. Soc.*, **17**, 55–60. (Cited on page 131)

Sundberg, R. 1972. *Maximum likelihood theory and applications for distributions generated when observing a function of an exponential family variable*. Ph.D. thesis, Inst. of Mathem. Statistics, Stockholm University. (Cited on pages 152, 153, and 155)

Sundberg, R. (1974a). Maximum likelihood theory for incomplete data from an exponential family. *Scand. J. Statist.*, **1**, 49–58. (Cited on pages 143, 150, and 155)

Sundberg, R. (1974b). On estimation and testing for the folded normal distribution. *Comm. Statist.*, **3**, 55–72. (Cited on pages 150 and 155)

Sundberg, R. (1976). An iterative method for solution of the likelihood equations for incomplete data from exponential families. *Comm. Statist. – Sim. Comp. B*, **5**, 55–64. (Cited on pages 152 and 153)

Sundberg, R. (2001). Comparison of confidence procedures for type I censored exponential lifetimes. *Lifetime Data Analysis*, **7**, 393–413. (Cited on page 126)

Sundberg, R. (2002). The convergence rate of the TM algorithm of Edwards and Lauritzen. *Biometrika*, **89**, 478–483. (Cited on page 46)

Sundberg, R. (2003). Conditional statistical inference and quantification of relevance. *J. Roy. Statist. Soc. Ser. B*, **65**, 299–315. (Cited on page 190)

Sundberg, R. (2010). Flat and multimodal likelihoods and model lack of fit in curved exponential families. *Scand. J. Statist.*, **37**, 632–643. (Cited on pages 131, 132, 133, and 135)

Sundberg, R. (2018). A note on "shaved dice" inference. *The American Statistician*, **72**, 155–157. (Cited on page 144)

Tjur, T. (1982). A connection between Rasch's item analysis model and a multiplicative Poisson model. *Scand. J. Statist.*, **9**, 23–30. (Cited on pages 233, 236, and 239)

Turner, H., and Firth, D. (2012). Bradley–Terry models in R: The BradleyTerry2 Package. *J. Statist. Software*, **48**, 1–21. (Cited on page 116)

Urbakh, V.Yu. (1967). Statistical testing of differences in causal behaviour of two morphologically indistinguishable objects. *Biometrics*, **23**, 137–143. (Cited on page 147)

Vardi, Y., Shepp, L.A., and Kaufman, L. (1985). A statistical model for positron emission tomography (with discussion). *J. Amer. Statist. Assoc.*, **80**, 8–37. (Cited on page 264)

Wald, A. (1943). Tests of statistical hypotheses concerning several parameters when the number of observations is large. *Trans. Amer. Math. Soc.*, **54**, 426–482. (Cited on page 92)

Wei, B.-C. (1998). *Exponential Family Nonlinear Models*. Springer. (Cited on page 168)

Whitehead, J. (1980). Fitting Cox's regression model to survival data using GLIM. *J. Roy. Statist. Soc. Ser. C*, **29**, 268–275. (Cited on page 186)

Whittaker, J. (1990). *Graphical Models in Applied Multivariate Statistics*. Wiley. (Cited on page 192)

Zellner, A. (1962). An efficient method of estimating seemingly unrelated regressions and tests for aggregation bias. *J. Amer. Statist. Assoc.*, **57**, 348–368. (Cited on page 120)

Zucchini, W., MacDonald, I.L., and Langrock, R. (2016). *Hidden Markov Models for Time Series. An Introduction Using R*. 2nd edn. CRC Press. (Cited on pages 159, 160, and 161)

Index